MULTICRITERIA OPTIMIZATION AND ENGINEERING

MULTICRITERIA OPTIMIZATION AND ENGINEERING

Roman B. Statnikov

Joseph B. Matusov

Mechanical Engineering Research Institute
of the Russian Academy of Sciences in Moscow

CHAPMAN & HALL

I(T)P An International Thomson Publishing Company

ew York • Albany • Bonn • Boston • Cincinnati • Detroit • London • Madrid • Melbourne • Mexico City
Pacific Grove • Paris • San Francisco • Singapore • Tokyo • Toronto • Washington

Printed in the United States of America

For more information, contact:

Chapman & Hall
One Penn Plaza
New York, NY 10119

International Thomson Publishing
Berkshire House 168-173
High Holborn
London WC1V 7AA
England

Thomas Nelson Australia
102 Dodds Street
South Melbourne, 3205
Victoria, Australia

Nelson Canada
1120 Birchmount Road
Scarborough, Ontario
Canada M1K 5G4

International Thomson Editores
Campos Eliseos 385, Piso 7
Col. Polanco
11560 Mexico D.F. Mexico

International Thomson Publishing Gmbh
Königwinterer Strasse 418
53228 Bonn
Germany

International Thomson Publishing Asia
221 Henderson Road #05-10
Henderson Building
Singapore 0315

International Thomson Publishing - Japan
Hirakawacho-cho Kyowa Building, 3F
1-2-1 Hirakawacho-cho
Chiyoda-ku, 102 Tokyo
Japan

1 2 3 4 5 6 7 8 9 10 XXX 01 00 99 98 97 96 95

Library of Congress Cataloguing-in-Publication Data
Statnikov, R. B. (Roman Bentsionovich)
Multicriteria optimization and engineering
Roman B. Statnikov and Joseph Matusov.
p. cm.
Includes bibliographical references and index.
ISBN 0-412-99231-0
1. Mathematical optimization. 2. Engineering mathematics .
I. Matusov, Joseph, 1948- . II. Title.
QA402.5.S7125 1995 94-25876
621.8 ' 15' 015193--dc20 CIP

British Library Cataloguing in Publication Data available

Please send your order for this or any other Chapman & Hall book to
Chapman & Hall, 29 West 35th Street, New York, NY 10001, Attn: Customer Service Department.
You many also call our Order Department at 1-212-244-3336 or fax you purchase order to 1-800-248-47

For a complete listing of Chapman & Hall's titles, send your request to
Chapman & Hall, Dept. BC, One Penn Plaza, New York, NY 10119.

Contents

Preface

In January—February 1991 I had an opportunity to deliver lectures on multicriteria optimization at a number of American companies and universities. The contacts with the people working for the companies, as well as discussions with renowned experts in the field (Dr. W. Stadler, Dr. V. Ozernoy, Dr. E. Lieberman, et al.) convinced me that it would be worthwhile to write a book for the American audience. Although optimization has been dealt with in numerous books and papers of varying excellence. I was dissatisfied that the results of solving engineering optimization problems were not much more impressive. The more so, since there exist quite a number of first-rate optimization methods.

In solving optimization problems it is usually assumed that the problem has already been formulated, and one has only to find its solution. Actually, this is not the case for the majority of engineering problems. Even if one has at his disposal an adequate mathematical model, which is a rare occasion, this does not guarantee success. In problems involving conflicting criteria, which are most typical for engineering applications, the designer encounters objective difficulties in formulating constraints imposed on design variables and performance criteria. However, these constraints are just those determining to a considerable extent a set of feasible solutions that satisfy all necessary requirements to the object under design. Without constructing this set all further efforts to optimize the solution to a real problem often prove to be futile.

Both the book and the problems considered in it have been brought to life by the practical significance of the problems under consideration, which form a considerable part of the book. Though quite diverse, all the problems have one feature in common—to solve them one must first find the feasible solutions set.

Central to the book is the parameter (design-variable) space investigation (PSI) method, which has been brought to life by the necessity of correct statement and solution of engineering problems of optimization.

The multicriteria approach allows us to interprete in a new fashion many well-known and important problems such as identification and operational development of prototypes designed for batch and mass production, and to develop new methods for their solution.

The problems considered in this book are perennial problems for engineers, designers, and analysts engaged in creating various machines, mechanisms, structures, and devices.

At present, the PSI method is efficiently used in many areas of human activity. This book acquaints the Western reader with some practical results of its application.

We live in the vast world of real multicriteria problems. Previously, we were taught to see it in the single-criterion dimension, and, hence, in a distorted form. However, it is no exaggeration to state that the world must be seen as it really is since this is very important for our lives.

Long ago I was invited to Central Asia to deliver popular lectures. When speaking in Tashkent, Uzbekistan, I noticed a dozing old man very much like a traditional Eastern sage. He wore a turban and sat completely motionless with shut eyes. I was speaking about decision-making problems and presented an example. "Suppose five brides are presented to you and you are asked to choose the best one," said I. Imagine my surprise when the man approached me after the lecture and said: "I think nobody, save Allah, knows how to do this, since otherwise people wouldn't behave so foolishly. But the very question is why do you think that one has to choose among the presented women, not among some others?" In a single sentence the man managed to condense the extreme importance of the problem of obtaining feasible solutions.

Before us my coauthor and I have the attentive, interested, and benevolent audience. We were eager to write this book for you and hope that it will not be read in vain, but will help designers in their work.

R. Statnikov

Introduction

Optimization in Engineering Problems

The majority of engineering problems are essentially multicriteria. In designing machine tools, airplanes, automobiles, ships, and locomotives we do our best to increase their productivity, strength, reliability, longevity, efficiency, and utilization factor. At the same time we try to decrease vibration and noise, production and maintenance costs, the number of failures, material and fuel consumption, overall dimensions, etc. As a rule, different performance criteria of an engineering system are conflicting in the sense that improvement in some of them results in deterioration in some others.

At present, the annual world production reaches dozens of millions of diverse machines, mechanisms, structures, robots and manipulators, automatic transfer lines, as well as unique expensive objects, such as nuclear power stations and spacecrafts.

In order to create competitive objects one has to use up-to-date technology, materials, equipment, microprocessors, etc. However, the work still starts with the design, which is one of the most important links in the tedious process of creation of modern machines and machine systems. Clearly a superior machine cannot be created on the basis of a second-rate design. Also, since the fleet of machines is to be renovated in no more than five to seven years, a design must be not only optimal but accomplished in the shortest time measured in several months. However, a preliminary design often foresees excessive material consumption, dangerous noise levels, high vibration activity, low reliability, inadequate longevity and strength, all resulting in premature failures, emergency situations, excessive energy consumption, unacceptable pollution of the environment, and rapid exhaustion of natural resources. This is due to the fact that optimization has not yet become a technical policy. To confirm the validity of this statement it suffices to say that optimization of structural parameters of batch and mass-production machines may result in decreasing the energy and material consumption by no less than 15%, and lowering the cost by 20%. This makes optimization, considered as a technical policy, an objective necessity caused by a dramatic sharp increase in machinery production in the last 40–50 years.

About 20–30 years ago the methods of nonlinear programming suddenly became very popular. Many works were published in which the problems of optimal design were essentially reduced to traditional problems of optimization. The hope was that the epoch of total optimization had finally come, and plants would soon start manufacturing optimal automobiles and machine tools. However; soon the hopes faded and were replaced by disillusion, since the results of using mathematical methods of optimization in solving engineering problems proved to be ridiculously insignificant: "the mountain brought forth a mouse," despite the fact that plenty of first-rate methods of optimization have been developed.

A careful analysis of the majority of solved engineering problems has shown that, considered as problems of optimization, they have been ill-posed. In order to treat a patient one must first diagnose the disease correctly. This is why one of the main issues to be discussed in this book is the correct formulation of multicriteria optimization problems.

Traditionally, any problem is divided into two phases: formulation and solution. First, one poses a problem and then solves it with the help of a computer. However, for engineering optimization problems this primitive scheme is improper, and the designer cannot, as a rule, formulate a problem correctly prior to its solution. Actually, he solves the problem, analyzes the results, corrects the formulation, and solves the problem again, his way to the truth being a complicated spiral line. This is a multiply repeated, cyclic process of "formulation-solution-analysis-correction-. . ." typical for the majority of engineering problems of optimization.

Note that ill-posedness of an engineering optimization problem may be caused by more than just the use of an inadequate mathematical model. Quite often an optimization problem proves to be ill-posed though the mathematical model is all right. Also, though designers usually pay considerable attention to constructing an adequate mathematical model, the issues of formulating the problem of optimization (which actually lie in the "boundary layer" between the traditional spheres of interest of pure and applied mathematicians) are presently the least investigated ones.

Broad experience in solving problems has shown that the time needed to formulate a problem makes up 70–85% of the total time required for a complete treatment, from the formulation to results. Often the initial formulation has little in common with the final one, which is followed by the search for an optimal solution.

The life cycle of a complex technological system such as a machine includes the following stages: the development of the request for proposal and specification, design (subdivided into several phases), manufacture, the tests and operational development of a prototype, quantity production, and exploitation.

At each stage one encounters many diverse problems. Accordingly, we consider the following two extensive classes of problems.

1. **Multicriteria optimization of complex objects**. Successful optimization depends, in turn, on solving the following problems.

   Determination of feasible solutions set and Pareto optimal set. Let us ask ourselves how many design solutions of a machining center, an automobile, or a ship are considered before choosing the single one that is to be put into quantity production? The answer is: not many. The result is that we have to seek the optimal solution among a few candidates that often are far from being the best ones. In reality, there exists the so-called feasible solutions set which comprise all solutions meeting all the requirements to the future machine. Determination of such a set is one of the major problems of optimal design because nobody can guarantee that even a talented and experienced designer will be able to find the best solution, operating with a small number of candidates only, and without determining the feasible solutions set. Hence, the traditional approach does not guarantee obtaining the optimal design. Thus, to create competitive machines one must be able to construct the feasible solutions set. The problem is how to help the designer do this. In solving multivariable problems with conflicting criteria, the construction of the set proves sometimes to be a difficult task even for an experienced and highly skilled designer.

   A feasible solutions set incorporates a subset of unimprovable, or the so-called Pareto optimal[1], design solutions which cannot be improved in all the performance criteria simultaneously. Clearly, the ultimate design solution must necessarily be Pareto optimal. That is why it is so important to be able to construct and analyze the Pareto optimal set. Especially difficult is the task of approximately constructing the feasible solutions set and Pareto optimal set to a given accuracy. Though the problem is under study for a rather long time, the complete solution has not yet been obtained. In this book, we propose solutions based on sufficiently simple assumptions concerning the properties of performance criteria.

2. **Problems of multicriteria identification**. Usually, identification of an object is defined as the construction of its mathematical model and determination of the latter's design variables, based on the analysis of the object's responses to known external disturbances. In contrast to conventional (scalar) identification, we use the ensemble (vector) of proximity (closeness, adequacy) criteria, characterizing the discrepancies

---

[1]Wilfredo Pareto (1848–1923) was a well-known Italian economist and sociologist.

between the corresponding characteristics of a mathematical model and the full-scale experiment.

Multicriteria identification is an important fundamental and applied problem one has to deal with in any area where theoretical and experimental results have to be matched.

When optimizing the parameters of some model, we tacitly suppose that the model is adequate and the results obtained on its basis are reliable. If the optimization results turn out to be not of practical significance, one possible cause may be the inadequacy of the mathematical model. In this connection the Eastern proverb "One can't pour out of a jug more than it contains" should be kept in mind. Hence, it is important that the models under consideration adequately describe real objects. One must be aware of the mathematical models' advantages and drawbacks.

A solution to the problem of multicriteria identification must allow determination of the "sphere of applicability" of the mathematical model, evaluation of expediency of its further development, accuracy, completeness, and trustworthiness of the results, as well as correction of the variable boundaries and verification of the list of performance criteria for further solution of optimization problems.

The method of muticriteria identification proposed in this work allows solution of an important applied problem of operational development (improvement) of prototypes. The significance of the problem is stressed by the fact that the cost of operational development is often commensurate with that of the creation of a new machine.

The problems of operational development are solved in two stages: First, the problem of multicriteria identification is solved, and then the problem of optimizing the performance criteria of the object subjected to improvement, is considered.

In the late 1960s, when it became clear that the vast number of optimization methods had practically no effect on the quality of designed objects, we started the development of the conception, methods, and algorithms for formulating and solving the problems of optimization of complex technological objects.

The efforts were crowned by the creation of the parameter (design-variable) space investigation (PSI) method, which is central to this book. The PSI method was created by Sobol' and Statnikov (see, e.g., Statnikov (1978) and Sobol' and Statnikov (1981)). Primarily, the method is aimed at the formulation and solution of the problem of determination of the feasible solutions set. In this sense, the method has no analogue. In creating the method we did our best to take into account the specific features of designers' thinking and behavior. Of course, the optimal solution to a highly complicated multicriteria problem cannot be found in the automatic mode. In the case under consideration the search scenario is based on the designer-computer dialogues. Later, the PSI method was used as a basis for developing the methods for approximating the feasible solutions set and Pareto optimal set, multicriteria identification, decomposition and aggregation of large-scale systems, and the estimation of criteria sensitivity to design variable alteration.

Practical results obtained on the basis of the PSI method have been tested at a number of major enterprises.

We were aware of the uniqueness and large scale of this extensive experiment and thought that it must be within the reach of the designers engaged in the search for optimal solutions of engineering problems. This is demonstrated by numerous examples whose authors have kindly proposed them for publication in this book.

The PSI method finds numerous applications in design practice in Russia and former USSR republics. It was efficiently used in geophysics and pharmacy, fiber and nonlinear optics, nuclear power and technology, petrophysics, and other fields in which complex multicriteria problems were present. At present, the boundaries of the sphere of its application may hardly be drawn.

The book generalizes our personal and professional experience, and we hope it will be helpful to the reader.

Today, as well as long ago, many designers rely, for some reasons, only on their personal experience, intuition, and luck. Of course, all means are good for attaining good objectives, the more so, "victors are not judged." However, we could give numerous, notorious examples when even gifted, acclaimed designers have failed to find the best solutions without using the methods of multicriteria optimization. What, then, is to be said about rank-and-file designers?

In solving an engineering problem, of special interest is the designer's thought flow, reasoning, and use of the PSI method.

That is why in considering a number of examples we tried, as much as possible, to convey those considerations that resulted in an alteration of the initial problem formulation and substantiation of a new one.

In selecting the examples for the book we tried to find those that are instructive from the viewpoint of methodology.

Special attention is paid to multicriteria optimization of objects by using finite element models. This is done not only because the problems are of extreme practical importance and their solution guarantees huge economic gains, but primarily because the PSI method allows, for the first time, revelation and evaluation of the entire diversity of geometrical shapes of the object under study (or being designed). In turn, this allows approaching the solution of problems with unformalizable criteria for the choice of the best production technology.

In brief, we wished to show life in its genuine form, and this has predetermined the form of the book. We aimed to describe the process of formulation and solution of the problem of the feasible solutions set determination, despite the diverse nature of problems under consideration. However, the major objective was to demonstrate the single conception of analysis on the basis of the PSI method. We shall be thankful for the reader's patience in getting acquainted with all the examples presented in this book, which are predominantly of methodological character. By presenting so much factual material we wanted, as much as possible, to make our concept of multicriteria analysis accessible and understandable for everyone who is going to solve analogous problems.

Alongside with introducing the new approach to finding optimal solutions at numerous enterprises, we delivered the "Multicriteria Machines Design" course of lectures in many countries. This book is addressed to a wide audience, from undergraduate students to researchers and engineers; actually, to everyone engaged in solving engineering optimization problems.

Since this is one of the first works on the study of engineering multicriteria problems, we are well aware that it has not yet acquired its final perfect form, and we tried, therefore, to avoid making categorical conclusions.

In mass and batch manufacture of machines and mechanisms that involve enormous material resources and where the cost of an error (the loss of markets, incompetitiveness of products, premature failures, and emergency situations) is rather high, multicriteria optimization grows into an objective necessity.

The PSI method has been realized in the form of the multicriteria optimization and vector (multicriteria) identification (MOVI) programs package created with the invaluable contribution of Mr. Y. Y. Uzvolok. This is the package that has been used for solving numerous problems of multicriteria optimization.

We thank Mr. Mikhail M. Tsipenyuk, who has done much favoring the publication of this book. We appreciate Mrs. Nelya B. Statnikova, who helped us in our contacts with the Publishers and Mr. LeRoy M. Lefkowitz who organized Dr. R. B. Statnikov's tour in the US for delivering lectures on the PSI method.

Especially we are grateful to Dr. Wolfram Stadler, Dr. Vladimir M. Ozernoy and Dr. Ralph Steuer for their kind attitude and valuable advice, which added much to the book.

We are thankful to all our colleagues who helped us when preparing the manuscript of this monograph. We want to mention here Dr. I.S. Yenyukov and Dr. L.Y. Banach, who participated in the work on Sections 5–1 and 5–4, respectively, as well as Mr. G.I. Firsov, Dr. E.M. Stolyarova, and Dr. N.N. Bolotnik who discussed with us different issues related to the scope of the book. Especially helpful were Mr. Y.Y. Uzvolok, Mr. V.S. Shenfeld, Mr. Y.S. Yuzhakov, Mr. A.A. Pozhalostin, and Mrs. O.A. Frolova, our colleagues from the Laboratory of Theory and Methods of Optimal Design of Russian Academy of Sciences of the Mechanical Engineering Research Institute.

Different people look at optimization from different points of view: mathematical, philosophical, political, pragmatic, etc. Optimization has many faces, but it is always aimed at reaching perfection. That is why we consider this book a path to finding sound engineering solutions.

The history of science shows that the paths to the truth are multitudinous. Here we have described one of them. Said Montenne: "The truth is so great a thing that we must not ignore any way leading to it." We agree.

MULTICRITERIA OPTIMIZATION AND ENGINEERING

1

Multicriteria Optimization and the Parameter Space Investigation Method

1-1. Engineering Optimization Problems: Their Features and Formulation

First of all, we will outline a class of problems to be solved. In doing this, we will rely on the experience accumulated while solving numerous engineering optimization problems, the problems of design included. This enables us to cover a sufficiently wide class of problems encountered in different applications, especially in engineering. There are a number of features inherent in the class of the problems under consideration that predetermine both their formulation and approaches to their solution.

Let us enumerate some basic features of the problems to be considered.

1. The problems are essentially multicriteria. As a rule, attempts are made to reduce multicriteria problems to single-criterion ones. For example, productivity of a machine is undoubtedly an important index. However, should one always try to make it maximum? Besides, the single-criterion formulation of a problem ignores such questions of paramount significance as: What is the cost of the maximum productivity? How much does it deteriorate other performance criteria? Why is one criterion preferred over other ones?

   Numerous attempts to construct a generalized criterion in the form of combination of particular criteria proved to be fruitless.

   By cramming a multicriteria problem into the Procrustean bed of a single-criterion one, we replace the initial problem with a different one that has little in common with the original problem. Obviously, one should always try to take into account all basic performance criteria simultaneously.

2. The determination of the feasible solutions set is one of the fundamental issues of the analysis of engineering problems. The construction of this set is an important step in the formulation and solution of such problems.
3. The problem formulation and solution make up a single process. Customarily the designer first formulates a problem and then a computer is employed to solve it. However, in the case under consideration this approach is unsuitable because only in rare cases can one formulate a problem completely and correctly before its solution. The feasible solutions set may be obtained only in the process of solution, therefore the problems should be formulated and solved in the interactive mode.
4. As a rule, mathematical models are complicated systems of equations (including differential equations) that may be linear and nonlinear, deterministic and stochastic, with distributed and lumped parameters.
5. Usually the parameters of a model are continuous. The feasible solutions set can be multiply connected, and its volume may be several orders of magnitude smaller than that of the domain within which the optimal solution is sought.
6. Both the feasible solutions and Pareto optimal sets are nonconvex. As a rule, the information about smoothness of goal functions is absent. Usually these functions are nonlinear and continuous, however they may be nondifferentiable. Almost always, there are many various constraints, and the dimensionality of the design variables and criteria vectors reaches many dozens.
7. Very often, designers encounter serious difficulties neither in analyzing the feasible solutions and Pareto optimal sets nor in choosing the most preferred solution. They have a sufficiently well-defined system of preferences. Besides, the aforementioned sets usually contain a small number of elements.

As mentioned in the Introduction, to formulate and solve engineering optimization problems, the method of parameter space investigation (PSI) has been developed. Statnikov (1978) and Artobolevskii et al. (1974) have been among the first to discuss the PSI method. A systematic and comprehensive description of the method can be found in Sobol[1] and Statnikov (1977, 1981, 1982) and Genkin and Statnikov (1987). In what follows, the material of these works is used to a considerable extent.

Formulation of Multicriteria Optimization Problems

We discuss here the formulation of the mathematical problem and methods of its solution that can be applied to the majority of engineering optimization problems.

Let us consider an object (mechanical, biological, social, etc.) whose operation is described by a system of equations (differential, algebraic, etc.) or whose performance criteria may be directly calculated. We assume that the system depends on r design variables $\alpha_1,\ldots,\alpha_r$ representing a point $\boldsymbol{\alpha}=(\alpha_1,\ldots,\alpha_r)$ of an r-dimensional space. Commonly, $\boldsymbol{\alpha}$ appears in the aforementioned equations. In this book, when considering optimization problems, the design-variable vector (vector of design variables), $\boldsymbol{\alpha}=(\alpha_1,\ldots\alpha_r)$, is also referred to as solution or model, whereas the components of this vector are referred to as design variables or simply variables.

In the general case, when designing a machine, one has to take into account design-variable, functional, and criteria constraints.

Design-variable constraints (constraints on design variables) have the form

$$\alpha_j^*\leq\alpha_j\leq\alpha_j^{**},\quad j=1,\ldots,r. \tag{1-1}$$

In the case of mechanical systems, α_j represent the stiffness coefficients, moments of inertia, masses, damping factors, geometric dimensions, etc.

Functional constraints may be written as follows

$$C_l^*\leq f_l(\boldsymbol{\alpha})\leq C_l^{**},\quad l=1,\ldots,t \tag{1-2}$$

where the functional dependences $f_l(\boldsymbol{\alpha})$ may be either functionals depending on the integral curves of the differential equations mentioned previously or explicit functions of $\boldsymbol{\alpha}$ (not related to the equations); and C_l^* and C_l^{**} are constraints such as the allowable stresses in structural elements, the track gauge, etc.

Also, there exist particular performance criteria such as productivity, the material consumption, and efficiency. It is desired that, other things being equal, these criteria, denoted by $\Phi_\nu(\boldsymbol{\alpha})$, $\nu=1,\ldots,k$, would have the extreme values. For simplicity we suppose that $\Phi_\nu(\boldsymbol{\alpha})$ are to be minimized.

Obviously, constraints (1-1) single out a parallelepiped Π in the r-dimensional design-variable space (space of design variables). In turn, constraints (1-2) define a certain subset G in Π whose volume may be assumed to be positive without loss of generality.

In order to avoid situations in which the designer regards the values of some criteria as unacceptable, we introduce criteria constraints

$$\Phi_\nu(\boldsymbol{\alpha})\leq\Phi_\nu^{**},\quad \nu=1,\ldots,k \tag{1-3}$$

where Φ_ν^{**} is the worst value of criterion $\Phi_\nu(\boldsymbol{\alpha})$ the designer may comply with. (The choice of Φ_ν^{**} is discussed in Section 1-3.)

Criteria constraints differ from functional ones in that the former are determined when solving a problem and, as a rule, are repeatedly revised. Hence, unlike C_l^* and C_j^{**}, reasonable values of Φ_ν^{**} cannot be chosen before the problem solving.

Constraints (1-1)–(1-3) define the feasible solutions set D, i.e., the set of design solutions $\boldsymbol{\alpha}^i$ that satisfy the constraints, and hence, $D \subset G \subset \Pi$.

If functions $f_l(\boldsymbol{\alpha})$ and $\Phi_\nu(\boldsymbol{\alpha})$ are continuous in Π then the sets G and D are closed.

Let us formulate one of the basic problems of multicriteria optimization. It is necessary to find such a set $P \subset D$ for which

$$\Phi(P) = \min_{\boldsymbol{\alpha} \in D} \boldsymbol{\Phi}(\boldsymbol{\alpha}) \tag{1-4}$$

where $\boldsymbol{\Phi}(\boldsymbol{\alpha}) = (\Phi_1(\boldsymbol{\alpha}), \ldots, \Phi_k(\boldsymbol{\alpha}))$ is the criteria vector; and P is the Pareto optimal set.

We mean that $\boldsymbol{\Phi}(\boldsymbol{\alpha}) < \boldsymbol{\Phi}(\boldsymbol{\beta})$ if for all $\nu = 1, \ldots, k$, $\Phi_\nu(\boldsymbol{\alpha}) \leq \Phi_\nu(\boldsymbol{\beta})$ and at least for one $\nu_0 \in \{1, \ldots, k\}$, $\Phi_{\nu_0}(\boldsymbol{\alpha}) < \Phi_{\nu_0}(\boldsymbol{\beta})$.

Upon solving the problem one has to determine design-variable vector $\boldsymbol{\alpha}^0 \in P$, which is the most preferred among the vectors belonging to set P. However, if not all performance criteria can be formalized, then the optimal solution should be sought over the entire set D.

Let us give an alternative definition of the Pareto optimal set.

Definition. A point $\boldsymbol{\alpha}^0 \in D$, is called a Pareto optimal point, if there exists no point $\boldsymbol{\alpha} \in D$ such that $\Phi_\nu(\boldsymbol{\alpha}) \leq \Phi_\nu(\boldsymbol{\alpha}^0)$ for all $\nu = 1, \ldots, k$ and $\Phi_{\nu_0}(\boldsymbol{\alpha}) < \Phi_{\nu_0}(\boldsymbol{\alpha}^0)$ for at least one $\nu_0 \in \{1, \ldots, k\}$. A set $P \subset D$ is called Pareto optimal if it consists of Pareto optimal points.

The Pareto optimal set plays an important role in vector optimization problems, because (1) It can be analyzed easier than the feasible solutions set; and (2) the optimal vector always belongs to the Pareto optimal set, irrespective of the system of preferences used by the designer for comparing vectors belonging to the feasible solutions set. The importance of this set is determined to a great extent by the well-known theorem formulated, for example, in Sobol′ and Statnikov (1981).

Theorem. If feasible solutions set D is closed, and criteria $\Phi_\nu(\boldsymbol{\alpha})$ are continuous, then the Pareto optimal set is nonempty.

Thus, when solving a multicriteria optimization problem, one always has to find the set of Pareto optimal solutions.

Although these arguments in favor of the problem formulation are rather obvious, some alternative formulations are often used in practice. Next we analyze three such formulations and point out their drawbacks.

A. Substitution of a multitude of criteria by a single one

As a rule, this approach fails to provide acceptable results. For instance, sometimes it is wise to choose $\beta_\nu \geq 0$ (usually, $\beta_1 + \ldots + \beta_k = 1$) so that the function

$$\Phi=\beta_1\Phi_1(\boldsymbol{\alpha})+\ldots+\beta_k\Phi_k(\boldsymbol{\alpha}).$$

integrates all requirements of the criteria $\Phi_1,\ldots,\Phi_k$, and to consider $\Phi(\boldsymbol{\alpha})$ as the only performance criterion. The coefficient β_ν reflects the relative "importance" of the criterion Φ_ν, $\nu=1,\ldots,k$.

In practice, the "true" values of β_ν are usually unknown beforehand, especially if these criteria are of different natures and reflect different aspects of the system behavior. Moreover, it is clear that in engineering problems, the "importance" of different criteria depends on their values, and it seems reasonable to choose different β_ν for different parts of the set D.

In practice, the designer usually starts with choosing some values of $\beta_1,\ldots,\beta_k$, and then finds the best point $\boldsymbol{\alpha}'$ corresponding to the minimum value of $\Phi(\boldsymbol{\alpha})$ for $\boldsymbol{\alpha}\in D$. If some values of $\Phi_\nu(\boldsymbol{\alpha}')$ prove to be unsatisfactory, then the designer chooses $\beta_1,\ldots,\beta_k$, again. Clearly, such a procedure cannot be called *optimization* in the strict sense of the word; rather, this is a kind of exhaustive search whose completeness is not guaranteed.

B. Optimization of the most important criterion

In this case, the criterion considered by the designer to be the most important is retained, while all the others are replaced by constraints.

Let $\Phi_1(\boldsymbol{\alpha})$ be the basic criterion. Then we have to choose constraints $\Phi_2^{**},\ldots,\Phi_k^{**}$ and consider the problem of finding the minimum

$$\Phi_1(\boldsymbol{\alpha})\to\min$$

under the following constraints:

$$\begin{aligned}\alpha_j^*\le\alpha_j\le\alpha_j^{**},&\quad j=1,\ldots,r,\\ C_l^*\le f_l(\boldsymbol{\alpha})\le C_l^{**},&\quad l=1,\ldots,t,\\ \Phi_\nu(\boldsymbol{\alpha})\le\Phi_\nu^{**},&\quad \nu=2,\ldots,k.\end{aligned}$$

It is clear that in this case we also face the problem of choosing criteria constraints Φ_ν^{**} that cannot be reasonably solved without special calculations. If, however, there exists a reliable method for choosing Φ_ν^{**}, $\nu=2,\ldots,k$, then by using this method one can also select Φ_1^{**}, thus determining the set of feasible points D. In principle, it is possible to search for the best point in D taking into account only one criterion. However, as a rule, this way is not the most effective.

Besides, the majority of engineering problems contain several meaningful criteria, some of them conflicting. This is a feature of design problems.

C. Consecutive optimization of all criteria

There are several algorithms allowing consecutive improvement of all criteria. Here we consider an approach that is often called the method of successive concessions.

At the first step, we determine the minimum of $\Phi_1(\boldsymbol{\alpha})$ for $\boldsymbol{\alpha}\in D$. Let us denote this minimum by min Φ_1. Then, a "concession" h_1 is chosen for the criterion Φ_1 and the corresponding criterion constraint is specified:

$$\Phi_1^{**}=\min \Phi_1+h_1.$$

At the second step, the minimum value of $\Phi_2(\boldsymbol{\alpha})$ is determined for $\boldsymbol{\alpha}\in D$, under the constraint $\Phi_1(\boldsymbol{\alpha})\leq\Phi_1^{**}$. Upon calculating the minimum of Φ_2 and choosing a "concession" h_2, we specify the second criterion constraint

$$\Phi_2^{**}=\min \Phi_2+h_2.$$

At the third step, the minimum value of $\Phi_3(\boldsymbol{\alpha})$ is determined for $\boldsymbol{\alpha}\in D$, $\Phi_1(\boldsymbol{\alpha})\leq\Phi_1^{**}$ and $\Phi_2(\boldsymbol{\alpha})\leq\Phi_2^{**}$, and so on.

Finally, the minimum value of $\Phi_k(\boldsymbol{\alpha})$ is found for $\boldsymbol{\alpha}\in D$, $\Phi_1(\boldsymbol{\alpha})\leq\Phi_1^{**},\ldots,$ $\Phi_{k-1}(\boldsymbol{\alpha})\leq\Phi_{k-1}^{**}$. If min Φ_k is attained at some point $\boldsymbol{\alpha}'$ then this point is considered to be the best.

It is clear that the point $\boldsymbol{\alpha}'$ depends on both the order in which the criteria are enumerated, and on the choice of $h_1,\ldots,h_{k-1}$. Besides, doubt always persists that by making a concession somewhat larger one could have improved the values of the rest of the criteria considerably.

The Choice of a Single Criterion

The issue of mathematical construction of a single (determining) criterion Φ is dealt with by decision-making theory (Larichev 1987; Fishburn 1970; Keeney 1972). In the general case, the problem is reduced to the induction of a partial order on the set D or to the construction of a value (utility) function $U(\Phi_1,\ldots,\Phi_k)$. This function must reflect the designer's system of preferences, i.e.

$$U(\Phi_1'',\ldots,\Phi_k'')>U(\Phi_1',\ldots,\Phi_k').$$

if and only if the designer considers the point $\boldsymbol{\alpha}''$ corresponding to the values $\Phi_\nu(\boldsymbol{\alpha}'')=\Phi_\nu''$ as being preferred to the point $\boldsymbol{\alpha}'$ that yields the values $\Phi_\nu(\boldsymbol{\alpha}')=\Phi_\nu'$ to the performance criteria, $\nu=1,\ldots,k$. If such a function U has been constructed (Matusov and Statnikov 1981), then the problem of choosing the best point reduces to minimizing the value function.

However, even in those cases where the mathematical conditions of the existence of the function $U(\Phi_1,\ldots,\Phi_k)$ are satisfied, its construction is a very serious problem, since it requires much more information than the designer usually possesses. However, in the problems of design, the best solutions can be found comparatively easily by searching over the set of Pareto optimal solutions.

1-2. Systematic Search in Multidimensional Domains by Using Uniformly Distributed Sequences

The features of the problems under consideration make it necessary to represent vectors $\boldsymbol{\alpha}$ by points of uniformly distributed sequences in the space of design variables (Sobol' and Statnikov 1981). In the following we consider this issue in brief.

For many applied problems the following situation is typical. There exists a multidimensional domain in which a function (or a system of functions) is considered whose values may be calculated at certain points. Suppose we wish to get some information on the behavior of the function in the entire domain or in any subdomain. Then, in the absence of any additional information about the function, it is natural to wish that the points at which the function is calculated would be uniformly distributed within the domain. However, the question arises: What meaning should be assigned to the notion of a uniform distribution? This concept is quite evident only in the case of a single variable. By dividing the range of the variable into N equal parts and locating a point within each of the parts, we arrive at a sequence of N points (a net) uniformly distributed over the domain under consideration. Unfortunately, in the case of several variables the concept of uniformity is not so evident. If for each of the variables we make a partition similar to that done in the case of a single variable, then for n variables we get N^n points (a cubic net). However, the concept of uniformity should be independent of the number of points, and, besides, the use of nets containing so many points seriously complicates the solution of practical problems.

Weyl was the first to give the definition of the uniformity.

Let us consider a sequence of points $\mathbf{P}_1, \mathbf{P}_2, \ldots, \mathbf{P}_i, \ldots$ belonging to a unit r-dimensional cube K^r. By G we denote an arbitrary domain in K^r, and by $S_N(G)$ the number of points $\mathbf{P}_i$ belonging to G ($1 \leq i \leq N$). A sequence $\mathbf{P}_i$ is called uniformly distributed in K^r, if

$$\lim_{N\to\infty} \frac{S_N(G)}{N} = V(G) \qquad (1\text{-}5)$$

where $V(G)$ is the volume of the r-dimensional domain G. (If, instead of the unit cube, a parallelepiped Π is considered, then the right-hand side of (1-5) transforms into $V(G)/V(\Pi)$.)

The meaning of the definition is quite clear: For large values of N, the number of points of a given sequence belonging to an arbitrary domain G is proportional to volume $V(G)$:

$$S_N(G) \sim NV(G).$$

Figures 1-1 and 1-2 demonstrate different uniformly distributed sequences in the cubic net and the P_0-net discussed in the Addendum.

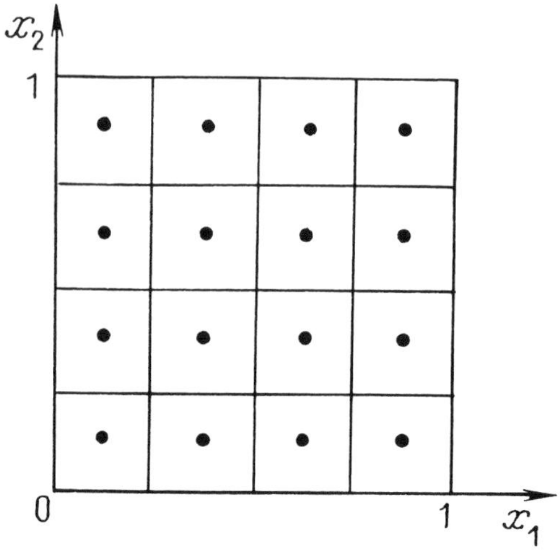

Figure 1–1 Cubic net for $n=2$ ($N=16$).

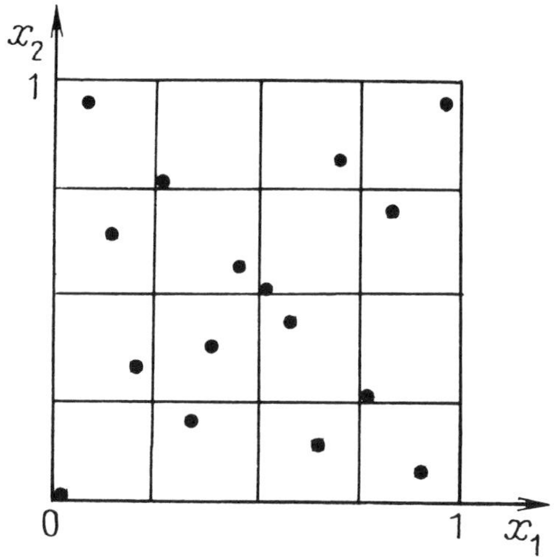

Figure 1–2 Improved net for $n=2$ ($N=16$).

In solving engineering problems one must commonly deal not with K^r, but with a certain parallelepiped Π, and, hence, transit from the coordinates of the points uniformly distributed in K^r to those in Π. Let us formulate the following statements (Sobol' and Statnikov 1981).

Lemma 1. If points $\mathbf{Q}_i$ with Cartesian coordinates $(q_{i1},\ldots,q_{ir})$ form a uniformly distributed sequence in K^r, then points $\boldsymbol{\alpha}^i$ with Cartesian coordinates $(\alpha_1^i,\ldots,\alpha_r^i)$ where

$$\alpha_j^i=\alpha_j^*+q_{ij}(\alpha_j^{**}-\alpha_j^*),\quad j=1,2,\ldots,r, \tag{1-6}$$

form a uniformly distributed sequence in parallelepiped Π consisting of points $(\alpha_1,\ldots,\alpha_r)$ whose coordinates satisfy the inequalities $\alpha_j^*\leq\alpha_j\leq\alpha_j^{**}$.

Proof. Let us choose an arbitrary parallelepiped $\Pi_0\subset\Pi$ specified by inequalities $\overline{\alpha_j}\leq\alpha_j\leq\overline{\overline{\alpha_j}}$. According to (1-6), there exists a one-to-one correspondence between the points of Π_0 and those of parallelepiped $\tilde{\Pi}\subset K^r$, which is specified by inequalities

$$\frac{\overline{\alpha_j}-\alpha_j^*}{\alpha_j^{**}-\alpha_j^*}\leq\alpha_j\leq\frac{\overline{\overline{\alpha_j}}-\alpha_j^*}{\alpha_j^{**}-\alpha_j^*}.$$

Hence, the number of points $\boldsymbol{\alpha}^i\in\Pi_0$, denoted by $S_N(\Pi_0)$, is equal to the number of points $\mathbf{Q}_i\in\tilde{\Pi}$. The latter is denoted by $\tilde{S}_N(\Pi_0)$. Since the volumes of the parallelepipeds are equal to

$$V(\Pi_0)=\prod_{j=1}^{r}(\overline{\overline{\alpha_j}}-\overline{\alpha_j}),\; V(\tilde{\Pi})=\prod_{j=1}^{r}\frac{\overline{\overline{\alpha_j}}-\overline{\alpha_j}}{\alpha_j^{**}-\alpha_j^*}=\frac{V(\Pi_0)}{V(\Pi)}$$

respectively,

$$\lim_{N\to\infty}\frac{S_N(\Pi_0)}{N}=\lim_{N\to\infty}\frac{\tilde{S}_N(\tilde{\Pi})}{N}=V(\tilde{\Pi})=\frac{V(\Pi_0)}{V(\Pi)}.$$

This completes the proof of the lemma.

If among the points $\boldsymbol{\alpha}^1,\ldots,\boldsymbol{\alpha}^i,\ldots$ forming a uniformly distributed sequence in Π, we choose all the points belonging to a certain domain $G\subset\Pi$ we obtain a sequence of points uniformly distributed in G. Let us prove this formulation.

Lemma 2. Let $\boldsymbol{\alpha}^1,\ldots,\boldsymbol{\alpha}^i,\ldots$ be a sequence of points uniformly distributed in Π, and $G\subset\Pi$ be an arbitrary domain whose volume is $V(G)>0$. If among the points $\boldsymbol{\alpha}^i$, one chooses all the points belonging to G, then he arrives at the sequence of points uniformly distributed in G.

Proof. Let $\boldsymbol{\alpha}^{i_1},\ldots,\boldsymbol{\alpha}^{i_N}$ be the first N selected points. If the number of the last point is N' (i.e., $\boldsymbol{\alpha}^{i_N}\equiv\boldsymbol{\alpha}^{N'}$) then $S_{N'}(G)=N$.

Let us choose an arbitrary parallelepiped $\Pi_0\subset G$ and denote by $\tilde{S}_N(\Pi_0)$ the number of the points from $\boldsymbol{\alpha}^{i_1},\ldots,\boldsymbol{\alpha}^{i_N}$ belonging to Π_0. Then $\tilde{S}_N(\Pi_0)=S_{N'}(\Pi_0)$, since those of the points $\boldsymbol{\alpha}^1,\ldots,\boldsymbol{\alpha}^N$ that do not belong to G, cannot belong to Π_0. Hence, as N, and hence, N', tends to infinity, we have

$$\frac{\tilde{S}_N(\Pi_0)}{N}=\frac{S_{N'}(\Pi_0)}{N'}\cdot\frac{N'}{N}=\frac{S_{N'}(\Pi_0)}{N'}\cdot\frac{N'}{S_{N'}(G)}\to\frac{V(\Pi_0)}{V(G)}.$$

This completes the proof of the lemma.

The extent of uniformity of a sequence may be estimated using the known characteristics presented in the Addendum, which also cites some additional requirements to the uniformity of the distribution of the first N points of a sequence. The requirements are of considerable importance, since these N points are used in practice. It is desirable that N not be too large, since otherwise too much computer time is needed.

The practical advantages of using more uniform sequences/nets are as follows. If we wish to solve a problem (for instance, to obtain the Pareto optimal and feasible solutions sets) with a prescribed accuracy, then the use of a more uniform sequence assures a higher convergence rate. However, if the time allowed for solving the problem is very short, and hence, N is small, then the problem cannot be solved in this way. Nevertheless, using more uniform sequences one may distribute the points in such a way that they would represent the whole domain G satisfactorily. As a result, the designer would have sufficiently reliable information about the problem under consideration.

In the following discussion, we consider two different classes of uniform sequences whose uniformity characteristics are among the best presently known. These are the so-called LP_τ-sequences and the novel P_τ-nets. The necessary definitions, descriptions of properties, and the methods for calculating the coordinates of the points of LP_τ-sequences, are presented in the Addendum.

1-3. Parameter Space Investigation (PSI) Method

In Section 1-1 we formulated the problem of multicriteria optimization and defined the feasible solutions set D, which is constructed using the values of Φ_ν^{**}, $\nu=1,\ldots,k$, and some other constraints. Now we proceed by describing the parameter space investigation method allowing correct determination of Φ_ν^{**} and, hence, of the feasible solutions too.

The parameter (design-variable) space investigation method involves the following three stages, see Fig. 1-3.

Stage 1. Compilation of the test tables with the help of a computer.

First, one chooses N trial points $\boldsymbol{\alpha}^1,\ldots,\boldsymbol{\alpha}^N$ from G, see Section 1-2. Then, all the particular criteria $\Phi_\nu(\boldsymbol{\alpha}^i)$ are calculated at each of the points $\boldsymbol{\alpha}^i$, and for each of the criteria a test table[2] is compiled so that the values of $\Phi_\nu(\boldsymbol{\alpha}^1),\ldots,\Phi_\nu(\boldsymbol{\alpha}^N)$ are arranged in the increasing order, i.e.

$$\Phi_\nu(\boldsymbol{\alpha}^{i_1})\leq\Phi_\nu(\boldsymbol{\alpha}^{i_2})\leq\ldots\leq\Phi_\nu(\boldsymbol{\alpha}^{i_N}),\ \nu=1,\ldots,k \qquad (1\text{-}7)$$

[2]Sometimes it is called an ordered test table. In an unordered table the columns are formed of the values of $\Phi_\nu(\boldsymbol{\alpha}^i)$, $i=1,\ldots,N$, $\nu=1,\ldots,k$. For example, see Table 2-1.

where $i_1, i_2, \ldots, i_N$ are the numbers of trials (a separate set for each ν). Taken together, the k tables form a complete test table. In the following discussion, the latter is called the test table[3].

Stage 2. Preliminary selection of criteria constraints.

This stage envisages the interference of the designer. By analyzing tables in Equation (1-7) in turn, the designer specifies criteria constraints Φ_ν^{**}. (It should be noted that the described method is convenient for a designer in practice. Actually, the designer has to consider one criterion at a time and specify the respective constraints.)

All Φ_ν^{**} are the maximum values of criteria $\Phi_\nu(\boldsymbol{\alpha})$, which guarantee an acceptable level of the object's operation. If the selected values of Φ_ν^{**} are not maximum, then many interesting solutions may be lost, since some of the criteria are contradictory. As a rule, the designer may put Φ_ν^{**} equal to a criterion value $\Phi_\nu(\overline{\boldsymbol{\alpha}})$ whose feasibility is beyond doubt. However, if he starts by determining the maximum possible value of Φ_ν^{**} then he has to pass to Stage 3.

Stage 3. Verification of solvability of problem (1-4) with the help of a computer.

Let us fix a criterion, say $\Phi_{\nu_1}(\boldsymbol{\alpha})$, and consider the corresponding table (Eq. 1-7), and let S_1 be the number of the values in the table satisfying the selected criteria constraint:

$$\Phi_{\nu_1}(\boldsymbol{\alpha}^{i_1}) \leq \ldots \leq \Phi_{\nu_1}(\boldsymbol{\alpha}^{i_{S_1}}) \leq \Phi_{\nu_1}^{**} = \Phi_{\nu_1}(\overline{\boldsymbol{\alpha}}). \tag{1-8}$$

One should choose the criterion Φ_{ν_1} for which S_1 is minimum among the analogous numbers calculated for each of criteria Φ_ν.

Then criterion Φ_{ν_2} is selected in analogy to Φ_{ν_1} and the values of $\Phi_{\nu_2}(\boldsymbol{\alpha}^{i_1}), \ldots, \Phi_{\nu_2}(\boldsymbol{\alpha}^{i_{S_1}})$ of Φ_{ν_2} in the test table are considered. Let the table contain $S_2 \leq S_1$ values such that $\Phi_{\nu_2}(\boldsymbol{\alpha}^{i_j}) \leq \Phi_{\nu_2}^{**}$, $1 \leq j \leq S_2$. Similar procedures are carried out for each of the criteria. Then, if at least one point can be found for which all inequalities (1-3) are valid simultaneously, then the set D defined by inequalities (1-1)–(1-3), is nonempty, and problem (1-4) is solvable. Otherwise, one should return to Stage 2 and ask the designer to make certain concessions in the specification of Φ_ν^{**}. However, if the concessions are highly undesirable then one may return to Stage 1 and increase the number of points in order to repeat Stage 2 using extended test tables.

The procedure is to be continued until D proves to be nonempty. Then, the Pareto optimal set is constructed in accordance with the definition presented in

[3]Fragments of the test tables are presented in Sections 1-4 and 6-1.

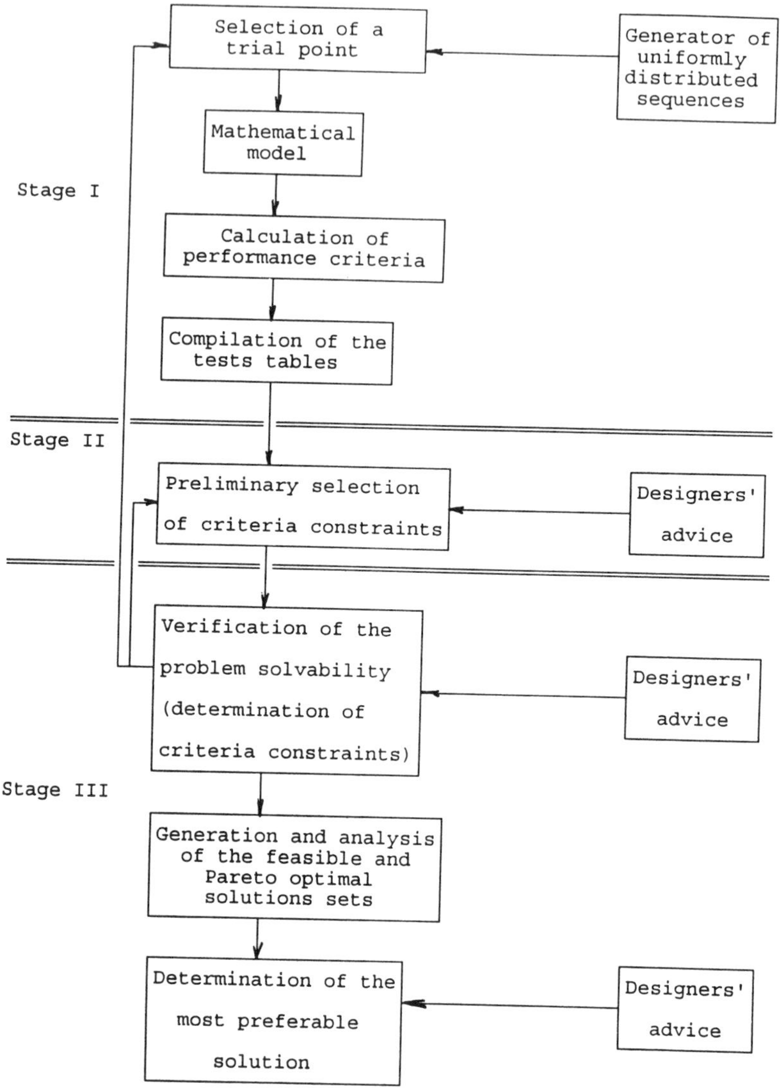

Figure 1–3 Flowchart of the algorithm.

Section 1-1. This is done by removing those feasible points that can be improved with respect to all the criteria simultaneously.

Let us consider the case where it is difficult to decide whether the value of a Φ_ν^{**} is maximum. Commonly, one is not sure whether the values of $\Phi_\nu(\boldsymbol{\alpha})$ from the interval $\Phi_\nu(\overline{\boldsymbol{\alpha}}) \leq \Phi_\nu(\boldsymbol{\alpha}) \leq \tilde{\Phi}_\nu^{**}$ are feasible. (Here $\tilde{\Phi}_\nu^{**}$ is the value of the νth criterion for which the values $\Phi_\nu(\boldsymbol{\alpha}) > \tilde{\Phi}_\nu^{**}$ are known to be unacceptable.) In such a case one has to go to Stage 3 and construct the feasible solutions set D, under the constraints $\Phi_\nu^{**}=\Phi_\nu(\overline{\alpha})$, and the corresponding Pareto optimal set P. Further, the set $\tilde{D}$ is constructed under the constraints $\tilde{\Phi}_{\nu'}^{**}$ $\nu=1,\ldots,k$, as well as the corresponding Pareto optimal set $\tilde{P}$. Let us compare $\Phi(P)$ and $\Phi(\tilde{P})$.

If the vectors belonging to $\Phi(\tilde{P})$ do not improve the value of the vectors from $\Phi(P)$ substantially, then one may put $\Phi_\nu^{**}=\Phi_\nu(\overline{\alpha})$. Otherwise, if the improvement is significant then the values of the criteria constraints may be set equal to $\tilde{\Phi}_\nu^{**}$. In this case, one has to make sure that the optimal solution thus obtained is feasible[4]. If the designer is unable to do this, then the criteria constraints are put equal to their previous values, $\Phi_\nu^{**}=\Phi_\nu(\overline{\alpha})$. This scheme can be used for all possible values of $\Phi_\nu(\overline{\alpha})$ and Φ_ν^{**}.

The Selection of Trial Points

In all the examples presented in this book, as well as in solving other problems, points $Q_1, Q_2,\ldots,Q_i,\ldots$ of the LP_τ-sequence were used.

According to Lemma 1 from Section 1-2, the Cartesian coordinates of a point $Q_i=(q_{i1}, q_{i2},\ldots,q_{ir})$ are used to calculate from Equation (1-6) the coordinates of a point $\boldsymbol{\alpha}^i=(\alpha_1^i,\ldots,\alpha_r^i)$ belonging to parallelepiped Π:

$$\alpha_j^i=\alpha_j^*+q_{ij}(\alpha_j^{**}-\alpha_j^*),\ j=1,\ldots,r,\ i=1,\ldots,N.$$

When using the points of the LP_τ-sequence, one should refer to Table A-1 presented in the Addendum. The table allows for solving problems with the number of design variables $r\leq 20$ and the number of trials $N\leq 2^{16}$. Sobol' and Statnikov (1981) contains a table where $r\leq 51$ and $N\leq 2^{20}$. Table A-6 corresponds to the novel P_τ-net where $r\leq 20$ and $N\leq 2^{12}$.

According to Lemma 2 from Section 1-2, these trial points form a sequence uniformly distributed in G, as $N\to\infty$.

At Stage 3 we find q points belonging to D where $q\leq N$. The method for constructing and selecting these points (see Lemma 2, Section 1-2) guarantees that q tends to infinity as N tends to infinity, and the sequence of the points is uniformly distributed within D.

Remark. Besides the LP_τ-sequence and the P_τ-nets, there exist some other useful sequences and nets, several of which are discussed in the Addendum. Prior to solving a concrete problem one cannot say with certainty which of them is most suitable. *Much depends on the behavior of criteria, the form of functional and design-variable constraints, and the feasible solutions set geometry. Hence, for the scheme presented in Figure 1-3, other sequences (nets) can be successfully used too.*

---

[4]To do this the designer will possibly have to analyze the mathematical model anew or, if necessary, conduct additional experimental studies.

Examples of the Feasible Solutions Set Construction

Suppose two design variables, α_1 and α_2, may be varied and the quality of an object is evaluated by criteria Φ_1 and Φ_2 depending on the design variables (see Fig. 1-4). It is required to minimize the criteria. We also suppose that a sufficiently large number of design solutions $\boldsymbol{\alpha}^i$ and $\boldsymbol{\Phi}(\boldsymbol{\alpha}^i)$, $i=1,\ldots, N$, represented in Figure 1-4*a* and *b* by dots may be generated by computer ($\Phi(P)$ in Figure 1-4*b* is the set of Pareto optimal solutions in the criteria space). Owing to the presence of the three functional constraints $C^{**}_{1,\alpha}$, $C^{**}_{2,\alpha}$, and $C^{**}_{3,\alpha}$ (Fig. 1-4c) the initial set of solutions reduces. The figure shows domain $G \subset \Pi$ satisfying the functional constraints. Within the criteria space shown in Figure 1-4*d*, $\Phi(G)$ is an image of G, so that $C^{**}_{i,\Phi}=\Phi(C^{**}_{i,\alpha})$, $i=1, 2, 3$. Having determined G, the designer seeks the set of feasible solutions D. Figure 1-4*f* illustrates three dialogues. The first one is represented by $\Phi^{**}_{1,1}$ and $\Phi^{**}_{2,1}$ where the second subscript indicates the number of the dialogue, and $D_1=\varnothing$. At this stage the designer makes a concession. The second dialogue is represented by $\Phi^{**}_{1,2}$ and $\Phi^{**}_{2,2}$, and $D_2=\varnothing$ again. The third dialogue is represented by $\Phi^{**}_{1,3}$ and $\Phi^{**}_{2,3}$; here $D_3 \neq \varnothing$, $D_3 \subset G$. In Fig. 1-4*e* $\hat{\Phi}^{**}_1$and $\hat{\Phi}^{**}_2$are inverse images of Φ^{**}_1 and Φ^{**}_2 in the design-variable space.

Figure 1-5 shows schematically three dialogues for another problem. The first one is represented by criteria constraints $\Phi^{**}_{1,1}$ and $\Phi^{**}_{2,1}$, which form the set $\Phi(D_1)$, whose inverse image in the design-variable space is D_1. The second and third dialogues are represented by criteria constraints $\Phi^{**}_{1,2}$ and $\Phi^{**}_{2,2}$ and $\Phi^{**}_{1,3}$ and $\Phi^{**}_{2,3}$ forming the sets $\Phi(D_2)$ and $\Phi(D_3)$, respectively: D_2 and D_3 are the inverse images of these sets. Upon analyzing D_3 the designer has decided that it may serve as a feasible solutions set, that is, $D_3=D$. Figure 1-5 shows the set of Pareto optimal solutions $\Phi(P)$ in the criteria space together with its inverse image P in the design-variable space.

In Sections 6-1 and 6-2, we present various dialogues together with the corresponding sets of feasible solutions.

Figure 1-6 shows a disconnected and nonconvex feasible solutions set often encountered in solving engineering problems.

The Complexity of Search

For sufficiently large values of N the property of uniform distribution of points implies that $\gamma=V(D)/V(\Pi)\approx N'/N$ where N is the number of points $\boldsymbol{\alpha}^i \in \Pi$, and N' is the number of points that have entered D. For many engineering problems $\gamma \leqslant 0.01$, and the search for the solution is like seeking a needle in a haystack. (In effect, γ characterizes the complexity of solution of the problems belonging to the class under consideration.)

"Soft" Functional Constraints and Pseudocriteria

For many practical problems, there can be found "good" solutions that lie slightly beyond the limits imposed by the constraints. If a designer is informed about

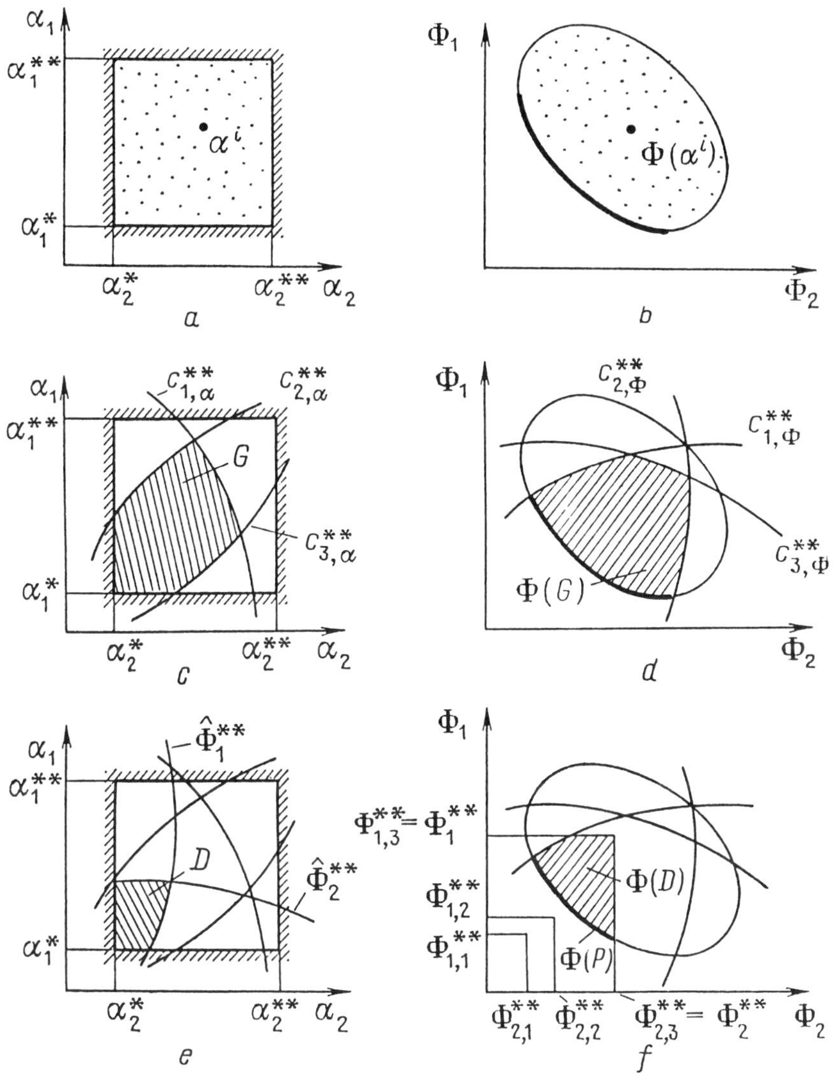

Figure 1–4 Procedure for determination of a feasible solutions set.

this, in some cases he will be ready to modify the constraints so that the "good" solutions would be found inside the feasible solutions set. The question is how to obtain such information.

Instead of the function $f_l(\boldsymbol{\alpha})$, whose constraints are not rigid (soft), we introduce an additional criterion $\Phi_{k+l}(\boldsymbol{\alpha})=f_l(\boldsymbol{\alpha})$, which we will call a pseudocriterion. However, to find the value of Φ_{k+l}^{**} one has to compile a test table containing $\Phi_{k+l}(\boldsymbol{\alpha})$. By using the aforementioned algorithm together with the new test

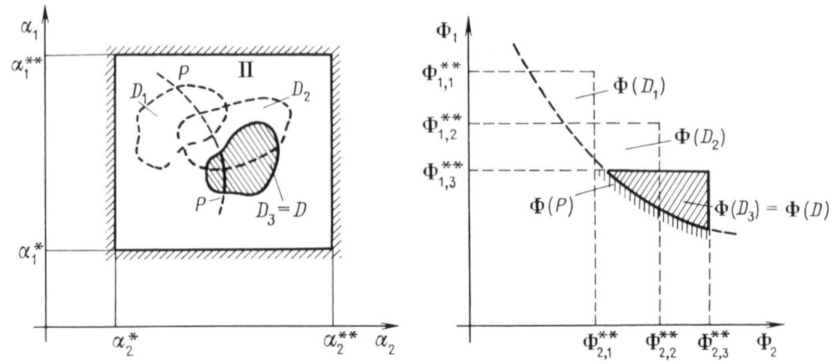

Figure 1–5 Procedure for constructing a feasible solutions set using the results of various dialogues.

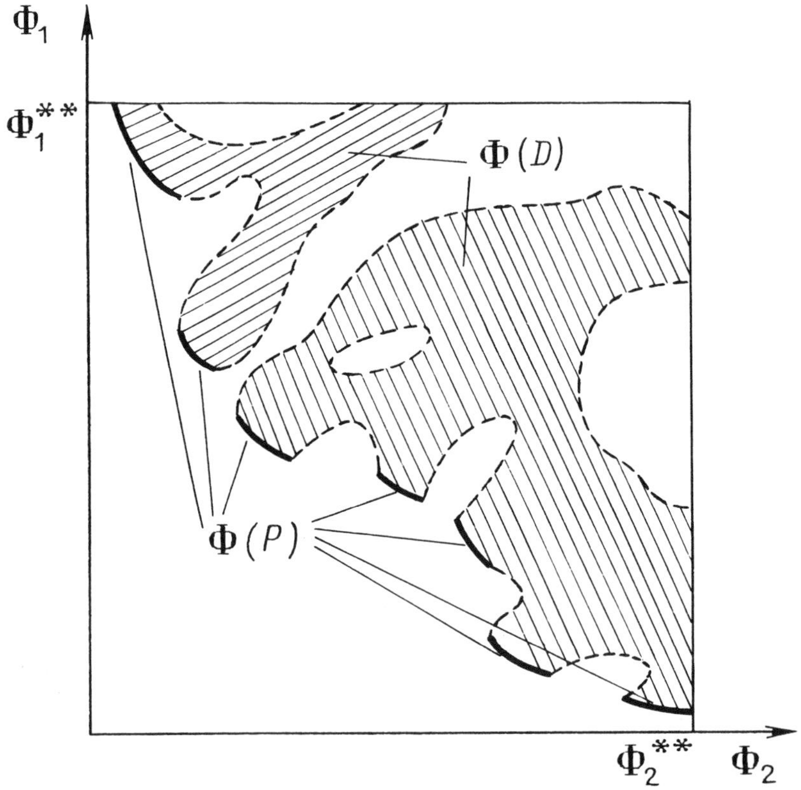

Figure 1–6 Disconnected feasible solutions set Φ (D).

table one can define Φ_{k+l}^{**} in a way preventing from the loss of interesting solutions.

Strange as it may seem, in solving engineering single-criterion problems involving soft functional constraints, one has to pass to multicriteria problems in order to find the feasible solutions set. This is due to the fact that Φ_ν^{**} may be determined correctly only upon analyzing the test table.

In the general case, in solving a problem with soft functional constraints, one has to find the set D taking all performance criteria into account, the functions $f_l(\boldsymbol{\alpha})$ being considered as pseudocriteria. In other words, one has to solve the problem with the constraints

$$\Phi_\nu(\boldsymbol{\alpha}) \leq \Phi_\nu^{**},\ \nu = 1,\ldots,k,k+1,\ldots,n.$$

It was already mentioned that in order to "avoid multicriteriality," attempts were made to transform all criteria except one into functional relationships with constraints of the form (1-2). It is clear that one cannot act in such a way because it can lead to a considerable reduction of the feasible solutions set. Whenever possible, the designer has to do just the opposite, viz, to transform the functional relationships into pseudocriteria and then reduce the problem solution to the analysis of the test table.

Investigation of Relations Between Criteria (Sobol' and Statnikov 1981)

The results of the parameter space investigation method may be used for constructing the correlation matrix $\| r_{\mu\nu} \|$ where $r_{\mu\nu}$ is the cross-correlation coefficient for criteria $\Phi_\nu(\boldsymbol{\alpha})$ and $\Phi_\mu(\boldsymbol{\alpha})$. The matrix allows estimating the extent of linear dependence between two criteria. For instance, if an element of the matrix $r_{\mu\nu} \approx 1$, $\mu \neq \nu$, then the criteria Φ_ν and Φ_μ are linearly related. Investigation of the matrix may be helpful for analyzing the feasible solutions set.

The Variations of the Design-Variable Constraints

In solving optimization problems one has to specify design-variable constraints $\alpha_j^{*(*)}$ correctly. However, this is not a simple matter as long as multivariable and multicriteria engineering problems of high dimensionality are considered. In Sections 1-4 and 1-6, we will show how this may be done.

Visualization of the Process of the Criteria and Design-Variable Spaces Investigation

This is an important process allowing the designer to grasp the very physical essence of a problem as well as to correct the mathematical model, the constraints, etc. In specifying boundaries $\alpha_j^{*(*)}$ it is useful to analyze the functions $\Phi_\nu(\boldsymbol{\alpha})$, since this allows us to decide whether the boundaries should be actually modified.

Let us consider the problem of optimal design of a four-cylinder automobile engine whose displacement volume is 1.6 *l*. Subjected to optimization were 18 criteria; 24 design variables were varied. Figures 1-7*a–c* show the values of criteria Φ_5, Φ_{14}, and Φ_{17} depending on α_3. Here α_3 is the pressure (in Kgf/cm^2) of the third piston ring on the cylinder wall; Φ_5 is the mean oil film thickness (in μm) under the third ring; Φ_{14} is mean heat conductivity (in W/m^2·deg) of the third ring; and Φ_{17} is the minimal oil film thickness (in μm) under the third ring. Criterion Φ_5 is to be minimized, while criteria Φ_{14} and Φ_{17} are to be maximized.

N=1024 trials were conducted, and 585 vectors were plotted satisfying the functional constraints. By analyzing the plots the following conclusions were drawn:

1. The system loses stability at $\alpha_3 \approx 4.40$. Therefore, in the subsequent analysis α_3^* was set equal to 5.0 (see Figs. 1-7*a* and *c*).
2. By analyzing the dependences shown in Figures 1-7*b* and *c* the designer was able to determine the value of α_3^{**}.

In more detail the correction of design-variable constraints is discussed in Sections 1-4, 1-6, etc.

The Required Number of Trials

As noted previously, unlike other optimization methods, the PSI method was devised not only to solve a problem, but also to help formulate it. Therefore, the number of trials N needed for constructing the feasible solutions and Pareto optimal sets depends to a great extent on how the problem is formulated.

Also, it should be noted that N depends on the class of functions subjected to optimization, the number of design variables being varied, the volume of the parallelepiped under investigation, and the functional and criteria constraints. In turn, the number of the functions may reach many dozens, and they may be differentiable, nondifferentiable, nonconvex, discrete, etc.

As a rule, the number of trials was determined on the basis of a nonformal analysis of the calculation results. Taken into account were the significance of the problem under consideration, the time available for obtaining the optimal solution, the quality of the mathematical model, the accuracy with which the criteria had to be calculated, etc.

The need for a large number of trials is predetermined by the following considerations. Since engineering problems are, as a rule, ill-posed, one has to correct the mathematical model, the initial boundaries of the design variables and the values of both functional and criteria constraints. Usually, 70–85% of the total number of trials are "spent" to formulate an optimization problem. After all the constraints have been determined, the optimal solution may be obtained

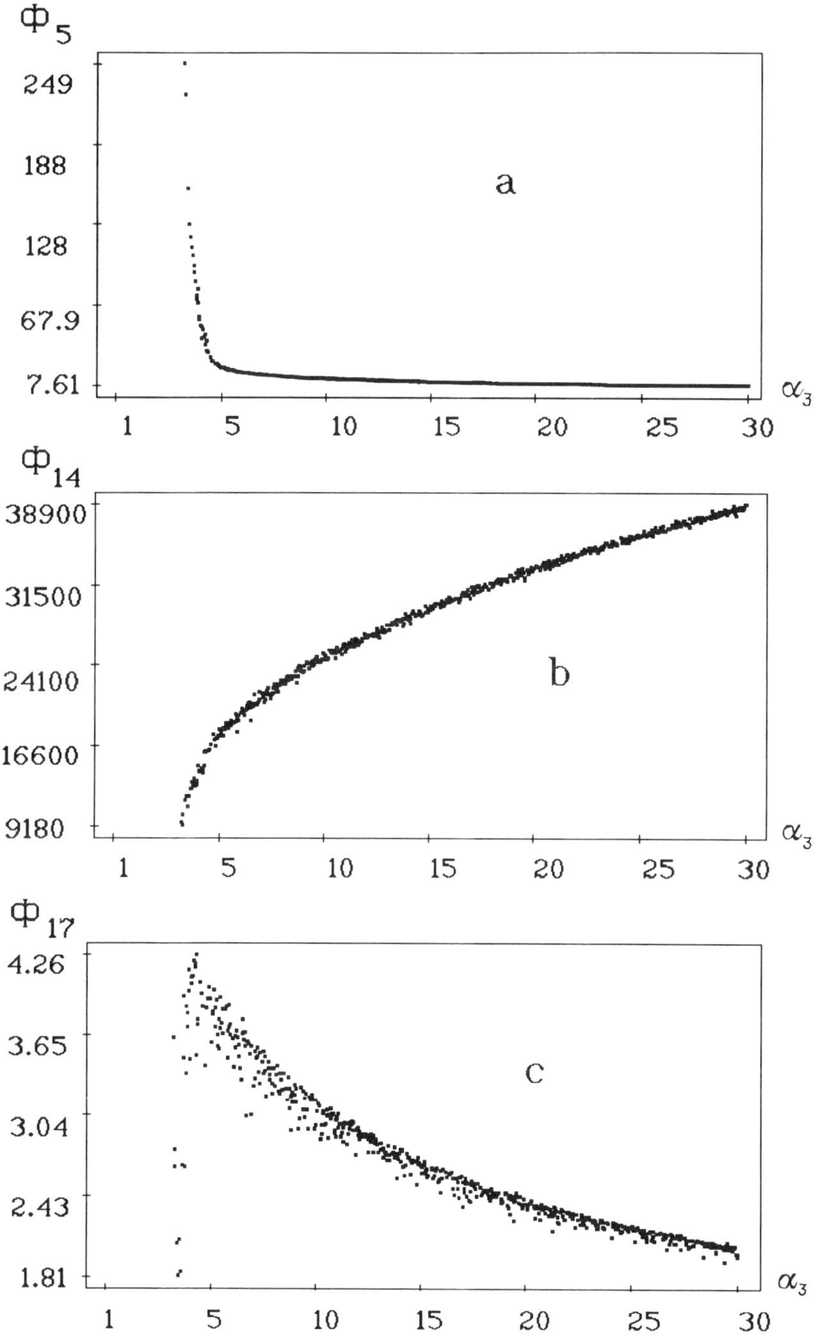

Figure 1–7 Dependences Φ_ν ($\boldsymbol{\alpha}$).

by running a comparatively small number of trials. Taking into account the importance of the problems to be solved and the expected effect of optimization, in the majority of cases the designer has to agree to a comparatively large number of trials, which sometimes may achieve several thousands.

This is the case primarily for batch and mass production of automobiles, machine tools, speed reducers, etc., which are manufactured in large quantities and for which the economy of metal and fuel as well as cost reduction are of paramount significance. In all these cases the efficiency of multicriteria optimization may be quite high, and it should be implemented with utmost care.

The experience accumulated in solving engineering problems shows that the time spent in formulating and solving an optimization problem is fully compensated by the results.

1-4. Example 1: The Choice of the Optimal Design Variables of an Oscillator

Let us consider a two-mass dynamic model (Statnikov and Uzvolok 1990) shown in Figure 1-8 where M_1 and M_2 are masses, K_1 and K_2 are stiffness coefficients, and C is a damping factor. Mass M_1 is acted upon by harmonic force $P\cos(\omega t)$ where P=2,000 N and ω=30 s^{-1}.

The equations of motion are given by

$$M_1X_1''+C(X_1'-X_2')+K_1X_1+K_2(X_1-X_2)=P\cos(\omega t),$$
$$M_2X_2''+C(X_2'-X_1')+K_2(X_2-X_1)=0.$$

The system contains five design variables, $\alpha_1=K_1$, $\alpha_2=K_2$, $\alpha_3=M_1$, $\alpha_4=M_2$, and $\alpha_5=C$. We specify upper and lower bounds for each of the design variables thus determining the parallelepiped Π_1:

$$1.1\cdot10^6\text{N/m}\leq\alpha_1\leq2.0\cdot10^6\text{N/m},$$
$$4.0\cdot10^4\text{N/m}\leq\alpha_2\leq5.0\cdot10^4\text{N/m},$$
$$950\text{ kg}\leq\alpha_3\leq1{,}050\text{ kg},$$
$$30\text{ kg}\leq\alpha_4\leq70\text{ kg},$$
$$80\text{ N}\cdot\text{s/m}\leq\alpha_5\leq120\text{ N}\cdot\text{s/m},$$

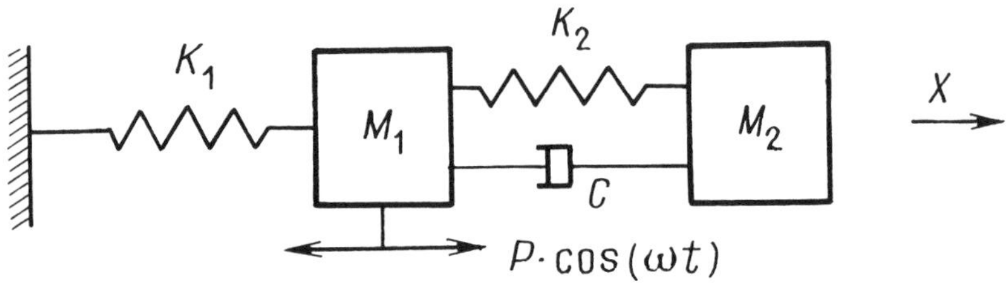

Figure 1–8 Two-mass dynamic model.

and three functional constraints (imposed on the total mass and partial frequencies):

$$f_1(\boldsymbol{\alpha})=\alpha_3+\alpha_4<1{,}100\ \text{kg},$$

$$33\ \text{s}^{-1}\leq f_2(\boldsymbol{\alpha})=p_1=\sqrt{\frac{\alpha_1}{\alpha_3}}\leq 42\text{s}^{-1},$$

$$27\ \text{s}^{-1}\leq f_3(\boldsymbol{\alpha})=p_2=\sqrt{\frac{\alpha_2}{\alpha_4}}\leq 32\text{s}^{-1}.$$

The following four performance criteria were optimized: the first mass oscillation amplitude X_{1d}, the total mass of the system M_1+M_2, and the dimensionless criteria X_{1d}/X_{1st}, and ω/p_1 characterizing certain dynamic properties of the system, where X_{1st} is the static deflection of mass M_1 caused by force P.

Both the constraints imposed on $f_1(\boldsymbol{\alpha})$ and the lower constraints on $f_2(\boldsymbol{\alpha})$ and $f_3(\boldsymbol{\alpha})$ are rigid. Conversely, the upper constraints on $f_2(\boldsymbol{\alpha})$ and $f_3(\boldsymbol{\alpha})$ may be slightly varied by the designer. Accordingly, functional dependences $f_2(\boldsymbol{\alpha})$ and $f_3(\boldsymbol{\alpha})$ should be transformed into pseudocriteria. They were denoted by Φ_1 and Φ_2, and the above performance criteria by Φ_3–Φ_6.

Analysis in Π_1

On the basis of the PSI method, 4,096 trials (N=4,096) were carried out in Π_1 (see Table 1-1 for a presentation of a fragment of the test table). The first portion embraces the 10 best models obtained for each of the six criteria, and the other portion corresponding to the end of the test table presents the three worst solutions for each of the criteria. For example, the 10 best solutions (models) in the first[5] criterion are presented in order of decreasing quality: 2,912; 3,072; 480; 1,216;...; 2,768; 1,280; the three worst ones being 1,791; 2,751; and 1,407. The best solution in the second criterion is 901. It is followed by 2,449; 2,402; 222;...; 3,555; the worst ones being 2,562; 3,007; 552; etc.

Dialogue 1: The following criteria constraints have been formulated: $\Phi_1^{**}=40$; $\Phi_2^{**}=32$; $\Phi_3^{**}=1.5$; $\Phi_4^{**}=1{,}030$; $\Phi_5^{**}=1.3$; $\Phi_6^{**}=0.75$.

Since no model has found itself in the feasible solutions set, three more dialogues were conducted.

Dialogue 2: $\Phi_1^{**}=41$; $\Phi_2^{**}=32$; $\Phi_3^{**}=2.0$; $\Phi_4^{**}=1{,}040$; $\Phi_5^{**}=1.5$; $\Phi_6^{**}=0.80$.

Thirty-five models have found themselves in the feasible solutions set.

Dialogue 3: $\Phi_1^{**}=42$; $\Phi_2^{**}=31$; $\Phi_3^{**}=1.9$; $\Phi_4^{**}=1{,}030$; $\Phi_5^{**}=1.45$; $\Phi_6^{**}=0.85$.

[5] Here representations $\boldsymbol{\alpha}^{2,912}$; $\boldsymbol{\alpha}^{3,072}$; $\boldsymbol{\alpha}^{480}$; $\boldsymbol{\alpha}^{1,216}$;...; $\boldsymbol{\alpha}^{2,768}$; $\boldsymbol{\alpha}^{1,280}$; and 2,912; 3,072; 480; 1,216;...; 2,768; 1,280 are equivalent.

Table 1–1

$\boldsymbol{\alpha}^i$	$\Phi_1(\boldsymbol{\alpha}^i)$	$\boldsymbol{\alpha}^i$	$\Phi_2(\boldsymbol{\alpha}^i)$	$\boldsymbol{\alpha}^i$	$\Phi_3(\boldsymbol{\alpha}^i)$	$\boldsymbol{\alpha}^i$	$\Phi_4(\boldsymbol{\alpha}^i)$	$\boldsymbol{\alpha}^i$	$\Phi_5(\boldsymbol{\alpha}^i)$	$\boldsymbol{\alpha}^i$	$\Phi_6(\boldsymbol{\alpha}^i)$
2,912	33.005	901	27.001	1,687	1.2366	3,003	982.11	1,120	1.1783	1,407	0.6552
3,072	33.013	2,449	27.014	2,703	1.2395	1,071	982.20	1,042	1.1845	2,751	0.6566
480	33.015	2,402	27.016	847	1.2489	3,843	982.42	1,687	1.1874	1,791	0.6574
1,216	33.024	222	27.020	2,231	1.2713	3,988	982.65	2,703	1.2075	3,199	0.6575
608	33.046	2,008	27.021	671	1.2798	3,185	984.00	3,203	1.2137	2,015	0.6577
1,664	33.048	1,481	27.021	1,503	1.2802	2,860	984.07	878	1.2142	63	0.6578
160	33.052	3,460	27.023	863	1.2924	2,142	984.36	280	2.2171	671	0.6582
2,192	33.075	3,447	27.024	167	1.3127	151	984.45	847	1.2198	3,423	0.6590
2,768	33.077	1,092	27.027	1,199	1.3159	1,208	984.46	1,425	1.2236	959	0.6594
1,280	33.107	3,555	27.027	3,743	1.3183	3,629	984.49	592	1.2280	2,367	0.6595
⋮	⋮	⋮	⋮	⋮	⋮	⋮	⋮	⋮	⋮	⋮	⋮
1,791	45.637	2,562	40.473	2,288	27.962	867	1,099.6	1,664	16.429	480	0.9087
2,751	45.691	3,607	40.489	1,664	29.739	1,176	1,099.8	1,592	16.745	3,072	0.9087
1,407	45.786	552	40.641	656	33.191	1,246	1,099.9	656	18.795	2,912	0.9090

Table 1–2

α^i	$\Phi_1(\alpha^i)$	$\Phi_2(\alpha^i)$	$\Phi_3(\alpha^i)$	$\Phi_4(\alpha^i)$	$\Phi_5(\alpha^i)$	$\Phi_6(\alpha^i)$
569	41.461	28.996	1.6263	1,018.4	1.3412	0.72358
1,425	40.488	29.238	1.5451	1,022.8	1.2236	0.74095
1,882	38.371	29.096	1.8115	1,022.6	1.2857	0.78184
2,325	41.942	28.485	1.6493	1,021.9	1.3958	0.71527
2,374	38.632	29.148	1.8256	1,024.9	1.3204	0.77655
2,753	40.520	29.101	1.6179	1,009.0	1.2633	0.74038
3,109	41.903	28.883	1.6530	1,012.0	1.3862	0.71593
3,361	40.559	29.605	1.6867	1,004.9	1.3211	0.73967

Twenty models have found themselves in the feasible solutions set.

Dialogue 4: $\Phi_1^{**}=42$; $\Phi_2^{**}=31$; $\Phi_3^{**}=1.85$; $\Phi_4^{**}=1,025$; $\Phi_5^{**}=1.4$; $\Phi_6^{**}=0.85$.

The feasible solutions set constructed using the PSI method contains eight models. Table 1-2 presents the values of the criteria for all eight feasible models, six of which are Pareto optimal. Models 1,425 and 3,361 were identified as the best ones (see Tables 1-3 and 1-4). Having analyzed the results the designer agreed to consider them as the final solution for the aforementioned parallelepiped Π_1.

One of the ways of correcting design-variable constraints requires the construction and analysis of histograms of the design-variables distribution over the

Table 1–3

	Results of investigations in Π_1		Results of investigations in Π_2	
Φ_ν	$\Phi(\alpha^{1,425})$	$\Phi(\alpha^{3,361})$	$\Phi(\alpha^{2,753})$	$\Phi(\alpha^{6,569})$
Φ_1	40.488	40.559	40.948	41.512
Φ_2	29.238	29.605	29.899	29.306
Φ_3	1.5451	1.6867	1.4539	1.3622
Φ_4	1,022.8	1,004.9	989.18	1,001.2
Φ_5	1.2236	1.3211	1.1352	1.1074
Φ_6	0.7410	0.7397	0.7326	0.7227

Table 1–4

	Results of investigations in Π_1		Results of investigations in Π_2	
α_j	$\alpha^{1,425}$	$\alpha^{3,361}$	$\alpha^{2,753}$	$\alpha^{6,569}$
α_1	$1.584\cdot10^6$	$1.567\cdot10^6$	$1.562\cdot10^6$	$1.626\cdot10^6$
α_2	$4.840\cdot10^4$	$4.613\cdot10^4$	$5.164\cdot10^4$	$4.955\cdot10^4$
α_3	966.16	952.27	931.38	943.52
α_4	56.621	52.627	57.803	57.690
α_5	82.012	91.104	80.974	70.775

ranges of their variation. For vectors of the feasible solutions set D, the histograms of distribution of their coordinates over the variation ranges of respective design variables are constructed. For each design variable, the range of its variation is divided into n equal parts (segments). If any of these segments contains at least one trial from the feasible solutions set, these segments are marked by a black rectangle. The domain of feasible values of the design variable α_j is denoted by $[\alpha^{*}_{D_j}; \alpha^{**}_{D_j}]$ (see Fig. 1-9).

Figure 1-9 demonstrates histograms for the distribution of the eight feasible models over the ranges of the design variables. The interval of variation of each design variable was divided into 10 equal segments. Those segments containing the feasible solutions $[\alpha^{*}_{D_j}; \alpha^{**}_{D_j}]$ are marked with black rectangles.

Figure 1-9 shows that the fourth, sixth, and seventh segments in $[\alpha^{*}_{1}, \alpha^{**}_{1}]$ incorporate models 1,882 and 2,374; 1,425, 2,753, and 3,361; and 569, 2,325, and 3,109, respectively. The remaining segments contain no solutions. Figure 1-9 presents the feasible models distribution in Π_1 for the remaining four design variables.

From Figure 1-9 it follows that the feasible models for the second, third, and fifth design variables lie near the boundaries of variation of α^{**}_{2}, α^{*}_{3}, and α^{*}_{5}, respectively. Intervals $[\alpha^{*}_{2}, \alpha^{*}_{D_2})$, $(\alpha^{**}_{D_3}, \alpha^{**}_{3}]$ are "holes" caused by functional and criteria constraints. As to the aforementioned design variables, the designer may agree to revise the original constraints α^{**}_{2}, α^{*}_{3}, and α^{*}_{5} if the concessions would result in improving the values of the criteria.

For example, it is important to know how much the feasible and Pareto optimal solutions would be improved in the basic particular performance criteria if the initial constraints $[\alpha^{*}_{j}, \alpha^{**}_{j}]$ are replaced by new ones $[\hat{\alpha}^{*}_{j}, \hat{\alpha}^{**}_{j}]$.

In the general case, substitution of Π_1 by new parallelepipeds may result in the disappearance or shift of the "holes" owing to the formation of new combinations of the design variables.

Let us consider the construction of a new parallelepiped Π_2 for which $\hat{\alpha}^{*}_{j} = \alpha^{*}_{j} - \delta^{*}_{j}$ and $\hat{\alpha}^{**}_{j} = \alpha^{**}_{j} + \delta^{**}_{j}$. Here δ^{*}_{j} and δ^{**}_{j} are the concessions the designer has made with respect to the jth design variable (the jth coordinate). As noted above, the final approval depends on the values of the performance criteria attained within the new parallelepiped as compared with the original one. The boundaries of the remaining design variables $\hat{\alpha}^{*(*)}_{j}$ stay the same as in Π_1.

For the parallelepiped Π_2, $\Pi_1 \subseteq \Pi_2$, we have:

$$
\begin{aligned}
&\hat{\alpha}^{*}_{1} = \alpha^{*}_{1}, && \hat{\alpha}^{**}_{1} = \alpha^{**}_{1};\\
&\hat{\alpha}^{*}_{2} = \alpha^{*}_{2}, && \hat{\alpha}^{**}_{2} = \alpha^{**}_{2} + \delta^{**}_{2};\\
&\hat{\alpha}^{*}_{3} = \alpha^{*}_{3} - \delta^{*}_{3}, && \hat{\alpha}^{**}_{3} = \alpha^{**}_{3};\\
&\hat{\alpha}^{*}_{4} = \alpha^{*}_{4}, && \hat{\alpha}^{**}_{4} = \alpha^{**}_{4};\\
&\hat{\alpha}^{*}_{5} = \alpha^{*}_{5} - \delta^{*}_{5}, && \hat{\alpha}^{**}_{5} = \alpha^{**}_{5}.
\end{aligned}
$$

The boundaries of parallelepiped Π_2 were specified by the inequalities:

$$1.1 \cdot 10^6 N/m \leq \alpha_1 \leq 2.0 \cdot 10^6 N/m;$$
$$4.0 \cdot 10^4 N/m \leq \alpha_2 \leq 5.3 \cdot 10^4 N/m;$$
$$930\, kg \leq \alpha_3 \leq 1{,}050\, kg;$$
$$30\, kg \leq \alpha_4 \leq 70\, kg;$$
$$70 N \cdot s/m \leq \alpha_5 \leq 120 N \cdot s/m.$$

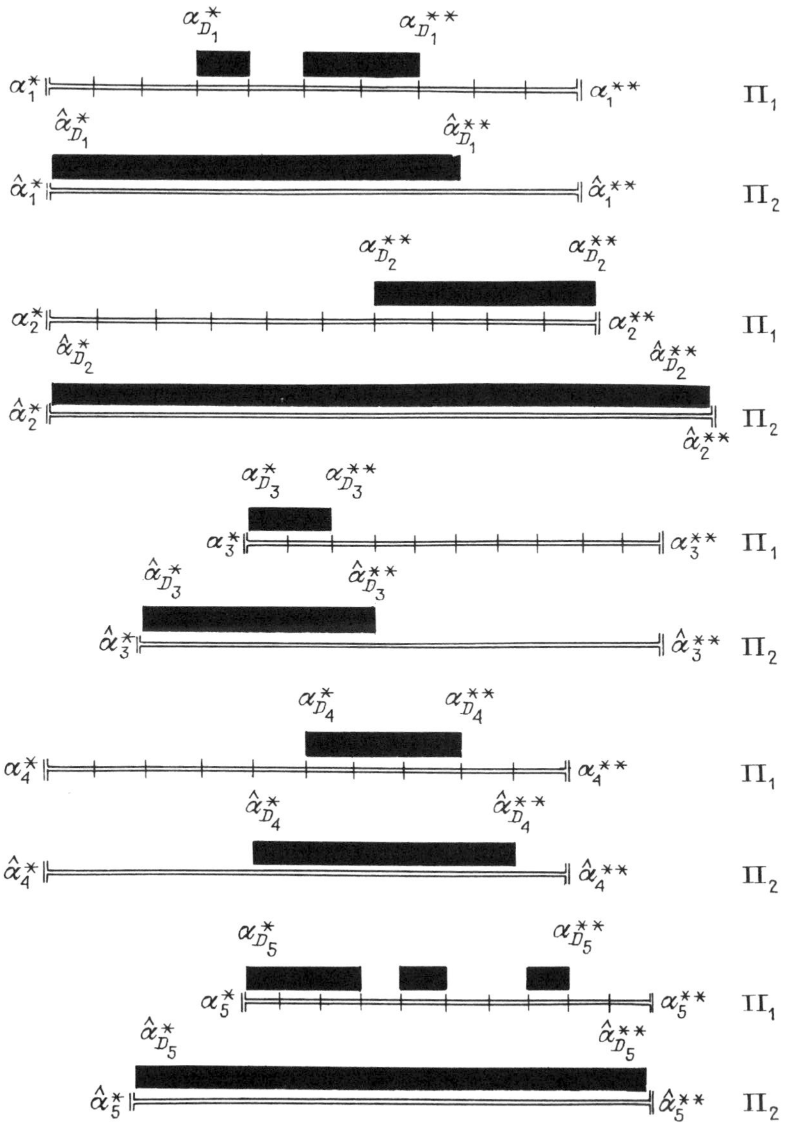

Figure 1–9 Histograms of the feasible solutions distribution.

Figure 1-9 shows the old, $\alpha_j^{*(*)}$, and the new $\hat{\alpha}_j^{*(*)}$, design-variable boundaries forming parallelepipeds Π_1 and Π_2; also shown are the boundaries of the feasible solutions $\hat{\alpha}_{D_j}^{*(*)}$ in Π_2. For the previous functional and criteria constraints (see Dialogue 4), the feasible solutions set contained 67 solutions, while Π_1 included only eight. In Π_2, 8,192 trials were conducted; the best results are presented in Tables 1-3 and 1-4.

The following conclusions may be made:

1. The analysis of the feasible solutions in parallelepiped Π_1 (see Table 1-2) has shown that for the initial functional constraints imposed on $f_1(\boldsymbol{\alpha})-f_3(\boldsymbol{\alpha})$, the feasible solutions set would contain only two vectors, 1,882 and 2,374. The remaining six vectors have found themselves in the feasible solutions set owing to the transformation of the functional dependence $f_2(\alpha)$ into pseudocriterion Φ_1.
2. The analysis of Tables 1-3 and 1-4 has shown that the results of optimization in Π_1 were improved by correcting the design-variable constraints for all four performance criteria $\Phi_3,\ldots,\Phi_6$. Thus, vectors 2,753 and 6,569 from Π_2 are undoubtedly better than solutions 1,425 and 3,361 from Π_1.

The design variables of the most preferred Pareto optimal vectors in Π_2 do not belong to Π_1.

The need for correcting both design-variable and functional constraints and for determining the criteria constraints in the interactive mode, is typical for the majority of applied optimization problems, especially those of optimum design. By using the PSI method one can readily solve the problems of finding the feasible solutions set.

1-5. Example 2: Automotive Valve Gear Design

The structural schematics of the valve gear used in the present-day automobile internal combustion engines with a camshaft in the cylinder block, are rather simple (see Fig. 1-10). Nevertheless, the choice of the mechanism's design variables is one of the most complicated problems one encounters in designing an automobile engine. Conventional design methods fail to satisfy all the conflicting requirements satisfactorily. As a result, the operational development of an engine takes more time and becomes more expensive. However, the problem may be solved efficiently using the PSI method (Genkin et al. 1983).

The motion of the links of a dynamic model used for estimating and choosing the design variables of the valve gear of the majority of modern automobiles is described by the equation of longitudinal oscillation of the valve spring coils (Korchemnyi 1981)

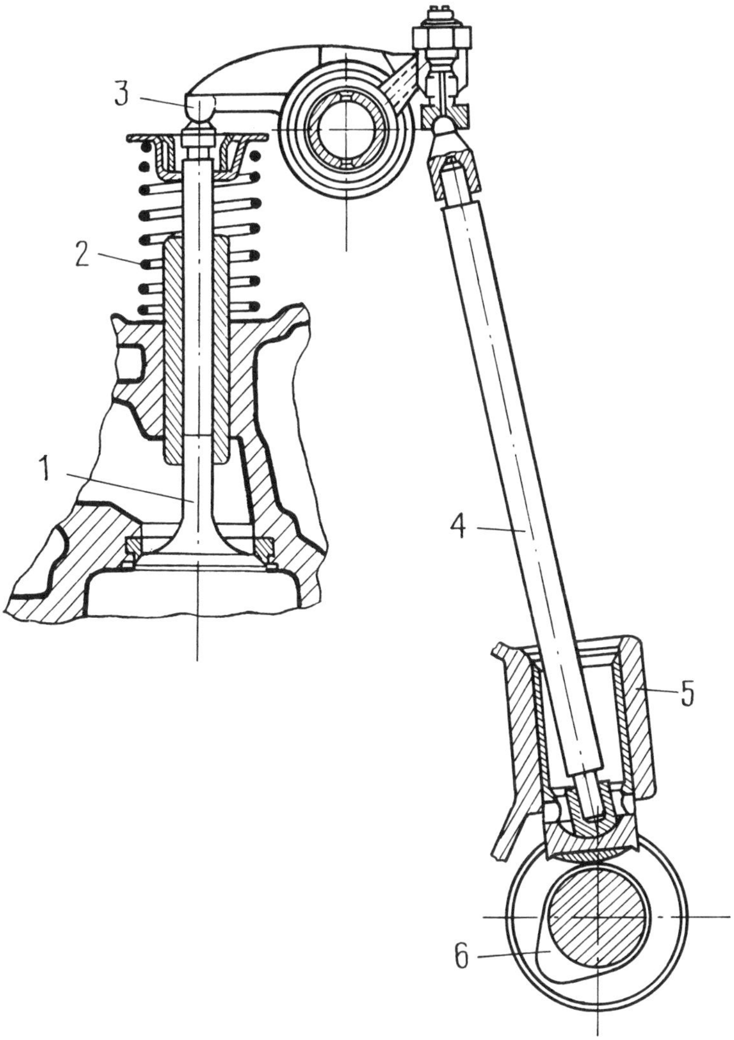

Figure 1–10 Automotive valve gear. (1) Valve; (2) spring; (3) rocker; (4) push rod; (5) tappet; and (6) cam.

$$\frac{\partial^2 u}{\partial \phi^2}+\frac{2\mu}{\omega}\frac{\partial u}{\partial \phi}+\frac{F}{\omega^2}\,\text{sign}\,\frac{\partial u}{\partial \phi}=\frac{a^2}{\omega^2}\frac{\partial^2 u}{\partial \xi^2}$$

subject to the following boundary conditions

$$u(0,\phi)=0,\; u(1,\phi)=y(\phi)$$

where $u(\xi,\phi)$ is the spring's displacement from the static-equilibrium state when the valve is shut; μ is the damping factor (due to viscous drag) of the oscillations of the spring; F is the dry friction force between the spring coils and the damper; a is the speed with which disturbances travel along the spring (along the ξ coordinate); ϕ is the rotation angle of the cam; ω is its angular velocity; and y is the valve displacement.

The spring vibration damper is made of a steel ribbon whose width is b and thickness is h and which is actually a spring mounted within the valve spring with interference δ.

To determine $y(\phi)$ one has to use the equation of motion of the reduced mass of the valve, M:

$$z''+\frac{b}{M\omega}z'+\frac{C}{M\omega^2}z=x''+\frac{F_0+F_r}{M\omega^2}+\frac{c}{M\omega^2}\frac{\partial u(1,\phi)}{\partial \xi}$$

where $z=x-y$ is the valve drive elastic deformation; $x(\phi)$ is the tappet displacement reduced to the valve; b is the valve gear conditional total viscous drag coefficient; F_0 is the valve spring preload; F_r is the force due to the cylinder gas pressure exerted onto the valve head; c is the valve spring stiffness; and C is the stiffness coefficient of the valve drive.

The valve is initially at rest: $y(0)=y'(0)=0$. The perfection of the valve gear design is estimated using the following performance criteria.

Criterion Φ_1 characterizes the maximum gas flow rate through the valve that is proportional to average lift y_m. The larger y_m, the higher the engine power and its economical operation. The maximum y_m is practically equivalent to the maximum tappet average lift, which may be found at the stage of the kinematic calculation of the mechanism. This quantity is used as the performance criterion

$$\Phi_1=x_m=\frac{1}{\phi_f-\phi_i}\int_{\phi_i}^{\phi_f} x\,d\phi$$

where ϕ_i and ϕ_f are the cam rotation angles corresponding to the beginning and termination of the theoretical valve lift (determined when ignoring the drive's elastic deformations). Note that the attempts to increase x_m may have an adverse effect on other criteria characterizing the possibilities of a practical realization of the mechanism and its operability.

Criterion Φ_2 is numerically equal to the minimum radius of the flat tappet for which the contact line length (equal to the cam width l) does not decrease for any relative position of the cam and the tappet:

$$\Phi_2 = \sqrt{(x'_{max})^2 + \left(\frac{l}{2} + e\right)^2}.$$

Here, e is the tappet axis displacement with respect to the cam center.

Criterion Φ_2 should be minimized, since this would allow decreasing the overall dimensions of the cam pair as well as its mass and inertia.

Criteria Φ_3 and Φ_4 are the extreme values of the tappet acceleration analogue[6]: $\Phi_3 = x''_{max}$, $\Phi_4 = |\ x''_{min}\ |$. Commonly, x''_{min} corresponds to the top of the cam. Since Φ_3 is approximately proportional to the maximum force of inertia acting onto the cam, and Φ_4 to the maximum force of inertia applied to the valve spring, both criteria should be minimized.

Criterion Φ_5 is equal to the maximum static contact stress at the cam top:

$$\Phi_5 = A\sqrt{\frac{(F_0 + cx_{max})i^2}{l(x_{max} + x''_{min} + r_0)}}$$

where A is a factor depending on the Young moduli of the cam and tappet materials; i is the valve rocker arm transmission ratio; and r_0 is the radius of the initial circle of the equivalent cam. In designing valve gears, Φ_5 is to be made as small as possible. For a plane flat tappet this quantity depends mainly on the cam profile.

Criteria Φ_6 and Φ_7 allow preliminary estimation of the correct choice of the valve gear design variables. Criterion $\Phi_6 = m_v + m_t + m_r$ where m_v, m_t, and m_r are the masses of the valve, the tappet, and the push rod respectively. Φ_6 should be made as small as possible.

Besides decreasing the specific quantity of metal, this helps to decrease the valve's reduced mass, thus affecting the dynamic properties of the valve gear. The effect may be evaluated with the help of criterion $\Phi_7 = (2\pi)^{-1}\sqrt{(C/M - (b^2/4M^2)}$, which is the valve gear natural frequency and should be maximized.

Criterion Φ_8 characterizes the valve spring fatigue safety margin. The larger Φ_8, the less the probability of the spring's failure.

Criterion Φ_9 represents the maximum elastic deformation of the valve drive z_{max}, which is proportional to the maximum force applied to the mechanism. The latter operates the better the smaller Φ_9.

Criterion Φ_{10} is equal to the absolute value of an analogue of the valve velocity

---

[6]Here, the acceleration analogue (velocity analogue) is the second (first) derivative of the tappet displacement with respect to angle of the cam rotation.

at the instant it hits the seat, $|y'_{im}|$. It defines both the impact intensity and the maximum stresses in colliding parts. To a considerable extent, this criterion also characterizes the possibility of repeated opening of the valve due to its bouncing after the strike against the seat. This unfavorably affects the valve's longevity. Criterion Φ_{10} should be minimized.

The longevity of a cam pair greatly depends on the lubrication condition, which is characterized by the derivative p' of the "lubricant number" p calculated at the cam rotation angle at which p changes sign. At this point, p' must satisfy the inequality $p' \geq 540/\pi$. Therefore, the following functional constraint was used:

$$\Phi_{11} = \frac{|x'(\phi) + 2x'''(\phi)|}{i} \geq \frac{540}{\pi} \text{ for } |r_0 + x(\phi) + 2x''(\phi)| = 0.$$

The functional constraint $\Phi_{12} = z_{\min} \geq 0$ assures absence of breakings in the kinematic chain of the valve gear.

Of greatest importance for the solution of optimization problems are the valve gear design variables defining the tappet's law of motion $x(\phi)$ (see Fig. 1-11). In constructing the law one should use piecewise-polynomial functions only, regarding as design variables the values of the corresponding function and its three derivatives at the points where either $x(\phi)$ or $x'(\phi)$ or $x''(\phi)$ attains extreme values. Curve $x(\phi)$ is composed of six arcs, three of which correspond to the valve lift and the remaining three to its downward travel. The aforementioned design variables specify the conditions for matching the arcs that form the $x(\phi)$ curve and assure that it is uniquely defined. If the curve is symmetric, then one has to specify 13 design variables (see Table 1-5).

It was found that not all the combinations of the design variables assure the desired behavior of the curve $x(\phi)$ and its derivatives. Therefore, a functional constraint has been introduced that requires that the sign of $x'''(\phi)$ remain constant within each portion of the $x(\phi)$ curve. Table 1-5 presents all the design variables chosen in formulating the optimization problem.

The problem was solved in several stages. At the stage of preliminary calculation, the expediency of transforming functional constraints Φ_{11} and Φ_{12} into pseudocriteria was revealed, and the validity of the functional constraints imposed on x_i''', $i=1, 2, 3$ was checked. Of the total number of 2,048 models (points of design-variable space) 40 were included into the test table. Thus, the efficiency of the tappet's law of motion specified in the form of a piecewise-polynomial function was demonstrated. The laws of motion corresponding to the models presented in the test table are characterized by more favorable profiles of the tappet velocity and acceleration curves as compared with the initial model (the prototype) $\boldsymbol{\alpha}^1$.

Optimization of the intake valve design variables was reduced to carrying out

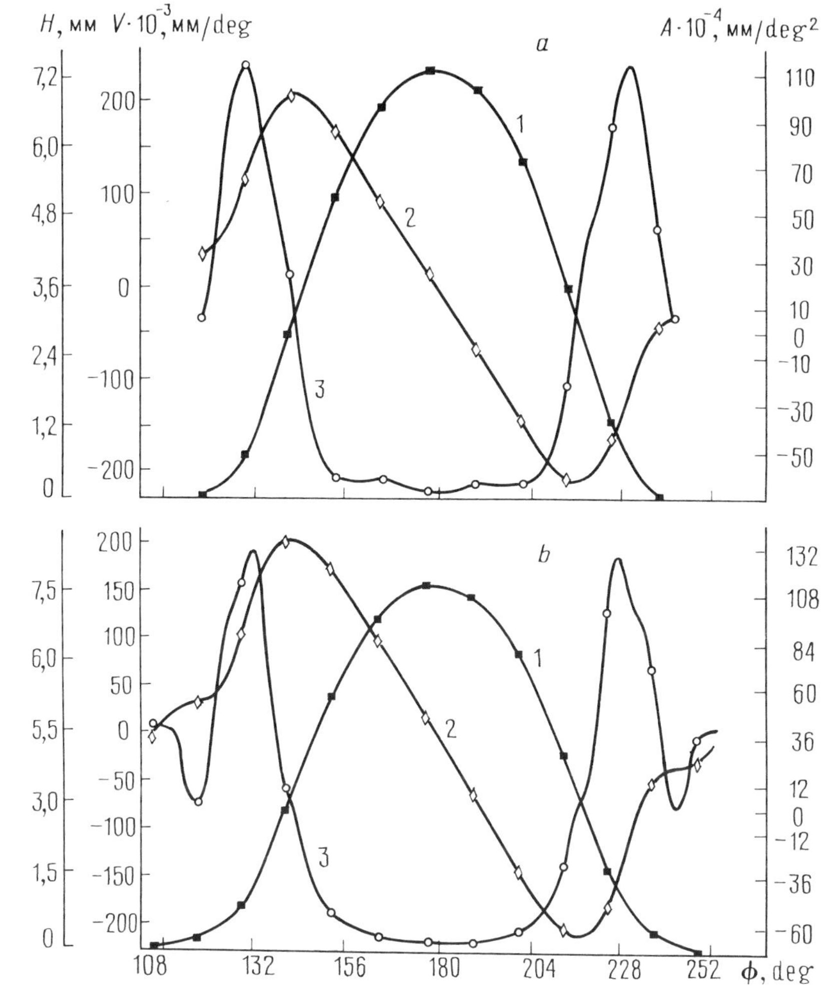

Figure 1–11 Law of tappet motion. (1) Curve of the tappet lift; (2) curve of the tappet velocity analogue; and (3) curve of the tappet acceleration analogue.

seven numerical experiments on a computer. These differed in the number of design variables being varied. In the first experiment α_{15}, α_{17}, α_{18}, and α_{23} were varied. In the second experiment, the values of these design variables remained unchanged and equal to the values of the corresponding design variables in the initial model $\boldsymbol{\alpha}^1$. In the seventh experiment α_1 was also kept constant.

For the design-variable values of the initial model, we took the design variables of the presently available intake valve drive of the valve gear of a V8, 180 hp

Table 1–5

Designation	Design variables	Dimension	Variation boundaries
α_1	Tappet lift at the start of the 1st part ($x=x_{max}$)	mm	7; 7.6
α_2	Analogue of tappet acceleration at the start of the first part	mm/deg^2	−0.0077; −0.0057
α_3	Angular extent of the first part	deg	36; 36
α_4	Tappet lift at the end of the first part ($x''=0$)	mm	2.9; 3.5
α_5	Analogue of tappet velocity at the end of the first part	mm/deg	−0.238; −0.176
α_6	Third derivative of the tappet lift with respect to the camshaft rotation angle at the end of the first part	mm/deg^3	0.00117; 0.00143
α_7	Tappet lift at the start of the second part	mm	0.84; 0.9
α_8	Analogue of tappet velocity at the start of the second part	mm/deg	0.106; 0.15
α_9	Analogue of tappet acceleration at the start of the second part ($x''=x''_{max}$)	mm/deg^2	0.0102; 0.0138
α_{10}	Angular extent of the second part	deg	13; 13
α_{11}	Analogue of tappet velocity at the end of the third part ($x=0$)	mm/deg	−0.0414; −0.0306
α_{12}	Third derivative of the tappet lift with respect to the camshaft rotation angle at the end of the third part	mm/deg^3	$-12\cdot10^{-6}$; $-8\cdot10^{-7}$
α_{13}	Angular extent of the third part	deg	13; 13
α_{14}	Valve mass	g	210; 210
α_{15}	Push rod mass	g	114; 154
α_{16}	Tappet mass	g	135; 135
α_{17}	Diameter of the valve spring wire	mm	4.5; 5.5
α_{18}	Average diameter of the valve spring	mm	34; 36
α_{19}	Radial fit between the spring and the damper	mm	1; 1.5
α_{20}	Damper spring ribbon width	mm	4; 9
α_{21}	Damper ribbon thickness	mm	0.5; 1.5
α_{22}	Preload force of the valve spring	kgf	29; 35
α_{23}	Number of effective coils of the valve spring	—	4.25; 5.25
α_{24}	Number of effective coils of the damper spring	—	3; 4

automobile diesel. It was necessary to construct the feasible solutions and Pareto optimal sets and find within the latter a model that would be better than the initial model $\boldsymbol{\alpha}^1$.

The first-experiment design-variable domain is presented in Table 1-5. A total of 2,048 trials were carried out. The performance criteria were calculated for the trials that were not discarded owing to the functional constraint in x_i'''. Of the 2,048 trials, 40 models have satisfied the constraint and were included into the test table (omitted here because of its large size). The criteria constraints were found using the PSI method. Those of the 40 models that satisfied the criteria constraints have formed the feasible solutions set D. There are six such models that belong to the Pareto optimal set. The fact that of the 2,048 trials only six appear in the feasible solutions set is explained by the rigid functional and criteria constraints, which, evidently, cut off comparatively small disconnected domains from the parallelepiped. Upon analyzing the Pareto optimal models table the designer could readily define the most preferred one. It proved to be model $\boldsymbol{\alpha}^{224}$. Being insignificantly worse than model $\boldsymbol{\alpha}^1$ in criteria Φ_1 and Φ_7, the model $\boldsymbol{\alpha}^{224}$ exceeds $\boldsymbol{\alpha}^1$ in all other criteria. Thus, the fatigue safety margin increased by 14%, the contact stresses at the cam top decreased by 10%, and the impact velocity decreased by a factor of almost 2.2 to become less than the theoretical value defined by the cam profile. Besides, the maximum positive and negative accelerations have decreased by 9% and 4%, respectively.

In the experts' opinion, model $\boldsymbol{\alpha}^{224}$ is, on the whole, undoubtedly better than the initial $\boldsymbol{\alpha}^1$ model. This conclusion was confirmed by comparing the calculated kinematics and dynamics of the mechanisms corresponding to models $\boldsymbol{\alpha}^1$ and $\boldsymbol{\alpha}^{224}$. The model $\boldsymbol{\alpha}^{224}$ acceleration curve is much smoother than that of model $\boldsymbol{\alpha}^1$ (compare Figs. 1-11*b* and *a*), the smoothness affecting the mechanism's operability favorably. The maximum stresses in the model $\boldsymbol{\alpha}^{224}$ valve spring are smaller by 20% than in the case of model $\boldsymbol{\alpha}^1$ (see Figs. 1-12*b* and *a*, respectively). The discontinuity of the kinematic chain due to negative tappet accelerations is practically absent, and the premature contact of the valve and the seat is eliminated.

In the second experiment, 15 design variables were varied. Similarly to the first experiment, 40 models entered the test table, and the feasible solutions and Pareto optimal sets contain practically the same models as in the first experiment. The values of all criteria except Φ_5 and Φ_{10} are almost equal for the first and second experiments. However, the values of criteria Φ_5 and Φ_{10} obtained in the first experiment are somewhat better.

By comparing the results of optimization in the two experiments we conclude that, if possible from the viewpoint of manufacture, the design of the push rod and the valve spring should be modified in accordance with model $\boldsymbol{\alpha}^{224}$ obtained in the first experiment. However, it is worth mentioning that the design was mostly improved due to the modification of the tappet's law of motion, which for model $\boldsymbol{\alpha}^{224}$ was the same in both experiments. Therefore, the use of model

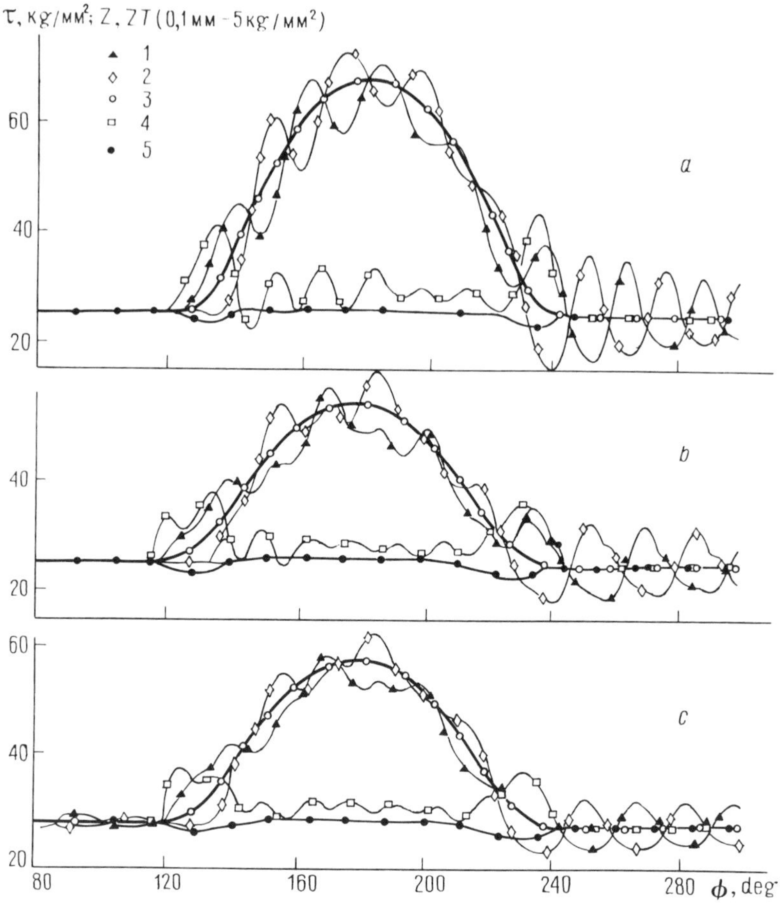

Figure 1–12 Dynamics of a valve gear. (1) Corresponds to the stresses (τ) in a moving coil; (2) corresponds to the stresses (τ) in a stationary coil (3) static stress (τ); (4) valve drive deformation (*Z*); and (5) conditional minimum allowable deformation of the valve drive (*ZT*).

$\boldsymbol{\alpha}^{224}$ constructed in the course of the second experiment (in which the specific features of mass production were taken into account to a greater extent) gives practically the same results.

The effect of the variation of design-variable constraints on the performance criteria was analyzed in the subsequent four experiments.

Table 1-6 shows the results of the intake and exhaust valves optimization.

Model $\boldsymbol{\alpha}^{224}$ surpasses model $\boldsymbol{\alpha}^{1}$ in all the criteria, except Φ_1, which, however, is one of the most important performance criteria. Therefore, the seventh experiment was conducted with the objective of improving Φ_1 without a considerable

Table 1–6

Valves	Experiment No.	Best models	Criteria Φ_1 (mm)	Φ_2 (mm)	Φ_3 (mm/deg^2)	Φ_4 (mm/deg^2)	Φ_5 (kgf/mm^2)	Φ_6 (g)	Φ_7 (Hz)	Φ_8	Φ_9 (mm)	Φ_{10} (mm/deg)
Intake valve	1	$\boldsymbol{\alpha}^{224}$	3.771	13.348	0.0110	0.0065	57.992	470	650.32	1.785	0.2055	0.1806
	2	$\boldsymbol{\alpha}^{224}$	3.771	13.348	0.0110	0.0065	60.155	479	665.96	1.789	0.1977	0.2127
	7	$\boldsymbol{\alpha}^{1663}$	3.938	14.002	0.0106	0.0065	59.380	479	665.96	1.551	0.2042	0.2547
	Initial model		3.908	14.033	0.0120	0.0067	64.445	479	665.96	1.566	0.2098	0.3901
Exhaust valve	8	$\boldsymbol{\alpha}^{1}$	4.062	13.936	0.0115	0.0066	60.429	459	689.37	1.634	0.1729	0.2173
	9	$\boldsymbol{\alpha}^{701}$	4.138	13.694	0.0124	0.0063	55.335	459	689.37	1.573	0.1989	0.1507
	Initial model		4.062	13.936	0.0115	0.0066	60.429	459	689.37	1.634	0.1729	0.2173

deterioration of the remaining criteria as compared with model $\boldsymbol{\alpha}^{224}$. The analysis of the test table has shown that all the criteria cannot be improved simultaneously, since, for instance, Φ_1 and Φ_8, Φ_9, Φ_{10} are conflicting criteria. However, the analysis of the Pareto optimal models has shown that the deterioration of Φ_1 in model $\boldsymbol{\alpha}^{224}$ as compared with model $\boldsymbol{\alpha}^1$ is due to the value of design variable α_1 in $\boldsymbol{\alpha}^{224}$ being smaller than in model $\boldsymbol{\alpha}^1$. It was decided to put $\alpha_1=7.3$ as in model $\boldsymbol{\alpha}^1$. After that, 2,048 trials were conducted, which allowed us to find a model possessing the desired properties. This is model $\boldsymbol{\alpha}^{1,663}$ (see Table 1-6 and Fig. 1-12*c*). Actually, this model improves the initial model $\boldsymbol{\alpha}^1$ in all criteria. Unlike models $\boldsymbol{\alpha}^1$ and $\boldsymbol{\alpha}^{224}$, its kinematic chain stays continuous. Besides, for model $\boldsymbol{\alpha}^{1,663}$ both the valve drive maximum deformation and the intensity of the valve spring coils vibration are smaller.

Thus, despite the fact that the initial models $\boldsymbol{\alpha}^1$ for both the intake and the exhaust valves corresponded to sufficiently good designs, and the boundaries of the design variables, taking the mass production specific features into account, were rather narrow, the use of the method of multicriteria optimization has allowed a substantial improvement in the kinematic and dynamic characteristics of the mechanism. Models $\boldsymbol{\alpha}^{224}$, $\boldsymbol{\alpha}^{1,663}$, and $\boldsymbol{\alpha}^{701}$ notably surpass the corresponding models $\boldsymbol{\alpha}^1$.

1-6. Specific Features of the Optimization Problems Formulation Using Finite Element Models

The finite element method (FEM) is widely used in numerous engineering problems of fluid mechanics, heat transfer, dynamics, strength, etc. However, the specific features of the problems make it necessary to modify the multicriteria formulation of optimization problems (see Section 1-1), since some basic criteria cannot be formalized. At the same time, without allowing for unformalizable criteria one cannot guarantee correct results. Usually, the criteria are related to optimal product manufacture technology, aesthetics, and similar aspects. As a rule, unformalizable criteria may be taken into account in analyzing the geometrical shapes of parts, units, structures, etc.

Problem Formulation and Its Specific Features (Statnikov et al. 1993)

Let us consider a finite element model of an object to be designed and a system of design-variable, functional, and criteria constraints (1-1)–(1-3). We define $\overline{D}$ as a set of vectors $\boldsymbol{\alpha}^i$ satisfying the constraints. Note that $\overline{D}$ is determined using formalizable criteria $\Phi_1,\ldots,\Phi_k$ and functional dependences.

Let the set $\overline{D}$ contain p elements for each of which the geometrical shape of the object under consideration may be generated. In visualizing and analyzing the set, the designer takes the remaining (unformalizable) criteria $\Phi_{k+1},\ldots,\Phi_{k+m}$ into account, that is by considering the geometry, he tries to find whether an element belonging to $\overline{D}$ is actually feasible. This results in the construction of a new feasible solutions set $D, D \subset \overline{D}$, which is often much smaller than $\overline{D}$.

Then the optimization problem reduces to finding the optimal vector $\boldsymbol{\alpha}^0$ on the feasible solutions set D. As is well-known, the optimal solution is commonly sought on the Pareto optimal set $\overline{P} \subset \overline{D}$.

In the problems under consideration the geometrical shape of an object is calculated using the design variables of each of the points $\boldsymbol{\alpha}^i$, $i=1,\ldots,N$. The finite element model must be modified accordingly. Since the number of trials is large, the model should be modified automatically. This may be done by using various methods for modifying the shapes of the curves and surfaces (Bezier 1987) as well as by using automatic finite element mesh generators.

An Example of Formulation and Solution of an Optimization Problem

Figure 1-13 shows the structure subjected to optimization. This is a plate rigidly fixed along the contour 1-2 and freely supported along the contour 3-4. The loads are applied along the contour 5-6 and are represented in the form of distributed forces whose intensity $q_x(x)$, $q_y(x)$, $q_z(x)$ is specified in such a way that the resultant forces and moments are not affected by a variation in the contour length.

The designers have defined the following performance criteria:

$$\Phi_1(\boldsymbol{\alpha})=\omega_1 \rightarrow \min,$$
$$\Phi_2(\boldsymbol{\alpha})=\omega_2 \rightarrow \max,$$
$$\Phi_3(\boldsymbol{\alpha})=\frac{|W_5+W_6|}{2} \rightarrow \max,$$
$$\Phi_4(\boldsymbol{\alpha})=\frac{|W_5-W_6|}{x_5} \rightarrow \min,$$
$$\Phi_5(\boldsymbol{\alpha})=m \rightarrow \min$$

where ω_1 and ω_2 are the first and second natural vibration frequencies respectively; W_5 and W_6 are the displacements of points 5 and 6 in the direction of z-axis; x_6 is the coordinate of point 6; Φ_3 and Φ_4 characterize the average linear and angular displacements of the line 5-6 points; and m is the plate mass. The stress-strained state of the structure is defined by the Mises maximum equivalent stress

$$f(\boldsymbol{\alpha})=\sigma_{\max} \leq [\sigma]$$

where $[\sigma]$ is the allowable stress.

Since the functional constraint $[\sigma]$ is not rigid (because the structure may be manufactured of different materials), $f(\boldsymbol{\alpha})$ should be represented in the form of pseudocriterion $\Phi_6(\boldsymbol{\alpha})=\sigma_{\max}$.

Thus, the criteria vector has the form

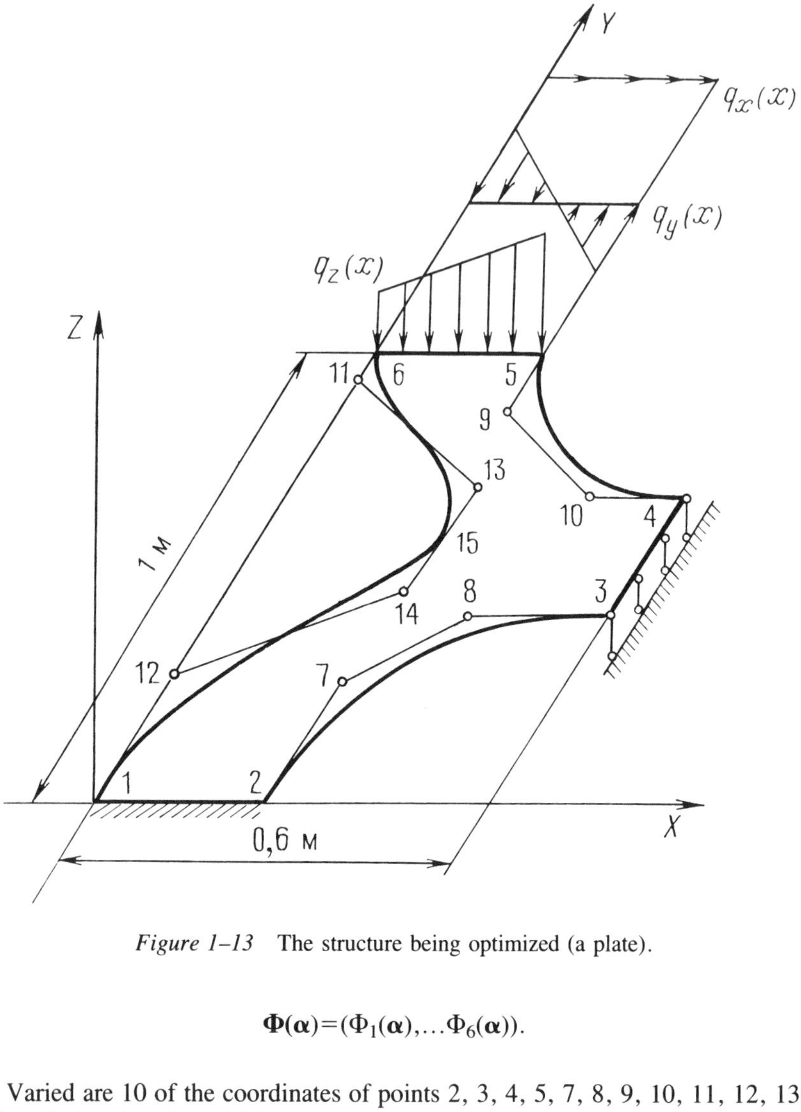

Figure 1–13 The structure being optimized (a plate).

$$\mathbf{\Phi}(\boldsymbol{\alpha})=(\Phi_1(\boldsymbol{\alpha}),\ldots\Phi_6(\boldsymbol{\alpha})).$$

Varied are 10 of the coordinates of points 2, 3, 4, 5, 7, 8, 9, 10, 11, 12, 13, 14 and also the plate thickness. Points 7–14 are used as reference points for obtaining curves 2-3, 4-5, 6-15, 1-15. The design-variable boundaries define the Π_1 parallelepiped.

Each point from the design-variable space corresponds to a separate finite element model.

Platelike four-node plane elements possessing both bending and membrane stiffness are used. Calculation of each solution $\boldsymbol{\alpha}^i$ is accompanied by automatic generation of a new finite element mesh.

Stage 1: Investigation of parallelepiped Π_1. In Π_1, 2,112 trials were carried out. The four dialogues implemented in investigating the design-variable space have provided the following criteria constraints $\mathbf{\Phi}^{**}=(11.2, 60.0, 0.03, 0.05, 18.0, 250.0)$. These data were used for constructing the feasible solutions set $\overline{D}(\Pi_1)$ containing six design-variable vectors, all of which are Pareto optimal. Figure 1-14 shows (against a black background) three most interesting solutions of the plate (white color), the difference being in geometrical shapes of the plate. These feasible solutions correspond to vectors 1,238, 387, and 1,353. Figure 1-14 presents the corresponding values of the performance criteria and design variable α_9 (the plate thickness). This design variable is rather important for analyzing the results. The complexity of the problem of determination of the set $\overline{D}$ is demonstrated by the fact that the ratio of the number of feasible vectors to N is a small quantity of the order of 0.003.

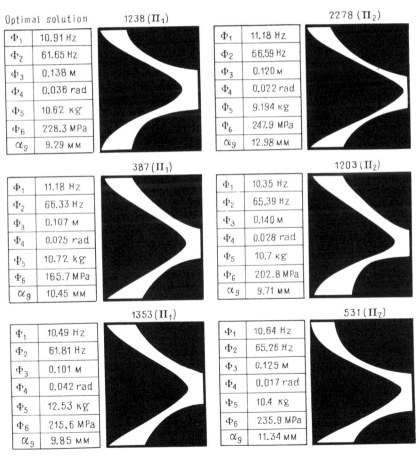

Optimal solution

1238 (Π_1)

Φ_1	10.91 Hz
Φ_2	61.65 Hz
Φ_3	0.138 м
Φ_4	0.036 rad
Φ_5	10.62 кg
Φ_6	228.3 MPa
α_9	9.29 мм

2278 (Π_2)

Φ_1	11.18 Hz
Φ_2	66.59 Hz
Φ_3	0.120 м
Φ_4	0.022 rad
Φ_5	9.194 кg
Φ_6	247.9 MPa
α_9	12.98 мм

387 (Π_1)

Φ_1	11.18 Hz
Φ_2	66.33 Hz
Φ_3	0.107 м
Φ_4	0.025 rad
Φ_5	10.72 кg
Φ_6	165.7 MPa
α_9	10.45 мм

1203 (Π_2)

Φ_1	10.35 Hz
Φ_2	65.39 Hz
Φ_3	0.140 м
Φ_4	0.028 rad
Φ_5	10.7 кg
Φ_6	202.8 MPa
α_9	9.71 мм

1353 (Π_1)

Φ_1	10.49 Hz
Φ_2	61.81 Hz
Φ_3	0.101 м
Φ_4	0.042 rad
Φ_5	12.53 кg
Φ_6	215.6 MPa
α_9	9.85 мм

531 (Π_2)

Φ_1	10.64 Hz
Φ_2	65.26 Hz
Φ_3	0.125 м
Φ_4	0.017 rad
Φ_5	10.4 кg
Φ_6	235.9 MPa
α_9	11.34 мм

Figure 1–14 A portion of the representations album.

Stage 2: Correction of design-variable constraints and construction of a new parallelepiped Π_2. Figure 1-15 shows the histograms for distribution of the six solutions in Π_1 over the ranges of the design variables α_1, α_2, α_3, α_8, and α_{10}. Having analyzed the histograms the designer was able to construct parallelepiped Π_2, within which 2,378 trials were conducted subject to the

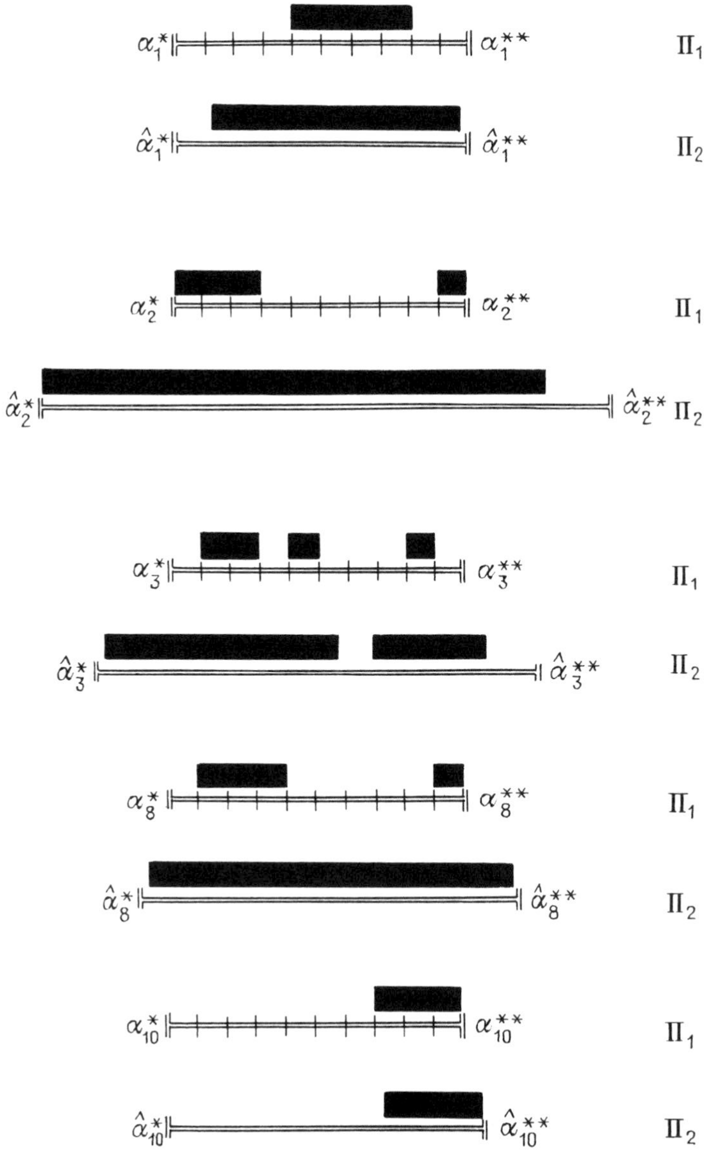

Figure 1–15 Histograms of the feasible solutions distribution.

aforementioned criteria constraints. The effectiveness of the correction of design-variable constraints is demonstrated by a substantial extension of the set $\bar{D}$ (Π_2), which now contains 30 vectors (including 11 Pareto optimal ones). The geometrical shapes of the three structures corresponding to vectors 2,278, 1,203, and 531 are shown in Figure 1-14. The histograms presented in Figure 1-15 demonstrate the feasible solutions distribution in Π_2. Note that for $\Phi_5<18$ kg, the mass of feasible structures has increased in parallelepipeds Π_1 and Π_2 from 10.62 kg to 12.558 kg and from 9.19 kg to 17.24 kg, respectively. However, Π_2 does contain three vectors, 531, 1,076, and 2,278, for which the structure's mass is less than 10.62 kg, viz. 10.4, 9.6, and 9.19 kg, respectively.

Stage 3: Visualization and analysis of the data. Thus, the investigations carried out in Π_1 and Π_2 have provided the set $\bar{D}=\bar{D}(\Pi_1)\cup\bar{D}(\Pi_2)$ containing 36 solutions. With respect to basic formalizable criteria Φ_1, Φ_2, and Φ_5, the previous three solutions in Π_2, 531, 1,076, and 2,278, were assumed to be the best ones. After the analysis of the geometrical shapes of the structures corresponding to vectors from $\bar{D}$ (taking into account the technological features of their manufacture), the feasible solutions set D appeared to contain eight solutions. The designers have preferred solution 1,238 from Π_1 (see Fig. 1-14). In such an important criterion as mass, this structure is inferior to the aforementioned solutions from Π_2, since its mass is 10.62 kg. Also, vector 1,238 is not Pareto optimal in $\bar{D}$ with respect to criteria Φ_1–Φ_5. This confirms the conclusion that the optimal solution must be sought not on the Pareto optimal set $\bar{P}$ but on $\bar{D}$. Hence, the majority of multicriteria optimization methods (Molodtsov and Fedorov 1979) are inefficient for the class of problems under consideration, since they do not allow construction of the set $\bar{D}$, and hence construction of the feasible solutions set D. Thus, we see that the PSI method should be used.

Conclusions

The use of the PSI method for optimizing various objects with the help of finite element models allows correct construction of the structure shapes set. The resulting solutions constitute the so-called album of the object representations or the album of visualization of an object. Figure 1-14 demonstrates a fragment of the album. In analyzing it one can take unformalizable criteria into consideration. As a result, feasible solutions set D is determined. By analyzing multiple solutions of a structure one can readily choose the best product manufacture process, including the optimal processing technique, equipment, tools, and devices.

The creation of the representations album should be considered one of the most important features of the class of problems under consideration. The album helps the designer to analyze formerly unknown geometrical shapes of the structure subjected to optimization and thus facilitates the search for innovative solutions.

The efficiency of the PSI method for optimizing finite element models was demonstrated by choosing the optimal design variables of the truck frame (see Section 6-2). As a result, the frame mass was reduced by 28 kg and some other performance criteria were improved (Velikhov et al. 1986).

Remark. In Statnikov and Matusov (1994), the search for optimal design variables of a platelike structure according to a multiple criterion is discussed. To solve the problem the programs implementing the finite element method and optimization techniques are used. The results of the comparison of the single-criterion and multicriteria approaches are presented.

In the first case, the well-known I-DEAS and ANSYS software packages with single-criterion optimization modules have been used. I-DEAS and ANSYS are general-purpose finite element analysis program packages that are used by engineers and designers around the world to analyze the stress, vibration, and heat transfer characteristics of structures and mechanical components. To this class of programs one should also relate such packages as MSC/NASTRAN, COSMOS, NISA, and others.

In the second case, the MOVI software package implementing the PSI method combined with the ANSYS finite element analysis program has been used.

The advantage of the multicriteria approach delivering important information about all Pareto optimal solutions to engineers and designers is proved.

Statnikov and Matusov (1994) conclude that to come to the best decision it is necessary to use multicriteria optimization in general-purpose finite element analysis programs.

The PSI method allows:

1. Determination of design-variable, functional, and criteria constraints
2. Taking into account the design-variables effect on criteria
3. Finding the criteria whose values remain practically constant and may thus be excluded from the further study
4. Singling out interdependent and conflicting criteria, etc.

The possibility of finding and evaluating the diversity of shapes of the object under consideration as well as its visualization and analysis allows the designer to take unformalizable criteria into account.

However, the determination of the feasible solutions and Pareto optimal sets is of paramount importance.

2

Approximation of Feasible Solutions and Pareto Optimal Sets

2-1. Approximation of a Feasible Solutions Set

We have introduced the notion of a feasible solution in the multicriteria optimization problem. The algorithm discussed in Section 1-3 allows simple and efficient identification and selection of feasible points from the design-variable space. However, the question arises: How can one use the algorithm for constructing a feasible solutions set D with a given accuracy? Since it is known that for the problems involving continuous design variables and criteria the set D is also continuous, the latter is constructed by singling out a subset of D that approaches any value of each criterion in region $\Phi(D)$ with a predetermined accuracy.

The possibility of approximating a feasible solutions set is illustrated by the following example (Sobol' and Statnikov 1981). Within the square

$$\Pi=\{-0.5\leq\alpha_1\leq0.5;\ 0\leq\alpha_2\leq1\}$$

criteria $\Phi_1=\alpha_1^2+4\alpha_2^2$ and $\Phi_2=(\alpha_1+1)^2+(\alpha_2-1)^2$ are specified, and are to be minimized taking the functional constraint $|\alpha_2-\alpha_1-0.375|\geq0.125$ into account. In this case, the set D is the square Π from which a strip has been cut out (see Fig. 2-1). The Pareto optimal set is composed of portions AA^1 and A^2A^3 of hyperbola $\alpha_2=-\alpha_1(3\alpha_1+4)^{-1}$, segment A^3A^4 of boundary $\alpha_1=-0.5$, and segment A^1A^5 of boundary $\alpha_2-\alpha_1=0.25$. The method for obtaining the set has been discussed in Bartel and Marks (1974). Also, the figure shows region $\Phi(D)$ on the criterion plane.

Points B^i on the criteria plane shown in Figure 2-1 are the images of points A^i. The exact trade-off curve (the Pareto optimal set) is shown in Figure 2-2, while Figure 2-3 presents the trial points in $\Phi(D)$. Judging by Figure 2-3*a* obtained for $N=64$, one cannot be sure that the set $\Phi(D)$ consists of two separate

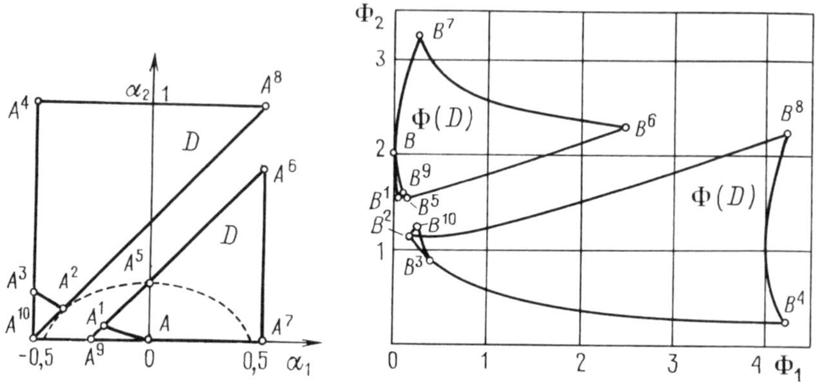

Figure 2–1 Feasible solutions sets in the design-variable space, D, and the criteria space, Φ (D). The set P consists of arcs AA^1A^5 and $A^2A^3A^4$, while Φ (P) consists of BB^1B^5 and $B^2B^3B^4$.

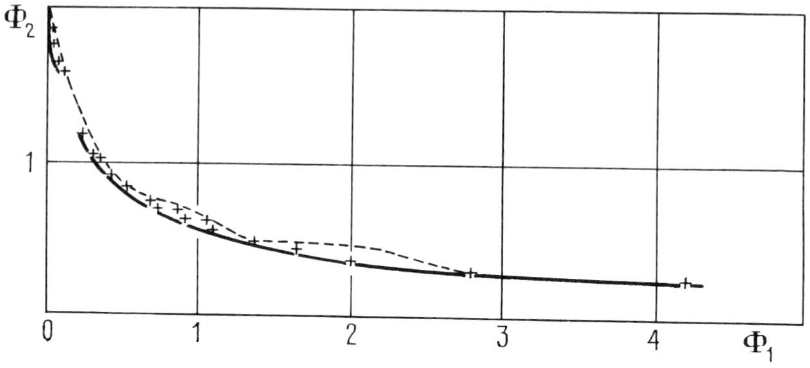

Figure 2–2 Exact and approximate (dashed line) Pareto optimal sets of $N=64$. The crosses indicate approximate Pareto optimal points for $N=512$.

parts; however, for $N=256$ this is quite clear (see Fig. 2-3*b*). We see that in the latter figure, and more so in Figure 2-3*c* plotted for $N=512$, the feasible region $\Phi(D)$ is approximated quite well.

Let ϵ_ν be an admissible (in the designer's opinion) error in criterion Φ_ν. By $\boldsymbol{\epsilon}$ we denote the errors set $\{\epsilon_\nu\}$, $\nu=1,\ldots,k$. We will say that region $\Phi(D)$ is approximated by a finite set $\Phi(D_{\boldsymbol{\epsilon}})$ with the accuracy up to the set $\boldsymbol{\epsilon}$, if for any vector $\boldsymbol{\alpha}\in D$, there can be found a vector $\boldsymbol{\beta}\in D_{\boldsymbol{\epsilon}}$ such that $|\Phi_\nu(\boldsymbol{\alpha})-\Phi_\nu(\boldsymbol{\beta})|\leq\epsilon_\nu$, $\nu=1,\ldots,k$.

Hence, for not too large values of ϵ_ν, region $\Phi(D)$, or D, may be constructed only if the number of points belonging to D is sufficiently large. The latter circumstance leads to a considerable consumption of computer time. It is clear,

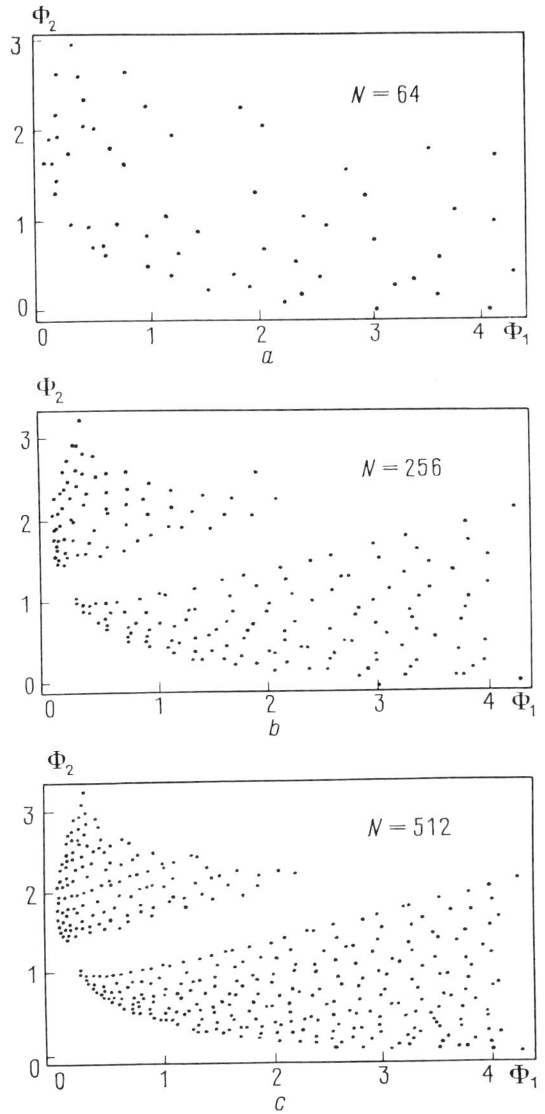

Figure 2–3 Trial points approximating the feasible solutions domain for N=64, 256, and 512.

however, that incomplete construction of the feasible solutions set may lead to results that are far from best.

We assume that the functions we shall be operating with are continuous and satisfy the Lipschitz condition (L) formulated as follows: For all vectors $\boldsymbol{\alpha}$ and $\boldsymbol{\beta}$ belonging to the domain of definition of the criterion Φ_ν, there exists a number L_ν such that

$$|\Phi_\nu(\boldsymbol{\alpha})-\Phi_\nu(\boldsymbol{\beta})|\leq L_\nu \max_j|\alpha_j-\beta_j|$$

In other words, there exists L'_ν such that

$$|\Phi_\nu(\boldsymbol{\alpha})-\Phi_\nu(\boldsymbol{\beta})|\leq L'_\nu \sum_{j=1}^{r} |\alpha_j-\beta_j|.$$

This is one of the least limiting conditions one encounters in the theory of optimization. In practice, its violation means that one has to deal with a "pathological" function. Fortunately, in engineering optimization problems such functions are extremely rare.

We will say that a function $\Phi_\nu(\boldsymbol{\alpha})$ satisfies the special Lipschitz condition (*SL*) if for all vectors $\boldsymbol{\alpha}$ and $\boldsymbol{\beta}$ there exist numbers L^j_ν , $j=1,\ldots,r$ such that

$$|\Phi_\nu(\boldsymbol{\alpha})-\Phi_\nu(\boldsymbol{\beta})|\leq \sum_{j=1}^{r} L^j_\nu\,|\alpha_j-\beta_j|$$

where at least some of L^j_ν are different.

The class of functions *SL* is of interest because:

1. Class *L* incorporates all the functions belonging to class *SL*. (In the majority of practical cases these classes coincide since the functions one encounters in solving engineering problems have different sensitivities with respect to design variables, and hence the constants L^j_ν are different too.)
2. For class *SL*, the convergence rate of the approximation process is greater than for the class with the Lipschitz condition (see Theorem 1 as follows).
3. The P_τ-nets (see Addendum) used for calculating criteria are optimal for the class *SL* of functions (Sobol' 1987).

Let us estimate the number of points of an *r*-dimensional P_τ-net, which is sufficient for approximating $\Phi(D)$ with a given accuracy for criteria $\Phi_\nu(\boldsymbol{\alpha})\epsilon L$ or $\Phi_\nu(\boldsymbol{\alpha})\epsilon SL$.

Theorem 1. If criteria $\Phi_\nu(\boldsymbol{\alpha})$ are continuous and satisfy either the Lipschitz condition or the special Lipschitz condition, then to approximate $\Phi(D)$ to an accuracy of $\boldsymbol{\epsilon}$ it is sufficient to have

$$\max_\nu 2^\tau\left(\frac{[L_\nu]}{[\epsilon_\nu]}\right)^r \text{ or } \max_\nu 2^\tau\left(\frac{\left[\sum_{j=1}^{r} L^j_\nu\right]}{[\epsilon_\nu]}\right)^r$$

points of the P_τ-net (Statnikov and Matusov 1985).

Proof. Let $[L_\nu]$ (or $\sum_{j=1}^{r} L_\nu^j$) be a dyadic rational number[7] exceeding L_ν (or $\sum_{j=1}^{r} L_\nu^j$) and sufficiently close to the latter, and let $[\epsilon_\nu]$ be the maximum dyadic rational number that is less than or equal to ϵ_ν and whose numerator is the same as that of $[L_\nu]$ (or $[\sum_{j=1}^{r} L_\nu^j]$). For any $\boldsymbol{\alpha}=(\alpha_1,\ldots,\alpha_r)$ we consider an r-dimensional cube $\mathscr{L}_\alpha^\nu$ with the edge length $[\epsilon_\nu]/[L_\nu]$, $\boldsymbol{\alpha}\in\mathscr{L}_\alpha^\nu$. The volume of the cube is given by $([\epsilon_\nu]/[L_\nu])^r$. Since this number is dyadic rational and its numerator is equal to unity, it may be represented in the form $([\epsilon_\nu]/[L_\nu])^r= 2^\tau/2^{\gamma_\nu}$ where $\gamma_\nu>\tau$ is unknown and τ is the subscript of the P_τ-net corresponding to the r-dimensional cube K^r. From the letter equality we get

$$2^{\gamma_\nu}=2^\tau[L_\nu]^r/[\epsilon_\nu]^r. \tag{2-1}$$

According to the definition of the P_τ-net, any binary parallelepiped of the cube K^r of volume $2^\tau/2^{\gamma_\nu}$ contains 2^τ points from 2^{γ_ν} points of the P_τ-net (Sobol' 1969). Hence, if γ_ν satisfies (2-1) then cube $\mathscr{L}_\alpha^\nu$ contains 2^τ points. By the Lipschitz condition and the definition of cube $\mathscr{L}_\alpha^\nu$, the inequality

$$|\Phi_\nu(\boldsymbol{\alpha})-\Phi_\nu(\boldsymbol{\beta})|\le\epsilon_\nu$$

is satisfied for any point $\boldsymbol{\beta}\epsilon\mathscr{L}_\alpha^\nu$ of the 2^τ points. Thus, an arbitrary value of $\Phi_\nu(\boldsymbol{\alpha})$ may be approximated to the accuracy of ϵ_ν by 2^{γ_ν} points of the P_τ-net. The value of τ may be calculated using the formulas presented in the Addendum.

If for some ν_i and ν_j $[\epsilon_{\nu_i}]/[L_{\nu_i}]<[\epsilon_{\nu_j}]/[L_{\nu_j}]$, then $\mathscr{L}_\alpha^{\nu_i}\subset\mathscr{L}_\alpha^{\nu_j}$. Hence, by choosing a value of n satisfying the equality $2^n=\max_\nu 2^{\gamma_\nu}$, $\nu=1,\ldots,k$, we get the finite ϵ-approximation $\Phi(D_\epsilon)$ of the set $\Phi(D)$. In this case the inequality $|\Phi_\nu(\boldsymbol{\alpha})-\Phi_\nu(\boldsymbol{\beta})|\le\epsilon_\nu$, $\nu=1,\ldots,k$, is satisfied where $\boldsymbol{\alpha}\in K^r$, and $\boldsymbol{\beta}$ is one of the 2^n points.

Remarks.

1. Generally speaking, the set of points approximating $\Phi(D)$ may not belong to $\Phi(D)$, since it can incorporate the points with coordinates $\Phi_\nu^{**}<\Phi_\nu(\boldsymbol{\alpha})\le\Phi_\nu^{**}+\epsilon_\nu$ as well as the points that are not feasible due to functional constraints. By transforming the functional dependences into pseudocriteria $\Phi_{k+1}(\boldsymbol{\alpha}),\ldots,\Phi_{k+p}(\boldsymbol{\alpha})$ we get, in analogy to what was

[7]A dyadic number is a number of the form $p/2^m$ where p and m are natural numbers.

proved above, the approximation of the feasible solutions set $\Phi(D)$ by the points $\boldsymbol{\Phi}(\boldsymbol{\alpha})$ whose $k + p$ coordinates satisfy the functional and criteria constraints to the accuracy of $\boldsymbol{\epsilon}$.

2. The estimate of the approximation process convergence rate, considered in the previous theorem, is of an a priori nature. In other words, having specified the admissible errors ϵ_ν of criteria Φ_ν, and knowing constants L_ν (or $\sum_{j=1}^{r} L_\nu^j$), one may approximate the whole of the feasible domain with a given accuracy for any function corresponding to these constants. To do so one has to calculate the criteria at the points of the P_τ-net the number of which is specified by the theorem. However, this is an estimate since it takes into account any, even the "worst", function of the class. Hence, for a concrete problem, the number of trials needed for the approximation is less than the one provided by the aforementioned estimate. Similar estimates for the problems of finding the absolute extremum of functions satisfying the Lipschitz condition have been obtained in a number of works by alternative methods. It is appropriate to mention here that an interesting estimate has been obtained in (Sobol' 1987).

The estimate of the convergence rate considered in Theorem 1 (as well as the majority of a priori estimates used in approximate methods) is generally applicable for the theoretical determination of the number of trials. However, it is, as a rule, inapplicable for solving engineering problems. The number of points needed for calculating the performance criteria may be so large that the speed of present-day computers may prove to be inadequate. This difficulty may be overcome by developing "fast" algorithms dealing not with an entire class of functions but taking into account the features of the functions of each concrete problem.

For approximating a feasible region $\Phi(D)$ such an algorithm may be constructed in the following way. (Although all subsequent considerations presume the satisfaction of the Lipschitz condition, they are valid as well for the special Lipschitz condition if constant L_ν is replaced by $\sum_{j=1}^{r} L_\nu^j$).

Let the Lipschitz constants L_ν, $\nu=1,\ldots,k$, be specified, and N_1 be the subset of the points from D that are either the Pareto optimal points or lie within the $\boldsymbol{\epsilon}$-neighborhood of a Pareto optimal point with respect to at least one criterion. In other words, $\Phi_\nu(\boldsymbol{\alpha}^0)\leq\Phi_\nu(\boldsymbol{\alpha})\leq\Phi_\nu(\boldsymbol{\alpha}^0)+\epsilon_\nu$ where $\boldsymbol{\alpha}^0\in P$, and P is the Pareto optimal set. Let also $N_2=D\backslash N_1$ and $\overline{\epsilon_\nu}>\epsilon_\nu$ where $\overline{\epsilon_\nu}$ is a certain number defined in proving Theorem 2.

Definition. A feasible solutions set $\Phi(D)$ is said to be normally approximated if any point of set N_1 is approximated to an accuracy of $\boldsymbol{\epsilon}$, and any point of set N_2 to an accuracy of $\overline{\boldsymbol{\epsilon}}$.

Theorem 2. There exists a normal approximation $\Phi(D_\epsilon)$ of a feasible solutions set $\Phi(D)$ (Statnikov and Matusov 1989).

Proof. Let the values of criteria be calculated at N points of the P_τ-net, from among which we single out the feasible solutions, D_N, and the Pareto optimal, P_N, sets. Also, let N_1' be a subset of the vectors of N specified points whose images are either Pareto optimal points or lie in the ϵ-neighborhood of a Pareto optimal point $\boldsymbol{\Phi}(\boldsymbol{\beta})$ with respect to at least one criterion. Besides, we put $N_2'=D_N\backslash N_1'$ and refer to D those of the N points that satisfy the functional and criteria constraints to an accuracy of ϵ.

Step 1. Consider an arbitrary point $\bar{\boldsymbol{\alpha}}\in N_2'$. Let $K_\nu^\beta=|\Phi_\nu(\bar{\boldsymbol{\alpha}})-\Phi_\nu(\boldsymbol{\beta})-\epsilon_\nu|$. (If $\bar{\boldsymbol{\alpha}}\notin D$ then $\Phi_\nu(\boldsymbol{\beta})$ is replaced by Φ_ν^{**}.) Let us place $\bar{\boldsymbol{\alpha}}$ at the center of cube $K_{\bar{\alpha}}$ whose edge length is $2K_\nu^\beta/L_\nu$. For any $\boldsymbol{\alpha}\in K_{\bar{\alpha}}$ we get $|\Phi_\nu(\bar{\boldsymbol{\alpha}})-\Phi_\nu(\boldsymbol{\alpha})|\leq L_\nu\max_j|\bar{\alpha}_j-\alpha_j|\leq K_\nu^\beta$. If the cube's edge length is $\min_\nu 2K_\nu^\beta/L_\nu$ then the latter inequality holds for all ν. Let us perform the operation for all points from P_N and choose the cube $K_{\bar{\alpha}}$ edge length $\min_{\beta\epsilon P_N}\min_\nu 2K_\nu^\beta/L_\nu$. Then we arrive at a cube with the center at $\bar{\boldsymbol{\alpha}}$ such that $K_{\bar{\alpha}}\cap N_1'=\emptyset$. Upon constructing the cube for any $\bar{\boldsymbol{\alpha}}\in N_2'$ and finding $K_1=\bigcup_{\bar{\boldsymbol{\alpha}}\in N_2'}K_{\bar{\boldsymbol{\alpha}}}$ we choose a point $\hat{\boldsymbol{\alpha}}\in N_1'$ and construct for it cube $K_{\hat{\boldsymbol{\alpha}}}$ with the center $\hat{\boldsymbol{\alpha}}$ and the edge length $\min_\nu 2\epsilon_\nu/L_\nu$. Then the inequality

$$|\Phi_\nu(\hat{\boldsymbol{\alpha}})-\Phi_\nu(\boldsymbol{\alpha})|\leq L_\nu\max_j|\hat{\alpha}_j-\alpha_j|\leq\epsilon_\nu,\ \ \nu=1,\ldots,k$$

is valid for any $\boldsymbol{\alpha}\in K_{\hat{\boldsymbol{\alpha}}}$. Let $K_2=\bigcup_{\hat{\boldsymbol{\alpha}}\in N_1'}K_{\hat{\boldsymbol{\alpha}}}$ and $K^1=K_1\cup K_2$. Consider the complement $K^r\backslash K^1$ where K^r is the initial cube/parallelepiped.

Step 2. Since K^1 is a union of cubes, $K^r\backslash K^1$ may be represented in the form of a finite number of nonintersecting parallelepipeds. By defining K_i^1 and $\bigcup_i K_i^1=K^2$ in a similar way for all of the previous parallelepipeds Π_i, we arrive at the region $K^1\cup K^2$, which is a union of a finite number of cubes. The most promising points of the region, belonging to N_1, are approximated to an accuracy of ϵ. The rest of the points are approximated to a worse accuracy $\bar{\epsilon}$ and are of no interest in constructing the Pareto optimal set. It should be noted that the Pareto optimal set on the second step must be chosen from the union of the Pareto optimal set obtained on the first step and the set of feasible points obtained during the second step. The mth step is performed in a similar way. After performing n steps and determining K^i, $i=1,\ldots,n$, we get

$K^r \setminus \bigcup_{i=1}^{n} K^i = \varnothing$, that is, cover the whole K^r with a union of cubes whose points are approximated with a desired accuracy. This completes the proof of the theorem.

It should be stressed that this algorithm is in many aspects analogous to the one proposed by several authors in considering single-criterion problems (Yevtushenko 1971).

However, here we can obtain a "faster" algorithm, since in passage to the class with the special Lipschitz condition, either $\sum_{j=1}^{r} L_\nu^j \le L_\nu'$ or $\sum_{j=1}^{r} L_\nu^j \le L_\nu \cdot r$. Hence the cubes covering the whole of K^r will be larger than in the case of the functions subjected to the Lipschitz condition, and the whole of the cube will be covered more economically. Besides, as noted, we are using highly uniform P_τ-nets. This also results in a "faster" approximation of the feasible region.

2-2. The Pareto Optimal Set Approximation

Since the Pareto optimal set is unstable, even slight errors in calculating criteria $\Phi_\nu(\boldsymbol{\alpha})$ may lead to a drastic change in the set. This implies that by approximating a feasible solutions set with a given accuracy we cannot guarantee an appropriate approximation of the Pareto optimal set.

Let us consider the example illustrated by Figure 2-4 where the feasible region is represented by a triangle. Here the vertex $\Phi(P)$ is the only Pareto optimal point and the approximation of $\Phi(D)$ is shown by dots and crosses. The Pareto optimal set of this approximation (shown by crosses) is seen to differ drastically from $\Phi(P)$.

This instability is one of the major reasons why the problem of approximating the Pareto optimal set proved to be rather complicated. Although the problem has been tackled since the 1950s, a complete solution acceptable for the majority of practical problems is still to be obtained. Nevertheless, promising methods have been proposed for some classes of functions (Stadler and Dauer 1992; Lieberman 1991; Ozernoy 1988; White 1990). Let us consider some of them in brief.

Linear Problems

In this case, the theorems about the Pareto optimal set structure, in particular the well-known Arrow-Barankin-Blackwell theorem, allow construction of the set P in a straightforward manner (Gass and Saaty 1955; Kornbluth 1974). However, in practice, methods generalizing the well-known simplex method of linear programming are used. Some interesting methods for solving the problem in question are suggested in (Steuer 1986; Dauer and Saleh 1992; Cohon et al.

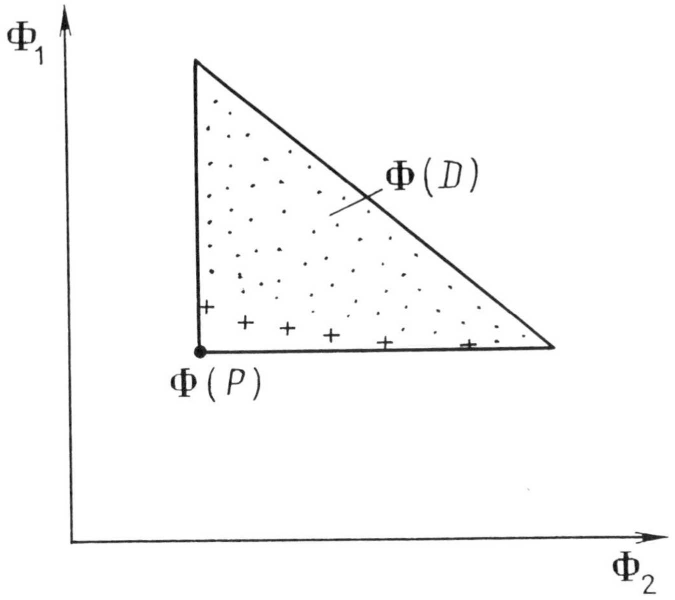

Figure 2–4 Instability of the Pareto optimal set.

1988; Benayoun et al. 1971). This issue is considered in the rather extensive literature (Isermann 1977; Zeleny 1974; Zionts and Wallenius 1980).

Concave Problems

These problems are commonly solved using the Karlin theorem stating that the Pareto optimal set coincides with the global minima set of the family of functions of the form $\sum_{i=1}^{k} \lambda_i \Phi_i(\boldsymbol{\alpha})$ where $\lambda_1+\ldots+\lambda_k=1$, $\lambda_i>0$ (see, e.g., Karlin (1959)).

The following result concerning the structure of the Pareto optimal set in the convex case is also used.

$$S_\Phi(\boldsymbol{\lambda})=\{\boldsymbol{\Phi}(\boldsymbol{\alpha})|\sum_{i=1}^{k} \lambda_i \Phi_i(\boldsymbol{\alpha})\leq \sum_{i=1}^{k} \lambda_i \Phi_i(\boldsymbol{\alpha}^0),\ \boldsymbol{\alpha}^0\in D\};$$

$$M=\bigcup_{\boldsymbol{\lambda}\in A} S_\Phi(\lambda),\ A=\{(\lambda_1,\ldots,\lambda_k)|\lambda_i>0,\ \sum_{i=1}^{k} \lambda_i=1\}.$$

Then, the Pareto optimal set is contained in the closure of the set M.

The proof of this theorem is given in many references (e.g., Dubov, Travkin and Yakimets (1986).

Some of the most important results in the theory of multicriteria optimization of concave functions are the formulations obtained by Kuhn et al. (Kuhn and Tucker 1951).

Other Problems

Here, the results that concern the structure of the Pareto optimal set have also been obtained (e.g., Da Cunha and Polak (1967)). The majority of methods for approximating the Pareto optimal set in problems that are neither linear nor concave are divided into the following two classes. The first class incorporates the methods based on minimization of various functions (Steuer and Choo 1983; Dyer et al. 1992; Benson 1992). Very often such functions are combinations of the criteria, such as $[\sum_{i=1}^{k} (\lambda_i\Phi_i)^s]^{1/s}$ where $s \geq 1$, $\lambda_i > 0$, $\sum_{i=1}^{k} \lambda_i = 1$ (Merkur'ev and Moldavskii 1979; Gearhart 1979). The combinations may be represented by the families of distances $d(\mathbf{x}^*, \mathbf{\Phi}(\boldsymbol{\alpha}))$ where $\mathbf{x}^*$ is an "ideal" vector, such as $\mathbf{x}^* = (0,\ldots,0)$, and $\mathbf{\Phi}(\boldsymbol{\alpha})$ is a point belonging to the feasible solutions set (Stadler 1988). Naturally, point $\mathbf{\Phi}(\boldsymbol{\alpha}^0)$ corresponding to the minimum distance d, is a Pareto optimal point. However, in the case under consideration, the set of points realizing the minima of the combinations does not form the whole of the Pareto optimal set. Therefore the major difficulty is related to finding the conditions assuring density of the points (Kelley 1957) in the Pareto optimal set. The density allows approximation of P. Thus, the approximation method discussed in Molodtsov and Fedorov (1979) is based on the use of a special kind of criteria combination. Summation of the conventional linear combination $\sum_{i=1}^{k} \lambda_i\Phi_i(\boldsymbol{\alpha})$ and a certain "additional" function assures density of the points corresponding to the minimum of the combinations in $\Phi(P)$. In Molodtsov and Fedorov (1979) the so-called ill-posed problem of the Pareto optimal set approximation is analyzed. The solution proposed in the work is obtained using the Hausdorff metric, which is discussed in the following. Similar approaches were employed in Popov (1981), Tanino and Sawaragi (1980), Dubov et al. (1986). The possibility of using the Hausdorff metric imposes certain constraints on the system of preferences of the decision maker. Besides, in using the previous methods one has to find the criteria combination global extremum to obtain a point belonging to P. Often this may require too much computer time.

Some interesting results related to applications of methods of the class under consideration are obtained in (Eschenauer 1988; Koski 1988; Ester 1987).

The other class comprises methods based on covering a feasible solutions set with subsets of a special shape: cubes, spheres, etc. Owing to the conditions imposed on the criteria, the cubes/spheres are chosen in such a way that all the points lying within them were approximated with a required accuracy. One can approximate a feasible solutions set $\Phi(D)$ by covering it with the cubes and then singling out the Pareto optimal points from the approximation of D. After that by performing necessary operations, taking into account the fact that the problem in question is ill-posed, one can construct the approximation of the Pareto optimal set. The methods of this class are more versatile (as far as the types of functions are concerned) than those of the first class. In Sukharev (1971) the problem of finding the optimal strategy for covering a set K with identical cubes was solved. It was also analyzed in a number of other works (see, e.g., Yevtushenko and Mazurik 1989).

Next we present a second-class method that has been developed without either using the Hausdorff metric or imposing any constraints on the designer's system of preferences. Besides, we use uniformly distributed sequences of points that allow us to hope that the resulting algorithms for approximating the Pareto optimal set are among the "fastest" ones. The only requirement is that the criteria are continuous and satisfy the Lipschitz conditions (Statnikov and Matusov 1989).

Let P be the Pareto optimal set in the design-variable space; $\Phi(P)$ be its image; and $\boldsymbol{\epsilon}$ be a set of admissible errors. It is desirable to construct a finite Pareto optimal set $\Phi(P_{\boldsymbol{\epsilon}})$ approximating $\Phi(P)$ to an accuracy of $\boldsymbol{\epsilon}$.

Let $\Phi(D_{\boldsymbol{\epsilon}})$ be the $\boldsymbol{\epsilon}$-approximation of $\Phi(D)$, and $P_{\boldsymbol{\epsilon}}$ be the Pareto optimal subset in $D_{\boldsymbol{\epsilon}}$. As has already been mentioned, the complexity of constructing a finite approximation of the Pareto optimal set results from the fact that approximating the feasible solutions set $\Phi(D)$ by a finite set $\Phi(D_{\boldsymbol{\epsilon}})$ to the accuracy of $\boldsymbol{\epsilon}$, in the general case one cannot achieve the approximation of $\Phi(P)$ with the same accuracy. This is due to the fact that the feasible point approximating a certain $\boldsymbol{\Phi}(\boldsymbol{\beta})\in\Phi(P)$ may be "knocked out" by another feasible point in selecting the Pareto optimal points from the $\boldsymbol{\epsilon}$-approximation of the feasible solutions set (see Fig. 2-5). As a result, $\boldsymbol{\Phi}(\boldsymbol{\beta})$ is not approximated by any of the selected Pareto optimal points. Such problems are said to be ill-posed in the sense of Tikhonov (Vasil'ev 1981). Although the latter notion is routinely used in numerical mathematics, let us recall it here.

Let P be a functional in space X, $P : X \rightarrow Y$. We suppose that there exists $y^* = \inf P(x)$, and $V_{\epsilon}(y^*)$ is the neighborhood of the required solution y^*. Let us single out an element x^* (or a set of elements) in space X and its δ-neighborhood $V_{\delta}(x^*)$ and call x_{δ}^{ϵ} a solution to the problem of finding the extremum of P if the solution satisfies simultaneously the conditions $x_{\delta}^{\epsilon}\in V_{\delta}(x^*)$ and $P(x_{\delta}^{\epsilon})\in V_{\epsilon}(y^*)$. If at least one of the conditions is not satisfied for arbitrary values of ϵ and δ then the problem is called ill-posed (in the sense of Tikhonov).

An analogous definition may be formulated for the case when P is an operator mapping space X into space Y. Let us set

$$X = \{\Phi(D_{\boldsymbol{\epsilon}}), \Phi(D)\}; \quad Y = \{\Phi(P_{\boldsymbol{\epsilon}}), \Phi(P)\}$$

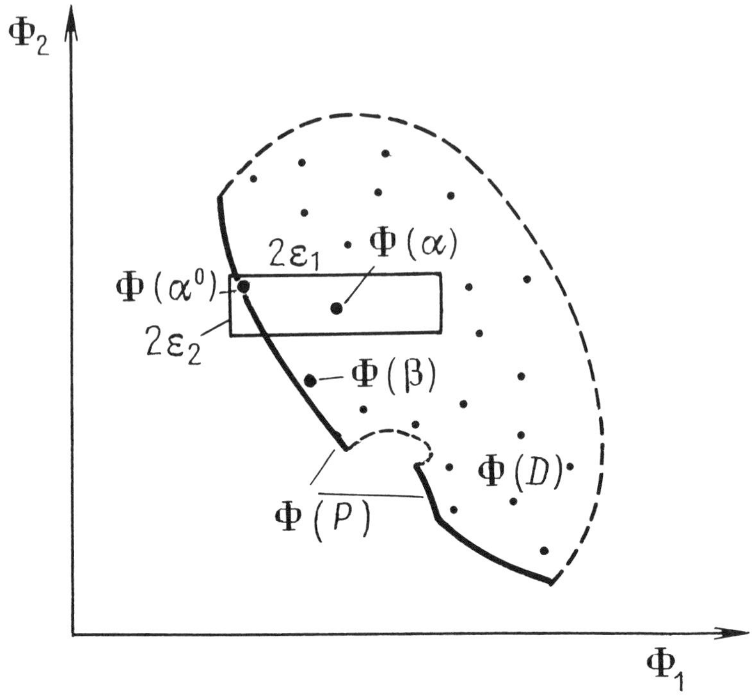

Figure 2–5 Non-Pareto point $\mathbf{\Phi}$ ($\boldsymbol{\alpha}$) looks suspicious. The truly Pareto point $\mathbf{\Phi}$ ($\boldsymbol{\alpha}^0$) lies in its ϵ-neighborhood.

where $\epsilon \to 0$, and let $P : X \to Y$ be an operator relating any element of X to its Pareto optimal subset. Then in accordance with what was said before, the problem of constructing sets $\Phi(D_\epsilon)$ and $\Phi(P_\epsilon)$ belonging simultaneously to the ϵ-neighborhoods of $\Phi(D)$ and $\Phi(P)$ respectively, is ill-posed. Of course, in spaces X and Y, the metric or topology (Kelley 1957) must be specified that corresponds to the system of preferences on $\Phi(D)$.

Let us define the V_ϵ-neighborhood of a point $\mathbf{\Phi}(\boldsymbol{\alpha}^0) \in \Phi(\Pi)$ as $V_\epsilon = \{\mathbf{\Phi}(\boldsymbol{\alpha}) \in \Phi(\Pi)\colon |\Phi_\nu(\boldsymbol{\alpha}^0) - \Phi_\nu(\boldsymbol{\alpha})| \le \epsilon_\nu,\ \nu = 1, \ldots, k\}$.

Next we have to construct a Pareto optimal set $\Phi(P_\epsilon)$ in which for any point $\mathbf{\Phi}(\boldsymbol{\alpha}^0) \in \Phi(P)$ and any of its ϵ-neighborhoods V_ϵ there may be found a point $\mathbf{\Phi}(\boldsymbol{\beta}) \in \Phi(P_\epsilon)$ belonging to V_ϵ. Conversely, in the ϵ-neighborhood of any point $\mathbf{\Phi}(\boldsymbol{\beta}) \in \Phi(P_\epsilon)$ there must exist a point $\mathbf{\Phi}(\boldsymbol{\alpha}^0) \in \Phi(P)$ (see Fig. 2-6). The set $\Phi(P_\epsilon)$ is called an approximation possessing property M.

An approximation $\Phi(P_\epsilon)$ will be said to possess the M_1-property if for any point $\mathbf{\Phi}(\boldsymbol{\alpha}^0) \in \Phi(P)$ and any its ϵ-neighborhood V_ϵ, there exists a point $\mathbf{\Phi}(\boldsymbol{\beta}) \in \Phi(P_\epsilon)$ belonging to V_ϵ.

Let there have been constructed $\Phi(D_\epsilon)$, an approximation of $\Phi(D)$.

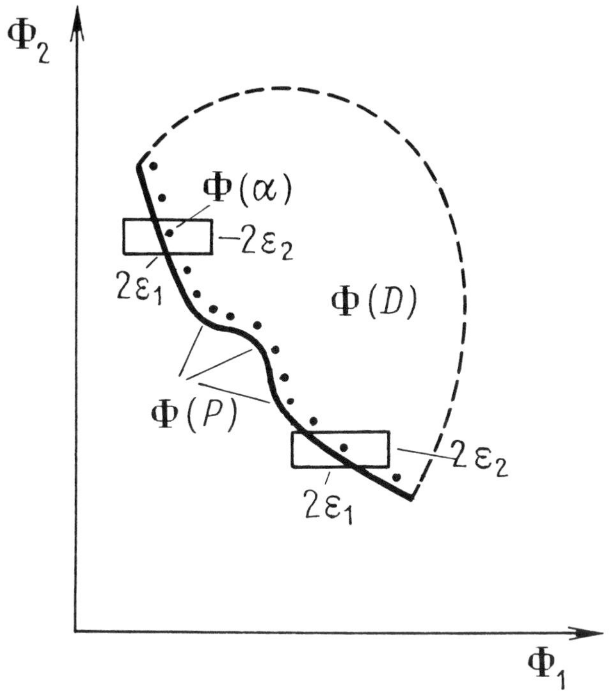

Figure 2–6 Approximation of Pareto optimal set Φ (P) by a finite set.

Lemma 1. If the conditions of Theorem 1 are satisfied, then there exists an approximation $\Phi(P'_\epsilon)$ possessing the M_1-property.

Proof. The lemma will be proved by analyzing the neighborhoods of the so-called "suspicious" points from $\Phi(D_\epsilon)$, that is, the points to whose neighborhoods the true Pareto optimal vectors may belong. If we find new Pareto optimal vectors in the neighborhoods of the "suspicious" points then these vectors may be added to $\Phi(P_\epsilon)$. Taken together with $\Phi(P_\epsilon)$, they form the ϵ-approximation of a Pareto optimal set (Matusov and Statnikov 1987).

Let us determine the set of "suspicious" points. Consider $\mathbf{\Phi}(\boldsymbol{\alpha})\in\Phi(P_\epsilon)$, and let

$$M_{\boldsymbol{\alpha}}^{+}=\{\mathbf{\Phi}(\boldsymbol{\beta})\in\Phi(D_\epsilon) : \forall\nu\ \Phi_\nu(\boldsymbol{\beta})\geq \Phi_\nu(\boldsymbol{\alpha})\},$$

$$M_{\boldsymbol{\alpha}}^{\epsilon}=\left\{\mathbf{\Phi}(\boldsymbol{\beta})\in M_{\boldsymbol{\alpha}}^{+} : \exists\nu\ \ \Phi_\nu(\boldsymbol{\beta})-\Phi_\nu(\boldsymbol{\alpha})\leq\frac{\epsilon_\nu}{2}\right\}.$$

(Here, the number $\epsilon_\nu/2$ has been taken arbitrarily. Instead of this, one can take any number less than ϵ_ν.) Let

$$M_{\alpha}^{-}=\Phi(D_{\epsilon})\backslash(M_{\alpha}^{+}\cup\Phi(P_{\epsilon})),\; M_{\alpha}=M_{\alpha}^{\epsilon}\cup M_{\alpha}^{-},\; M=\bigcup_{\alpha}M_{\alpha}.$$

Let us consider a point $\Phi(\alpha^0)\in M$. If $\Phi(P_\epsilon)$ contains no point $\Phi(\beta)$ such that $\Phi_\nu(\beta)\le\Phi_\nu(\alpha^0)-\epsilon_\nu$ for any ν, then $\Phi(\alpha^0)$ is a "suspicious" point. The set of suspicious points will be denoted by SM. It is easy to see that the points of $\Phi(P)$ that are not approximated by the set $\Phi(P_\epsilon)$ with the accuracy of ϵ may lie only in ϵ-neighborhoods of vectors from SM (see Fig. 2-5). Thus, if we construct a cube $\mathcal{L}_{\alpha^1}$ with the center at point α^1 and the edge length $\min_{\nu} 2\epsilon_\nu/L_\nu$, $\nu=1,\ldots,k$, for any point $\alpha^1\in D_\epsilon$ such that $\Phi(\alpha^1)\in SM$, then the cube may contain true Pareto optimal points from $\Phi(P)$, approximated with an accuracy of ϵ by no point from $\Phi(P_\epsilon)$.

Let ϵ'_ν be small errors that can be neglected. Let us approximate $\Phi(\mathcal{L}_{\alpha^1})\cap\Phi(D)$ by the P_τ-net points to an accuracy of ϵ', as before. (Since volume $\mathcal{L}_{\alpha^1}$ is quite small as compared with K^r, the number of points needed for the approximation is rather small.) At least one of the points of the P_τ-net in $\mathcal{L}_{\alpha^1}$ belongs to the neighborhood $V\subset\mathcal{L}_{\alpha^1}$ of a Pareto optimal point $\Phi(\alpha^0)$ if such a point does exist. Let us denote such a point from P_τ-net by $\Phi(\gamma)$. If $\Phi(\alpha^0)$ is a Pareto optimal point then $\Phi(\gamma)$ definitely improves the value of at least one criterion for an arbitrary point $\Phi(\alpha)\in\Phi(P_\epsilon)$. If such a point $\Phi(\gamma)$ exists it is added to $\Phi(P_\epsilon)$. Conversely, $\mathcal{L}_{\alpha^1}$ does not contain a point $\Phi(\alpha^0)\in\Phi(P)$, to an accuracy of ϵ'. The operation is repeated for all the vectors belonging to SM.

Let $\bigcup_i\Phi(\gamma^i)\cup\Phi(P_\epsilon)=\Phi(P'_\epsilon)$ and $\bigcup_i\gamma^i\cup D_\epsilon=D'_\epsilon$ where γ^i is a point obtained after performing the aforementioned procedure. Then $\Phi(P'_\epsilon)$ may contain points that are not Pareto optimal and are to be discarded. As a result, we arrive at the set $\Phi(P'_\epsilon)$, which is a Pareto optimal subset in $\Phi(\bigcup_i\gamma^i\cup D_\epsilon)$ and ϵ-approximation of $\Phi(P)$. This completes the proof of the lemma.

The approximation $\Phi(P'_\epsilon)$ thus obtained possesses the property M_1. However, the inverse formulation is invalid in the general case, since $\Phi(P'_\epsilon)$ may contain excessive points whose analysis would be fruitless.

The ϵ-approximation of Pareto optimal set $\Phi(P'_\epsilon)$ constructed in Lemma 1, is said to possess the property M_2 if there is a point $\Phi(\beta)\in\Phi(P)$ within the ϵ-neighborhood of any point $\Phi(\alpha)\in\Phi(P_\epsilon)$.

Lemma 2. There exists a subset $\Phi(P''_\epsilon)$ of the set $\Phi(P'_\epsilon)$, which possesses the property M_2.

Proof. Let $\Phi(\alpha)\in\Phi(P'_\epsilon)$; B be an arbitrary subset in $\{1,\ldots,k\}$; and $N_{\Phi(\alpha)}=\{\Phi(\beta)\in\Phi(P'_\epsilon):\forall\nu\in B\;\; \Phi_\nu(\alpha)\le\Phi_\nu(\beta)\le\Phi_\nu(\alpha)+\epsilon_\nu \;\vee\; \forall\nu\in\{1,\ldots,k\}\backslash B\;\; \Phi_\nu(\beta)\le\Phi_\nu(\alpha)-\epsilon_\nu\}$. As before, we start by investigating the neighborhoods of the points α and β for which $\Phi(\beta)\in N_{\Phi(\alpha)}$.

Let the condition $\forall\tilde{\alpha}\in\mathcal{L}_\alpha\cap D$ $\exists\gamma\in\mathcal{L}_\beta\cap D$ such that

$\Phi_\nu(\gamma) \leq \Phi_\nu(\tilde{\alpha}) + \epsilon'_\nu \ \forall \nu \in B$ where ϵ'_ν is small enough, be satisfied for all the P_τ-net points belonging to cubes $\mathcal{L}_\alpha$ and $\mathcal{L}_\beta$ and satisfying the previous requirement for the feasible solutions set approximation. Then point $\mathbf{\Phi}(\alpha)$ may be excluded from $\Phi(P'_\epsilon)$, since its ϵ-neighborhood will not contain any points from $\Phi(P)$. By carrying out a similar procedure for all $\mathbf{\Phi}(\alpha) \epsilon \Phi(P'_\epsilon)$ we arrive at the set $\Phi(P''_\epsilon) \subset \mathbf{\Phi}(P'_\epsilon)$ possessing the required property M_2.

It can be readily shown that having performed these procedures (see Lemmas 1 and 2) and obtained $\Phi(P''_\epsilon)$, we have actually proved the following

Theorem 3 (Matusov and Statnikov 1985). $\Phi(P''_\epsilon)$ is an approximation of the Pareto optimal set $\Phi(P)$, possessing the property M.

Thus, we have constructed the desired Pareto optimal set approximation[8] shown schematically in Figure 2-6. However, we have already mentioned that the problem of approximating the Pareto optimal set $\Phi(P)$ is ill-posed in the sense of Tikhonov. Also, we have pointed out that to solve an ill-posed problem one must specify a metric/topology in the spaces where solutions are sought. This metric/topology must reflect the system of preferences on $\Phi(D)$. In this connection, let us recall some definitions[9].

A space X is called metric if for any pair of its elements x and y there exists a function d, named the distance between the elements, possessing the following properties:

1. $d(x, y) = d(y, x)$.
2. $d(x, y) \geq 0$, $d(x, y) = 0$ if and only if $x = y$.
3. $d(x, y) \leq d(x, z) + d(z, y)$ for any x, y, and z.

Examples of Metric Spaces

Euclidean space for which $d(\mathbf{x}, \mathbf{y}) = \sqrt{\sum_{i=1}^{n} (x_i - y_i)^2}$ weighted Euclidean space with $d(\mathbf{x}, \mathbf{y}) = \sqrt{\sum_{i=1}^{n} (\rho_i (x_i - y_i))^2}$, and the Hemming space for which $d(\mathbf{x}, \mathbf{y}) = \sum_{i=1}^{n} |x_i - y_i|$.

A metric d is said to be adequate to the system of preferences defined on the pairs of vectors $(\mathbf{x}, \mathbf{y})$ from the space under consideration, if inequality

[8]If necessary, the computer time needed for implementing this method may be shortened by constructing only the set $\Phi(P'_\epsilon)$ possessing property M_1. Moreover, one may approximate set $\Phi(P)$ by the union $\Phi(P_\epsilon) \cup SM$ without obtaining $\Phi(P'_\epsilon)$.

[9]Since this material is of a rather theoretical nature, it may be skipped by the reader interested only in applications of multicriteria optimization.

$d(\mathbf{x}, \mathbf{y}) \leq d(\mathbf{x}, \mathbf{z})$ implies that $\mathbf{y}$ is no less preferred than $\mathbf{z}$. Here, $\mathbf{x}$ is the most preferred vector from the region under consideration, and $\mathbf{y}$ and $\mathbf{z}$ are arbitrary vectors.

Let us continue considering our problem. Clearly, for each concrete case one has to consider a metric corresponding to the physical nature of the problem. The metric is to be adequate to the system of preferences in the space under consideration. In our opinion, there exists no metric that would be adequate to the system of preferences on $\Phi(D)$ in the general case. Even if it occasionally does exist, its construction is more complicated than the Pareto optimal set approximation. However, one may introduce a topology corresponding to the system of preferences on $\Phi(D)$.

A topology on the space X is such a system of its subsets τ for which the following conditions are satisfied:

1. The union of any number of sets from τ also belongs to τ.
2. The intersection of a finite number of sets from τ also belongs to τ.
3. $X \in \tau$ (Kelley 1957).

All the aforementioned spaces are topological. However, there exist numerous topological spaces that cannot be made metric. Nevertheless, the topological spaces both generalize and inherit several basic properties of metric spaces such as closeness, the neighborhood properties, convergence, etc.

A topology may be specified at a given space in a variety of ways. Most often it is specified by a system of neighborhoods for any point from X.

Let us define a topology τ on X by specifying the neighborhood

$$W_{\epsilon}(x)=\{\Phi(D_{\epsilon}) : \forall \Phi(\boldsymbol{\alpha}) \in X \;\; \exists \Phi(\boldsymbol{\beta}) \in \Phi(D_{\epsilon}) : |\Phi_{\nu}(\boldsymbol{\alpha})-\Phi_{\nu}(\boldsymbol{\beta})| \leq \epsilon_{\nu} \vee$$
$$\forall \Phi(\boldsymbol{\gamma}) \in \Phi(D_{\epsilon}) \; \exists \; \Phi(\boldsymbol{\eta}) \in X : |\Phi_{\nu}(\boldsymbol{\gamma})-\Phi_{\nu}(\boldsymbol{\eta})| \leq \epsilon_{\nu}, \; \nu=1,\ldots,k\}$$

for any $\mathbf{x} \in X$.

The neighborhood $W_{\epsilon}(\mathbf{y})$ for an arbitrary $\mathbf{y} \in Y$, and hence the topology $\hat{\tau}$ on Y, $\Phi(D) \subset Y$, is specified in a similar way.

Clearly, the topology introduced here is a Hausdorff topology satisfying the second countability axiom (Kelley 1957). Hence, convergence in this topology may be described in terms of sequences.

As is well-known, solution of an ill-posed problem reduces to the construction of a regularizing sequence. In the present case this is a sequence of sets $\Phi(P_{\epsilon^j})$, $j=1,\ldots\infty$, such that for the corresponding sequence $\Phi(D_{\epsilon^j})$ and any $\boldsymbol{\epsilon}^j$-neighborhoods of sets $\Phi(P)$ and $\Phi(D)$, sets $\Phi(P_{\epsilon^j})$ and $\Phi(D_{\epsilon^j})$, starting from a certain j_0 belong to the respective neighborhoods.

Suppose that in accordance with Lemmas 1 and 2 sequences $\Phi(P''_{\epsilon^j})$ and $\Phi(D''_{\epsilon^j})$, $P''_{\epsilon^j} \subset D''_{\epsilon^j}$ are constructed for the sequence of sets $\boldsymbol{\epsilon}^j$, $j=1,\ldots\infty$. Then the following theorem can be proved.

Theorem 4. Sequence $\Phi(P''_{\epsilon^j})$ is regularizing.

Proof. From the formulated property M (which is valid for any term of sequence $\Phi(P''_{\epsilon^j})$) and the definition of the neighborhood $W_\epsilon(\Phi(P))$ $(W_\epsilon(\Phi(D)))$, it follows directly that the conditions of regularizability of the sequence are satisfied.

Thus the problem of constructing the Pareto optimal set is solved.

Remark. As mentioned previously, in a number of works the problem of the Pareto optimal set regularization is solved using the Hausdorff metric defined by the distance

$$d(A, B)=\max\{\sup_{\mathbf{a}\in A}\inf_{\mathbf{b}\in B}\rho(\mathbf{a}, \mathbf{b}), \sup_{\mathbf{b}\in B}\inf_{\mathbf{a}\in A}\rho(\mathbf{a}, \mathbf{b})\}$$

where $\rho(\mathbf{a}, \mathbf{b})=\max_\nu|a_\nu-b_\nu|$, and a_ν and b_ν are the coordinates of vectors **a** and **b**; $A, B \subset X$.

The class of the problems described by this metric is rather limited, since its utilization for a somewhat general situation results, as a rule, in distortion of the designer's system of preferences because of, for instance, different significance of the performance criteria. Besides, since a variation in ρ may affect convergence, the question arises: Why is the Hausdorff metric to be generated by the above or some other prespecified distance $\rho(\mathbf{a}, \mathbf{b})$? Therefore, in the general case one has to introduce a topology similarly to how it was done here. This topology is a generalization of the Hausdorff metric. Roughly speaking, it operates in the same way as the Hausdorff metric does without, however, distorting the designer's system of preferences.

In conclusion of this section, we would like to note that prospects of the methods similar to the one presented here are connected with the development of "fast" approximation algorithms. Such algorithms can be based, for instance, on considering the problems in which the criteria belong to a more "narrow" class of functions as compared with the one we have studied. Thus, for a class of sufficient number of times differentiable functions the convergence rate may increase. Besides, the convergence rate may be increased by using the decomposition and aggregation methods discussed in Chapter 3.

2-3. Example of Approximation of a Feasible Solutions Set

Let us analyze the problem of approximation of a feasible region by considering the following example (Sobol' and Statnikov 1982). The vibratory system shown in Figure 2-7 consists of two identical masses $m_1=m_2=m$ connected by springs whose stiffnesses are $k_1=k_2=k$ and k_0. Such a system depends on the three design variables k, k_0, and m. Suppose the design variables lie within the following specified limits:

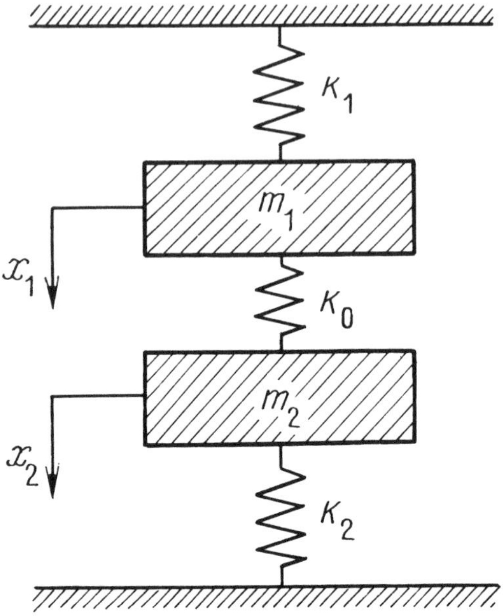

Figure 2–7 Oscillatory system.

$$k^* \leq k \leq k^{**}, \quad k_0^* \leq k_0 \leq k_0^{**}, \quad m^* \leq m \leq m^{**}. \tag{2-2}$$

The system can perform free harmonic oscillations with two natural frequencies ω_1 and ω_2. We will suppose that $0 \leq \omega_1 \leq \omega_2$. The formulas for the calculation of the natural frequencies are quite simple:

$$\omega_1 = \sqrt{\frac{k}{m}}, \quad \omega_2 = \sqrt{\frac{k+2k_0}{m}}. \tag{2-3}$$

Such dynamic systems are considered in many textbooks on the theory of oscillations, (see, e.g. Den Hartog (1956)).

In designing an oscillatory system one has to choose the natural frequencies in such a way as to avoid undesirable resonance phenomena. If the designer wishes to decrease ω_2 then he may introduce the criterion

$$\Phi_1 = \frac{k+2k_0}{m} \tag{2-4}$$

and assume that the smaller Φ_1 the more perfect the system is.

Suppose that alongside with decreasing frequency ω_2 the designer wishes to decrease the mass of the system $2m$. Then he may introduce another criterion

$$\Phi_2 = m \qquad (2\text{-}5)$$

and assume that the system is the better the smaller Φ_2.

Let us consider the criteria plane shown in Figure 2-8. To each set of design variables (k, k_0, m), there corresponds a pair of numbers Φ_1 and Φ_2 calculated from formulas (2-4) and (2-5), and hence, a point on the criteria plane. The same formulas allow one to find the set of points (Φ_1, Φ_2) on the criteria plane, which are calculated with the design variables (k, k_0, m) varying within the limits (2-2). The set is shown in Figure 2-8.

Consider the lower left-hand boundary of the set, formed by a segment of hyperbola

$$\Phi_1\Phi_2 = k^* + 2k_0. \qquad (2\text{-}6)$$

It can be readily shown that the points lying outside the hyperbola cannot correspond to the best solution.

To prove this statement we choose a point B (see Fig. 2-8). By drawing through the point a vertical and a horizontal line we get points B' and B'' belonging to the hyperbola. Since the abscissas of the points B' and B coincide (i.e., the values of Φ_1 at these points are equal) and the ordinate of B' is less than the ordinate of B (i.e., the value of Φ_2 corresponding to B' is smaller), the system corresponding to point B' is undoubtedly better than the one corresponding to point B.

Similarly, the system corresponding to point B'' (as well as to all points (except B) belonging to the curvilinear triangle $B'BB''$) may be shown to be undoubtedly better than the system corresponding to point B.

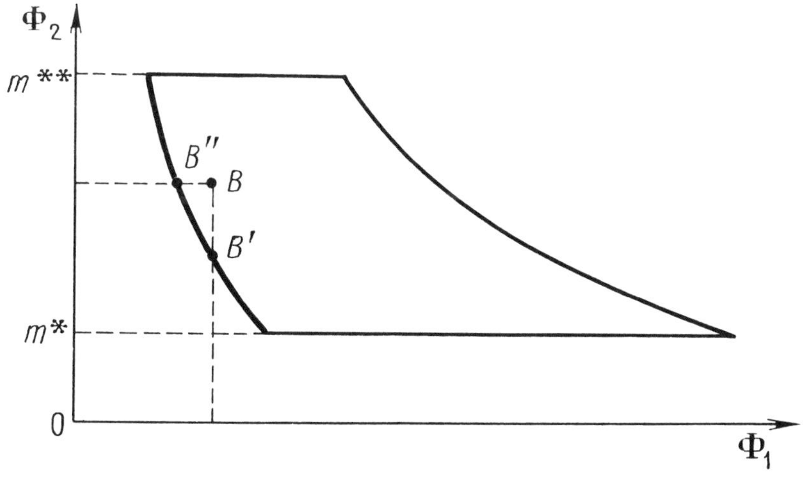

Figure 2–8 Criteria plane.

Hence, the best solutions should be sought among the systems corresponding to points belonging to a segment of hyperbola (Eq. (2-6)). This segment, shown in Figure 2-8 by a thick line, represents the Pareto optimal set.

Let constraints Φ_1^{**} and Φ_2^{**} be specified. In Figure 2-9 the set of points meeting these constraints is shaded. The set of design variables (k, k_0, m) satisfying conditions (2-2) is a parallelepiped Π in the three-dimensional design-variable space, see Figure 2-10. The set of points lying within parallelepiped Π and corresponding to curve (2-6) may readily be found. In fact, from Equations (2-4), (2-5), and (2-6) it follows that $k+2k_0=k^*+2k_0^*$, and since $k \geq k^*$ and $k_0 \geq k_0^*$ in Π we have $k=k^*$ and $k_0=k_0^*$.

Hence, the desired set of points is determined by conditions

$$k=k^*, \quad k_0=k_0^*, \quad m^* \leq m \leq m^{**}$$

and represents an edge of parallelepiped Π (see the thick line in Fig. 2-10). Naturally, point ($k=k^*$, $k_0=k_0^*$, $m=m^{**}$) lies on the edge.

Let us continue by finding the set of points in Π corresponding to the shaded region in Figure 2-9. From Equations (2-4) and (2-5) and inequalities $\Phi_1 \leq \Phi_1^{**}$ and $\Phi_2 \leq \Phi_2^{**}$ it follows that

$$m \leq \Phi_2^{**}, \quad k+2k_0 \leq \Phi_1^{**} \cdot m.$$

Together with (2-2), these equations define the portion of parallelepiped Π that is the feasible solutions set of points D (see Fig. 2-11).

Let k, k_0, and m be varying within the limits $2 \leq k \leq 6$, $1 \leq k_0 \leq 4$, $2 \leq m \leq 5$, and let us do the necessary calculations. In Table 2-1, the fragment

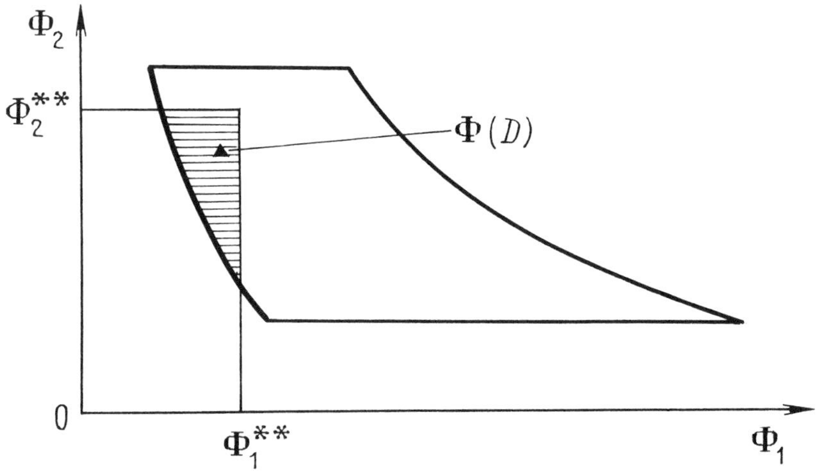

Figure 2–9 Feasible solutions set Φ (*D*) in the criteria plane.

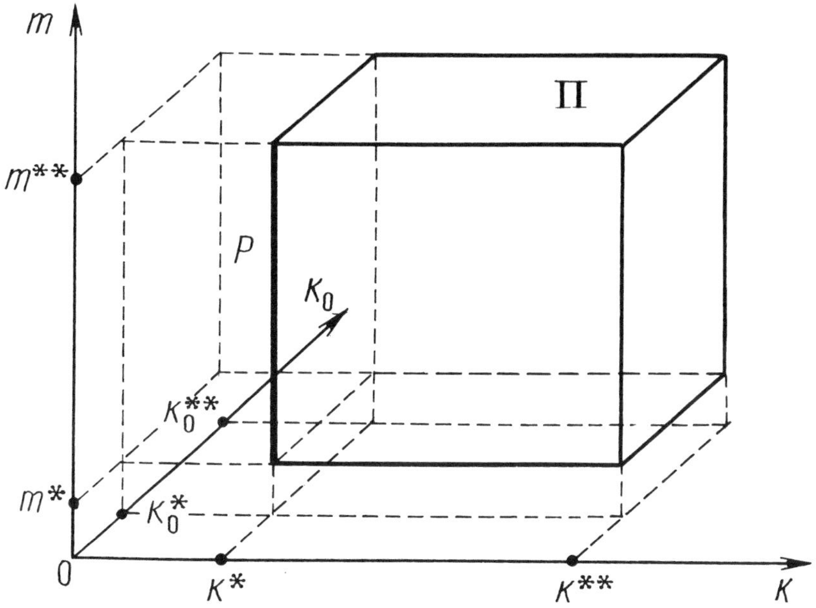

Figure 2–10 Pareto optimal set P in the design-variable space.

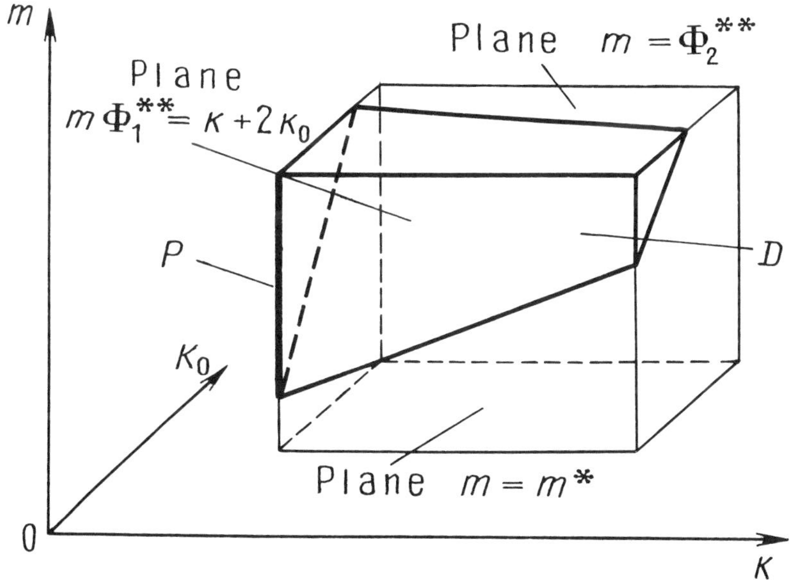

Figure 2–11 Feasible solutions set D in the design-variable space.

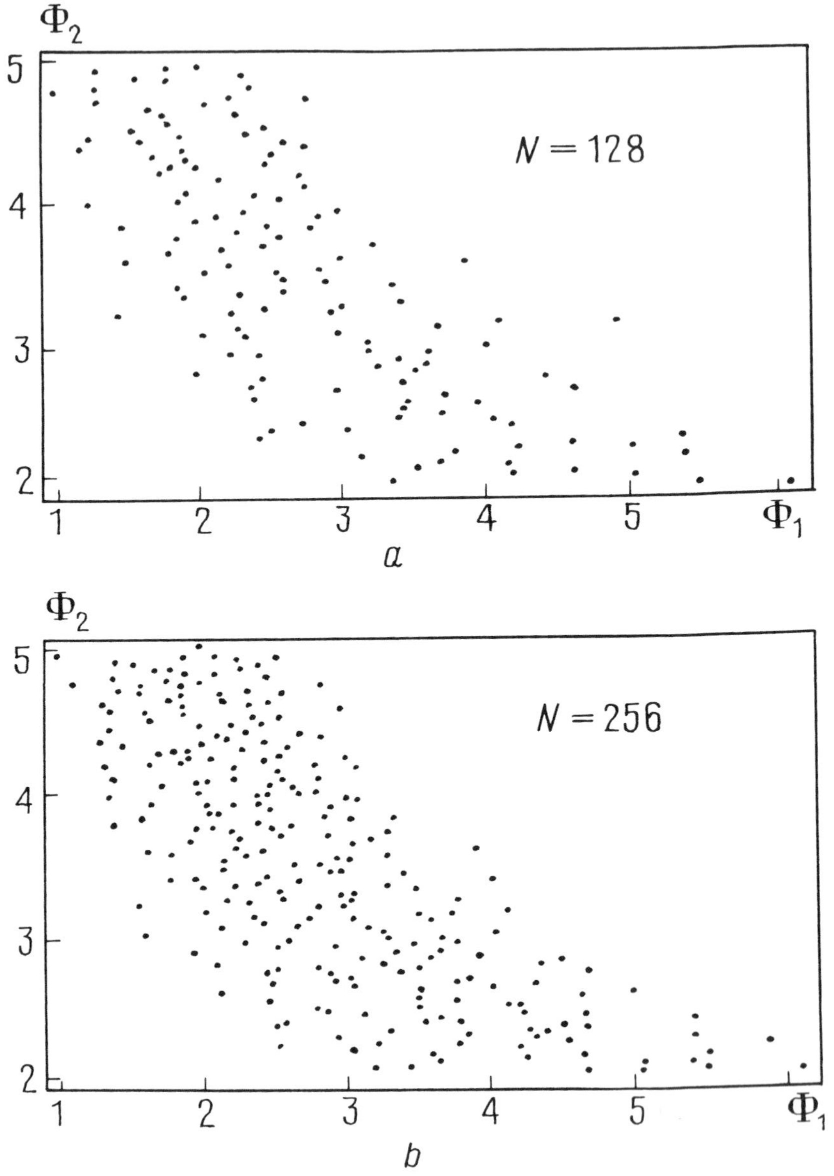

Figure 2–12 Approximation of Φ (D) for $N=128$ and 256.

of an unordered test table is given, the first 32 trials being represented. After 128 trials the criteria constraints Φ_1^{**} and Φ_2^{**} proved to be equal to 4.63 and 4.8 respectively. Figure 2-12*a* shows that 128 trials do not suffice to obtain a good approximation of the feasible solutions set in the criteria space. However, for $N=256$ the approximation is quite good, since $\epsilon_1=\epsilon_2=0.25$ (see Fig. 2-12b).

Table 2–1

α^i	$\Phi_1(\alpha^i)$	$\Phi_2(\alpha^i)$	α^i	$\Phi_1(\alpha^i)$	$\Phi_2(\alpha^i)$
1	2.5714	3.5	17	1.3377	4.7188
2	3.4545	2.75	18	2.7595	2.4688
3	2.0	4.75	19	2.9764	3.9688
4	1.7838	4.625	20	1.5478	3.5938
5	2.32	3.125	21	5.0448	2.0938
6	2.0	3.875	22	2.5468	4.3438
7	5.3684	2.375	23	3.5385	2.8438
8	2.4308	4.0625	24	1.5862	4.5313
9	3.4634	2.5625	25	4.0206	3.0313
10	1.3247	4.8125	26	2.562	3.7813
11	3.434	3.3125	27	3.8082	2.2813
12	2.2553	2.9375	28	3.7412	2.6563
13	2.6197	4.4375	29	2.1504	4.1563
14	4.1714	2.1875	30	1.8899	3.4063
15	2.2034	3.6875	31	2.3312	4.9063
16	2.2718	3.2188	32	1.8199	4.8594

3

Decomposition and Aggregation of Large-Scale Systems

3-1. Decomposition Methods

A large-scale system consists of a number of subsystems. For example, in a truck, one can separate the following subsystems: the frame, driver's cab, platform, engine, transmission, and steering system.

For the overwhelming majority of machines, there are rather many expensive units that can perform for a long time after the expiry of the normal period of the machine operation. In the case of production of millions of machines such as tractors, harvesters, motor cars, and machine tools, this leads to huge losses. This is caused by many factors, for example, drawbacks in the design. Very often, different departments of the design office engaged in creating a machine optimize their "own" subsystems ignoring others. The machine assembled from the "autonomously optimal" subsystems turns out to be far from perfect. A machine is a single whole. When improving one of its subsystems we can unwittingly worsen others. The subsystems are loaded in different ways and work in different conditions. It is desirable that the basic, most expensive units of a machine have equal durability and reliability indexes, be equally strong, etc. To meet this goal, we are to be able to find solutions hierarchically consistent with all subsystems. At present, such solutions are based mostly on the experience, intuition, and proficiency of a designer.

When designing machines, one has to deal with complicated mathematical models. Very often, these models have many hundreds of degrees of freedom, are described by high-order sets of equations, and the calculation of one solution can take an hour or more of computer time. This implies that it is not always possible to solve problems such as (1-1)–(1-4) directly (otherwise, we would have no problem with large-scale systems). One remedy may be to split (decompose) a large-scale system into subsystems that can be easily optimized, and then to aggregate the partial optimization results to obtain nearly optimal solutions for the whole system. This will allow a designer to determine requirements for the subsystems so as to make a machine optimal as a whole, and, by this, justify the proposals for designing different units of the machine.

The significance of using decomposition methods in solving optimal design problems for systems of high dimensionality consists in a considerable savings of computer time, whereas without using these methods many problems can not be solved practically at all.

At present, there are many exact decomposition methods that can be applied to solving problems of high dimensionality. Among them are Kron's methods (Kron 1963), the forces-and-displacements method (Goldman 1969), and the method of dynamic stiffnesses (compliances) (Craig and Bampton 1968).

Kron uses two sources of information: equations and graphs. When partitioning the system into subsystems, the graph of the system is first decomposed into comparatively large subgraphs for their subsequent unification. These methods are practiced on a large scale. In all these methods, when calculating each of the subsystems, the influence of the cut-off part of the system is replaced by its reaction. To obtain the solution for the whole system, we have to take into account the solutions of the equations of all subsystems, as well as the compatibility conditions for forces and displacements at the points of cutting the system.

Baranov, in the appendix to the Russian edition of the book by Kron (1963), shows that Kron's procedures are equivalent to matrix transformations such as the elimination of coordinates, introduction of additional unknowns, and permutation of rows and columns.

Application of Kron's methods to problems of mechanics has some features due to the multidimensionality of the graph branches and elastic interaction between inertial elements.

These methods are effective because they operate with subsystems matrices whose order is considerably less than the order of the matrix of the whole system. Note that Kron's methods are exact decomposition methods in the sense that they give the exact solution to the system. This solution can be obtained using direct methods without decomposition. However, the decomposition essentially facilitates the solving procedure. Apart from the exact methods, there are approximate techniques based on the decomposition. Next we consider some of them.

The Singular Perturbation Method

Consider a *k*th-order system of equations

$$\dot{\mathbf{x}}=\mathbf{A}\mathbf{x}+\mathbf{F}.$$

Within this system we separate *m* subsystems. Let us introduce a diagonal matrix $\boldsymbol{\epsilon}$ so that $\boldsymbol{\epsilon}\mathbf{A}=\bar{\mathbf{A}}$, with the elements of the matrix $\bar{\mathbf{A}}$ being close to unity. Then this system of equations can be represented by

$$\boldsymbol{\epsilon}_i \dot{\mathbf{x}}_i = \overline{\mathbf{A}}_i \mathbf{x}_i + \sum_{j=1\, j \neq i}^{m} \overline{\mathbf{A}}_{ij} \mathbf{x}_j + \overline{\mathbf{F}}_i, \quad i,j = \overline{1,m} \tag{3-1}$$

where $\boldsymbol{\epsilon}_i$ is the part of the matrix $\boldsymbol{\epsilon}$ corresponding to the ith subsystem. Small parameters are included here as factors of the derivatives. The change of $\boldsymbol{\epsilon}$ from a finite value to zero, or vice versa, leads to the change of the order of system (3-1). Such perturbations are called singular.

Within the matrix $\boldsymbol{\epsilon}$, let us separate its part $\overline{\boldsymbol{\epsilon}}$ with the elements being close to zero. Then (3-1) can be represented in the form

$$\begin{aligned} \dot{\mathbf{x}} &= f(\mathbf{x}, \mathbf{y}, t) \\ \overline{\boldsymbol{\epsilon}} \dot{y} &= g(x, y, t). \end{aligned} \tag{3-2}$$

The order of this system is $k = n + p$, where n and p are the dimensions of the vectors $\mathbf{x}$ and $\mathbf{y}$, respectively.

Having assumed $\overline{\boldsymbol{\epsilon}} = 0$, let us consider a more simple, degenerate system of the order n:

$$\begin{aligned} \dot{\bar{x}} &= f(\bar{x}, \bar{y}, t), \\ g(\bar{x}, \bar{y}, t) &= 0, \end{aligned} \tag{3-3}$$

where $\bar{x}(t)$ and $\bar{y}(t)$ are approximate solutions of system (3-2).

Tikhonov proved the theorem establishing the conditions that guarantee the convergence of the solution of the degenerate system to the solution of the full system, as $\boldsymbol{\epsilon} \rightarrow 0$ (Tikhonov 1952).

Aggregation of Variables

When studying systems with a large number of variables, we very often face the necessity of using amalgamated variables (aggregated variables, macrovariables) that are essentially less in number than the initial variables. In other words, the original system S_1 with n-dimensional state vector $\mathbf{x}$ is replaced by the system S_2 with l-dimensional state vector $\mathbf{z}$, l being considerably less than n. This replacement is done to mitigate the difficulties of the analysis of the system S_1 due to its high dimensionality. There are different interesting approaches to the variables aggregation (Lukyanov 1981).

Weakly Coupled Systems

In some cases, the behavior of a whole system can be described in terms of characteristics of its subsystems. In this connection, we encounter the problem of quantitative estimation of the presence of weakly coupled subsystems. Some of the estimates are given in Section 5-4. These problems are also studied in Tsurkov (1988) and Pel'tsverger (1984).

In Pervozanskii and Gaitsgori (1979), one can find some widespread decomposition methods for controlled systems.

Let us dwell on the possibility of using these methods to solve optimization problems. For linear systems, it is advisable to use exact decomposition methods, since in this case we obtain the same values of the performance criteria as when performing direct calculations. If this way does not lead to reducing the time necessary for the computations, one should use approximate methods. However, when doing this, it is necessary to make sure that these methods are applicable to the optimization problem, because even in case the approximate solutions of the system are obtained with sufficient accuracy, the error in calculating the performance criteria can appear to be unacceptably large. Therefore, we recommend estimating the solution accuracy that provides an acceptable error in calculating the criteria. If one meets difficulties in obtaining such estimates, it is necessary to calculate the values of the performance criteria for a restricted number of design-variable vectors by using the exact and approximate methods and, having compared the results, conclude about the applicability of the examined approximate method. A similar approach is described in Sections 5-2 and 5-3.

3-2. Construction of Hierarchically Consistent Solutions

We have dwelt on some methods of decomposition of large-scale systems. In a number of cases, after having applied these methods, we can optimize a large-scale system. However, for many systems, there are no effective methods similar to those described previously. There are a number of restrictions in using these methods. Even in the cases where these methods are applicable, the possibility of optimizing a large-scale system is not guaranteed yet. The reduction in time of calculating the system, though significant, can be insufficient for the optimization.

To solve this problem we can use another approach associated with considering the whole system as a hierarchical structure (Statnikov and Matusov 1989). The lower level of this structure comprises subsystems, whereas the higher level is the system as a whole. In many cases, the optimization can be done more simply at the lower level. Therefore, by using the results of the optimization at the lower level and, by this, reducing the number of competing solutions for the whole system, we can optimize the system in reasonable time. Such an approach was proposed comparatively recently, and only the first steps have been made in this direction (Krasnoshchokov et al. 1986). In particular, this is true for the methods proposed here. Nevertheless, the obtained results can be used for optimization of many large-scale systems.

Since the proposed approach is based on the optimization of the whole system through the optimization of its subsystems, we briefly describe the relation between the criteria for the system and subsystems. There are three possibilities for this relation.

1. Some of the criteria of the subsystem can implicitly affect the performance criteria of the system as a whole, and very often, such subsystem criteria are absent from the list of the performance criteria of the whole system. This situation is typical for the majority of complex engineering systems.
2. Some of the system criteria cannot be calculated at the subsystem level.
3. There are criteria that may be calculated both for the whole system and its subsystems.

The first two items are sufficiently simple. The third item is the most complicated.

To illustrate the last two items, let us consider the following example. In the structure of a slotting machine (a slotter), it is natural to separate three subsystems: the table, column, and hydraulic drive. The performance criteria for this system are the metal consumption, vibration resistance, processing accuracy, wear resistance of the guideways and tools, and dynamic forces in the junctions. All the criteria (apart from dynamic forces that can be determined only at the system level) can be calculated through the criteria of the subsystems: the table (the mass of the bed, the vibration amplitude of the workpiece, and wear resistance of the table guideways); the column (the mass of the column, the vibration amplitude of the cutting tool, and wear resistance of the ram guideways); and the hydraulic drive (the mass of the drive, the hydraulic cylinder diameter, and leak-proofness of the hydraulic drive).

In what follows, we give three ways (approaches, schemes) of searching for hierarchically consistent solutions in machine design problems. The first two schemes (A and B) are intended for optimization in comparatively simple cases, whereas the third scheme (C) is applicable to more complicated systems.

These three schemes do not cover all possible problems of machine design. However, by combining these basic approaches we can obtain other different methods for solving optimization problems for complicated systems.

The three methods have the following common features.

1. It is supposed that some of the mathematical models cannot be effectively optimized with respect to the whole criteria vector $\mathbf{\Phi}$, because it takes a great deal of computer time to formulate and solve problem (1-1)–(1-4). However, the calculation of the values of particular performance criteria Φ_ν needs a reasonable amount of computations.
2. The system is "partitioned" into subsystems. The couplings connecting the subsystems will be called external. To separate some of the subsys-

tems as autonomous, it is necessary to analyze the interaction of this subsystem with all other subsystems, as well as the external disturbances applied to the subsystem by environment. For example, in problems of dynamics, to determine the external disturbances they use the D'Alambert principle, the general dynamical equation of D'Alembert-Euler, and Lagrangian equations.

3. There are one or several criteria $\Phi_\nu(\boldsymbol{\alpha}^{(i)})$ of the ith subsystem that dominate corresponding criteria of other subsystems. This means that decreasing (increasing) the values of the criterion $\Phi_\nu(\boldsymbol{\alpha}^{(i)})$ (by no less than a certain amount $\epsilon_{\boldsymbol{\alpha}}$) entails decreasing (increasing) the value of the respective criterion $\Phi_\nu(\boldsymbol{\beta})$ for the whole system, compared with $\Phi_\nu(\boldsymbol{\alpha})$. Here, $\boldsymbol{\alpha}$ and $\boldsymbol{\beta}$ are the design-variable vectors of the system, and $\boldsymbol{\alpha}^{(i)}$ is the ith subsystem's vector of design variables corresponding to the vector $\boldsymbol{\alpha}$. This condition implies that the system contains one or several subsystems that determine the quality of the system in terms of the νth criterion.
4. It is supposed that the subsystems can be optimized by using methods of Chapter 1.
5. Let t be the total time of calculating the values of $\Phi_\nu(\boldsymbol{\alpha}^{(i)})$, $i=\overline{1,m}$, and T be the time of calculating the value of $\Phi_\nu(\boldsymbol{\alpha})$, where $\boldsymbol{\alpha}$ is the system design-variable vector corresponding to all $\boldsymbol{\alpha}^{(i)}$. Then the inequality $t \ll T$ is supposed to hold.

The idea of the optimization of the whole system consists in the following. First, when optimizing each (ith) subsystem, we obtain for this subsystem a pseudo-feasible solutions set $\overline{D}^i$, which, as a rule, is somewhat larger than the true feasible solutions set. After this, we compile the vectors for the whole system using the respective vectors from the sets $\overline{D}^i$. On the thus obtained domain, we check whether the criteria and functional constraints of the system are satisfied and, as a result, obtain the feasible solutions set D for the whole system. Finally, we search for the optimal solution over the set D.

The main point of this idea is the item 3. Let us consider it in more detail.

We will say that the pseudo-feasible solutions set $\overline{D}^i$ for the ith subsystem is dominant if the condition $\boldsymbol{\alpha}^{(i)} \notin \overline{D}^i$ entails $\boldsymbol{\alpha} \notin D$.

Assertion 1. In the systems satisfying the aforementioned conditions, there exist subsystems and criteria $\Phi_\nu(\boldsymbol{\alpha}^{(i)})$ such that the corresponding pseudo-feasible solutions sets $\overline{D}^i$ are dominant.

Proof. Let Φ_ν be a criterion satisfying the condition 3. The corresponding criterion constraint is given by $\Phi_\nu(\boldsymbol{\alpha}) = \Phi_\nu^{**}$. Here, as before, $\boldsymbol{\alpha}^{(i)}$ is the design-variable vector of the ith subsystem determined by the vector $\boldsymbol{\alpha}$. Denote by $\Phi_\nu^{i}{}^{**} = \Phi_\nu(\boldsymbol{\alpha}^{(i)})$ the constraint on the criterion Φ_ν for the ith subsystem. Let for some $\boldsymbol{\beta}^{(i)} \notin \overline{D}^i$, that is $\Phi_\nu(\boldsymbol{\beta}^{(i)}) > \Phi_\nu^{i}{}^{**}$, the inequality $\Phi_\nu(\boldsymbol{\beta}^{(i)}) \geq \Phi_\nu^{i}{}^{**} + \epsilon_{\boldsymbol{\alpha}}$ take place. Such $\boldsymbol{\beta}^{(i)}$ can always be chosen, since it is allowable to correct Φ_ν^{**} and hence, $\Phi_\nu^{i}{}^{**}$. Then, by the condition 3, for any vector $\boldsymbol{\beta}$, which the vector $\boldsymbol{\beta}^{(i)}$ of the ith subsystem corresponds to, we have $\Phi_\nu(\boldsymbol{\beta})>\Phi_\nu(\boldsymbol{\alpha})=\Phi_\nu^{**}$, that is, $\boldsymbol{\beta}\notin D$.

This assertion makes it possible to discard the design-variable vectors $\boldsymbol{\alpha}$ without calculation of the whole system, if the corresponding subvector $\boldsymbol{\alpha}^{(i)}$ violates the constraint $\Phi_\nu^{i}{}^{**}$. In other words, the optimization of the whole system is reduced, to a considerable extent, to the optimization of its subsystems. The schemes given next are based on this idea. These schemes are presented in the order of increasing their complication. We consider different relationships between the design variables of the system and its subsystems, discuss basic possibilities of simplifying the original model, the ways of determining external disturbances for subsystems, etc.

Scheme A

Let us have the mathematical models of subsystems that can be optimized (in reasonable time). We suppose that each component of the design-variable vector, $\boldsymbol{\alpha}=(\alpha_1,\ldots,\alpha_r)$, of the whole system is a component of at least one subsystem vector $\boldsymbol{\alpha}^{(i)}$ and, on the other hand, any component of the vector $\boldsymbol{\alpha}^{(i)}$ is a component of the vector $\boldsymbol{\alpha}$. Therefore, for each of the subsystems, the vector $\boldsymbol{\alpha}^{(i)}$ is uniquely determined by the vector $\boldsymbol{\alpha}$.

Taking this into account, we optimize the whole system. For each of the subsystems, we determine $\Phi_\nu^{i}{}^{**}$ satisfying condition 3. Regarding these constraints we construct pseudo-feasible solutions sets $\overline{D}^i$, see Chapter 1. For doing this, within the parallelepiped Π of the design variables for the whole system, N points are generated (in accordance with Section 1-2), and for each of these points, $\boldsymbol{\alpha}^j$, $j=\overline{1,N}$, the values $\Phi_\nu(\boldsymbol{\alpha}^{j(i)})$, $i=\overline{1,m}$, are calculated. The value $\Phi_\nu(\boldsymbol{\alpha}^{j(i+1)})$ is calculated only if $\boldsymbol{\alpha}^{j(i)}\in\overline{D}^i$. If it turns out that for any i and fixed j, $\boldsymbol{\alpha}^{j(i)}\in\overline{D}^i$, then, according to the condition 3, we will assume that $\boldsymbol{\alpha}^j$ belongs to a certain set $\overline{D}$ that is then used for determining the feasible solutions set D, $D\subset\overline{D}$. Otherwise, $\boldsymbol{\alpha}^i\notin\overline{D}$, and this vector is considered no longer.

For all $\boldsymbol{\alpha}^j\in\overline{D}$, we calculate the system as a whole, and also all $\Phi_\nu(\boldsymbol{\alpha}^j)$. Having done this, we determine Φ_ν^{**}, $\nu=\overline{1,k}$, and thereby, the feasible solutions set D. If the set D turns out to be empty ($D=\emptyset$), one should increase the number N of the points generated within the parallelepiped Π. After having found D in accordance with Chapter 1, we construct the Pareto optimal set P.

It should be noted that if it is possible to approximate the sets $\overline{D}^i$, the approximation of the feasible solutions set for the whole system can be constructed.

Example

Multicriteria optimization of cutter loaders (Dokukin et al. 1987). The mathematical model was taken that described the dynamics of cutter loaders with branching drive diagram and nonsymmetrical arrangement of cutting members.

This model is represented by a high-order system of nonlinear differential equations with stochastic functions on right-hand sides. The calculation of the model requires a lot of computer time, and for this reason, it is impossible to optimize the cutter loader to a sufficient extent by using conventional optimization techniques (the model contains 25 design variables and 20 criteria to be optimized). In this connection, we decomposed the system into three subsystems: drives, the arrangement of the cutter loader, and arrangement of the cutting tools on the machine.

The performance criteria have been taken as follows. The first criterion, Φ_1, reflects the unbalance of the actuator; the criteria Φ_2–Φ_5 characterize the nonuniformity of transmission loads when cutting the coal massif; Φ_6 and Φ_7 represent the nonuniformity of the driving torque of the electric motor; Φ_8 and Φ_9 give the life expectancy of the transmission with respect to the fatigue strength; Φ_{10} determines the probability of the electric motor reversal; Φ_{11} and Φ_{12} describe vertical displacements of the goaf side and face side of the machine housing and are used for estimating the stability. The criteria Φ_{13}–Φ_{16} characterize the variation of vertical displacements for each of the supports of the cutter loader, while the others describe the excess over the limiting level of the support displacements.

After the optimization of the first subsystem with respect to the criteria Φ_2–Φ_9 we obtained the pseudo-feasible solutions set that contained 14 models. For these models, we calculated the criteria Φ_{11}–Φ_{20} related to the second subsystem. The constraints for these criteria were satisfied by 9 of the 14 models. The third subsystem was optimized with respect to the force unbalance of the cutting member. Four of the nine models satisfied the constraint related to this criterion. According to Scheme A, for these four models, we calculated the criteria related to the whole system. After that we determined the optimal solution.

The optimization permitted us to reduce the dynamic load of the drive, to make the cutter loader more stable, and to improve other performance criteria.

It should be mentioned, however, that this procedure is effective only when applied to comparatively simple mechanisms, machines or their units. In more complicated cases, the assumption concerning the relationship between the vector of design variables $\boldsymbol{\alpha}$ of the whole system and respective vectors $\boldsymbol{\alpha}^{(i)}$ for subsystems is not valid, and we have to use Schemes B and C. Here, the situations are possible, when the design-variable vector of the whole system contains components that are absent from the subsystem level. This can take place, for example, if it is impossible to take into account correctly some external couplings when calculating the subsystem. Therefore, these couplings are usually ignored. Vice versa, among the subsystems design variables, there can be those weakly (if at all) affecting the performance criteria of the system to be optimized. As a rule, these design variables are not included in the list of design variables of the whole system.

There are also other ways for obtaining optimal solutions for the considered systems. For example, different criteria may require considerably different time for their calculations: say, the calculation of the ith criterion takes seconds, while the calculation of the pth criterion requires minutes. In this case, we divide the set of all criteria into n groups so that the time required for calculating criteria of some group significantly differs from the time corresponding to criteria of any other group. These groups are arranged in order of increasing computation time. Then we optimize the system with respect to criteria of the first group, and thus construct the pseudo-feasible solutions set $\overline{D}_1$. After this, for the models from $\overline{D}_1$, we calculate criteria of the second group, determine criteria and functional constraints, and construct the set $\overline{D}_2$. The process is repeated until the set $\overline{D}_n=D$ is constructed. Obviously, this procedure considerably reduces the time of optimizing the whole system.

Scheme B

Unlike Scheme A, we assume here, that the original model is simplified so that it becomes amenable to optimization. Here, external couplings between subsystems are retained, and the simplification is due to either aggregation of solutions for subsystems (this has been mentioned already) or aggregation of internal design variables of the subsystems. For example, if the subsystem contains masses $m_1,\ldots,m_n$, they (or part of them) can be replaced by the mass $M = \sum_{i=1}^{p} m_i$, $p \leq n$. This reduces the number of the subsystem design variables, the criteria, as a rule, being modified.

Let the optimization of the simplified system be carried out, and the corresponding feasible solutions set $\overline{\overline{D}}$ be constructed in accordance with Chapter 1. Consider the ith subsystem, $i=\overline{1,m}$. External disturbances acting on this subsystem are determined by taking into account design-variable vectors belonging to $\overline{\overline{D}}$.

Consider a vector $\boldsymbol{\gamma} \in \overline{\overline{D}}$ and separate the components of $\boldsymbol{\gamma}$ that are used when calculating external disturbances for the ith subsystem. These components form the vector of external couplings, $\tilde{\boldsymbol{\gamma}}$, for the given subsystem. Given the external disturbances, we calculate performance criteria $\Phi_\nu(\boldsymbol{\alpha}^{(i)})$ and determine constraints Φ_ν^{i**} and pseudo-feasible solutions sets $\overline{D}^i$ for each subsystem. Note that when optimizing the ith subsystem, its internal design variables are not aggregated but used in the form in which they are included in the original system.

By virtue of condition 3, the vectors $\boldsymbol{\alpha}$ for which $\boldsymbol{\alpha}^{(i)} \notin \overline{D}^i$ are excluded from further consideration.

The obtained sets $\overline{D}^i$ are unified to form a whole set. For simplicity, let us assume that there are only two subsystems. Consider the aforementioned coupling vector $\tilde{\boldsymbol{\gamma}}$. We separate all vectors $\boldsymbol{\alpha}$ of $\overline{D}^1$ and $\boldsymbol{\beta}$ of $\overline{D}^2$ that have been obtained when optimizing the subsystem under consideration by taking account of the vector $\tilde{\boldsymbol{\gamma}}$, and form augmented vectors $(\boldsymbol{\alpha}, \tilde{\boldsymbol{\gamma}}, \boldsymbol{\beta})$. This operation is repeated for all other vectors of external couplings similar to $\tilde{\boldsymbol{\gamma}}$. We call this operation the concatenation of subsystems. If there are more than two subsystems, their concatenation is made in a similar way. As a result, we obtain the set of vectors of the whole system $\tilde{D}$. At this point, we are to check whether the obtained vectors satisfy all the criteria and functional constraints that were determined when optimizing design variables corresponding to the simplified model. In the general case, the set $\tilde{D}$ reduces after this check, and we obtain the set $\overline{D} \subset \tilde{D}$ for which criteria of the original system are calculated, and the feasible solutions set D is found in accordance with Chapter 1.

It should be noted, however, that the set D can appear to be empty. This may take place if, when optimizing the ith subsystem, we cannot construct the whole feasible solutions set for this subsystem with acceptable accuracy. If D turns out to be empty, one should go back to the stage of the subsystems optimization, calculate a number of vectors $\boldsymbol{\alpha}^{(i)} \in \overline{D}^i$ and check whether D is nonempty. This has to be repeated until we obtain $D \neq \varnothing$. However, if we succeed in constructing approximations of the sets $\overline{D}^i$, $i=\overline{1,m}$, we can guarantee not only that D is nonempty but also that $\overline{D} \supset D$. The latter is established by the following assertion.

Assertion 2. The set $\overline{D}$ is an approximation of the feasible solutions set D, and $\overline{D} \supset D$.

Here we do not give the proof of this assertion. This proof is similar, with the exception of some minor details, to the proof of Assertion 3 given later.

Note that attempts to simplify the original model by means of aggregating internal design variables of subsystems, provided external couplings are conserved, are widely practiced in design. A modification of this scheme has been used for solving the problem from Section 3-3.

Scheme C

Suppose the simplification of the original model corresponding to Scheme B does not permit us to optimize subsystems in reasonable time. In this case, we can try to achieve success by ignoring some couplings between subsystems or aggregating these couplings as is done with internal design variables of subsystems in Scheme B. As a result, the number of criteria can reduce compared with the original model. The very criteria can appear to be altered too.

If such a simplification of the model permits us to optimize it, the solution reduces to the application of Scheme B. If this is not the case, we suppose that the system contains a sufficient number of design variables that influence criteria of the subsystem in which they are included, and do not affect criteria of other subsystems. By sufficiency we understand that each of the subsystems can be optimized, provided the previous condition is fulfilled. This condition is also necessary because, if it turns out that criteria of some subsystem depend on all or almost all design variables of the whole system, it will be difficult to optimize this subsystem as the whole system. If this condition is satisfied, we can optimize subsystems using the following two ways.

1. Suppose we can optimize the simplified system for fixed values of the design variables not influencing the ith subsystem $i=\overline{1,m}$. In other words, we can optimize the simplified system in reasonable time, having fixed the system design variables that do not influence criteria of the examined subsystem. External disturbances acting on the subsystem are determined as a result of computations related to the simplified model.
2. If the assumption of item 1 is not valid, the simplified model is not considered. In this case we construct simplified models for each of the subsystems. The simplification of the subsystem model is regarded as acceptable if at least one of the subsystem performance criteria can be calculated with sufficient accuracy and, in addition, the constraint related to this criterion permits us to exclude from consideration a sufficiently large number of design-variable vectors $\boldsymbol{\alpha}$. Having been considered separately, such models of subsystems are not of practical interest. However, provided we have a model of the whole system and the conditions defined above are satisfied, these models facilitate the optimization of the whole system. Note here, that external disturbances acting on subsystems are determined not from the model of the whole system, as it took place previously, but from the very subsystems models.

Both in the cases 1 and 2, the subsystem optimization differs from that in the previous procedures. Here, the design variables of external couplings of the subsystem are not fixed but vary simultaneously with design variables of other subsystems affecting the examined one.

So, let all subsystems be optimized, and for each of the subsystems, the pseudofeasible solutions set $\overline{D}^i$, $i=\overline{1,m}$, has been obtained according to Chapter 1.

We define the concatenation operation for the sets $\overline{D}^j$, $j=\overline{1,m}$, as follows. Denote by $\overline{D}_{1,2}$ the set consisting of vectors $\boldsymbol{\alpha}=(\boldsymbol{\alpha}^{(1)}, \boldsymbol{\alpha}^{(2)})$, $\boldsymbol{\alpha}^{(1)} \in \overline{D}^1$, $\boldsymbol{\alpha}^{(2)} \in \overline{D}^2$, such that common (i.e., influencing both subsystems) design variables included both in $\boldsymbol{\alpha}^{(1)}$ and $\boldsymbol{\alpha}^{(2)}$ assume equal values. (If some design variables, such as describing external couplings, have been omitted when calculating the subsystem, they are added to $\boldsymbol{\alpha}^{(1)}$ and $\boldsymbol{\alpha}^{(2)}$ when constructing vector $\boldsymbol{\alpha}$). We will denote the result of iterating this operation m times by $\overline{D}_{1,\ldots,m}=\tilde{D}$ and call the set $\tilde{D}$ the superstructure over the sets $\overline{D}^1,\ldots,\overline{D}^m$.

This definition allows us to aggregate different subsystems into the whole system by means of concatenation of their design-variable vectors.

Let the sets $\overline{D}^j$ be defined for all $j=\overline{1,m}$. The set $\overline{D}$ consisting of the design-variable vectors of the whole system $\boldsymbol{\alpha}$ such that $\boldsymbol{\alpha}^{(j)} \in \overline{D}^j$, $j=\overline{1,m}$, is called the pseudofeasible solutions set for this system.

Let us give the idea of the algorithm for constructing the feasible solutions set D. Let us take two subsystems of those obtained after partitioning the system.

Suppose, there are n common design variables influencing the criteria of both subsystems. We denote these design variables by $\alpha_1^{(j)},\ldots,\alpha_n^{(j)}$, $j=1,2$. Let us take an arbitrary vector $\boldsymbol{\alpha}^{(1)} \in \overline{D}^1$ and fix the values of the components $\alpha_1^{(1)},\ldots,\alpha_n^{(1)}$ of this vector. We assume that when probing design-variable spaces of the subsystems, we use the points of P_τ-nets for each of the subsystems. Then, since common design variables $\alpha_1^{(j)},\ldots,\alpha_n^{(j)}$ are first in each of the subsystems, they will assume the same values at all points with identical numbers (see Addendum). In $\overline{D}^2$, we find vectors $\boldsymbol{\alpha}^{(2)}$ whose first n components assume values equal (to the specified accuracy) to the values of the respective components of the vector $\boldsymbol{\alpha}^{(1)}$. After this, we concatenate the vectors $\boldsymbol{\alpha}^{(2)}$ with the vector $\boldsymbol{\alpha}^{(1)}$. As a result, we obtain vectors $\boldsymbol{\alpha}=(\boldsymbol{\alpha}^{(1)}, \boldsymbol{\alpha}^{(2)}) \in \overline{D}_{1,2}$. If we find no vector $\boldsymbol{\alpha}^{(2)} \in \overline{D}^2$ that can be concatenated with the vector $\boldsymbol{\alpha}^{(1)}$, the vector $\boldsymbol{\alpha}^{(1)}$ is considered no longer. Having done this operation with all vectors of $\overline{D}^1$, we obtain the superstructure $\overline{D}_{1,2}$.

If there are m subsystems, the process of constructing the superstructure $\tilde{D}$ is similar. We are only to provide that the concatenation condition is satisfied. After having constructed $\tilde{D}$ we calculate the system only at the points of this set.

Thus, the original model is calculated repeatedly. However, it is done only on the set $\tilde{D}$. If the number of elements in $\tilde{D}$ is not too large, the optimization of the whole system in a reasonable time becomes possible. After having introduced constraints Φ_ν^{**} we obtain the feasible solutions set D.

Note that here, as in Scheme B, the case is possible where D turns out to be empty. In this case one should repeat all the described operations until $D\neq\varnothing$. However, D cannot be empty if one succeeds in approximating the sets $\overline{D}^i$, $i=\overline{1,m}$. We denote these approximations by $\overline{\overline{D}}^i$. The following formulation is valid.

Assertion 3. The set $\tilde{D}$ being a superstructure over the sets $\overline{\overline{D}}^i$, $i=\overline{1,m}$, approximates the pseudofeasible solutions set, $\overline{D}$, of the whole system with a prescribed accuracy.

Proof. We will consider without loss of generality that there are only two subsystems ($m=2$). For $m>2$, the proof is analogous. Let $\boldsymbol{\alpha}^{(1)}=(\alpha_1^{(1)},\ldots, \alpha_p^{(1)}) \in \overline{\overline{D}}^1$ (more exactly, $\boldsymbol{\alpha}^{(1)}$ belongs to a neighborhood of a pseudofeasible solutions set of the first subsystem) and $\alpha_{c_1}^{(1)},\ldots,\alpha_{c_n}^{(1)}$ be the design variables of the first subsystem that influence the second subsystem criteria. The approximation of the set $\Phi(\overline{D}^i)$, $i=1,2$, is constructed following the algorithm given in Chapter 1, except for the following. In $\Phi(\overline{D}^1)$ and $\Phi(\overline{D}^2)$, all points must be approximated with the accuracy up to $\boldsymbol{\epsilon}^{(j)}=\{\epsilon_\nu^{(j)}\}$, $j=1,2$. These $\boldsymbol{\epsilon}^{(j)}$ are chosen so as to satisfy the condition that for any $\nu=\overline{1,k}$, $\boldsymbol{\alpha},\boldsymbol{\beta}\in\overline{D}$, and ϵ_ν the inequalities $|\Phi_\nu(\boldsymbol{\alpha}^{(j)}) - \Phi_\nu(\boldsymbol{\beta}^{(j)})|<\epsilon_\nu^{(j)}$, $j=1,2$, imply $|\Phi_\nu(\boldsymbol{\alpha})-\Phi_\nu(\boldsymbol{\beta})|<\epsilon_\nu$. Here $\overline{D}$ is the set of design variables of the whole system that satisfy the constraints $\Phi_\nu^j{}^{**}$, ϵ_ν is the admissible error for the criterion Φ_ν. It is easy to show that $\boldsymbol{\epsilon}^{(j)}$ always exist, due to the continuity of functions Φ_ν and closedness of the domain $\overline{D}$.

The vectors of the approximation of the set $\Phi(\overline{D}^2)$ must approximate also any values of the vector $(\alpha_{c_1},\ldots,\alpha_{c_n})$ with the accuracy up to $\delta_{c_1},\ldots,\delta_{c_n}$ respectively. Here, δ_{c_i} is the admissible error for the design variable α_{c_i}. To obtain this approximation it is sufficient to put $\Phi_{k+i}=\alpha_{c_i}$, $i=\overline{1,n}$. Instead of cubes covering the domain D (see the approximating algorithm in Section 2-1), one can take parallelepipeds with the edges corresponding to coordinates α_{c_i} being of the length δ_{c_i}. The lengths of the other edges are determined from the Lipschitz condition, as in the case of cubes. This additional property is necessary for the concatenation of design-variable vectors of different subsystems.

The concatenation results in the set of vectors $\overline{\boldsymbol{\alpha}}^j=(\boldsymbol{\alpha}^{(1)}, \boldsymbol{\alpha}_j^{(2)})$ with $\boldsymbol{\alpha}_j^{(2)} \in \overline{\overline{D}}^2$. Therefore $\overline{\boldsymbol{\alpha}}^j \in \tilde{D}$. Similar operation is to be done for all vectors $\boldsymbol{\alpha}^{(1)} \in \overline{\overline{D}}^1$. As a result, we obtain the set $\tilde{D}$ that approximates the pseudofeasible solutions set of the whole system with the prescribed accuracy. Indeed, let $\overline{\boldsymbol{\alpha}} \in \overline{D}$, $\overline{\boldsymbol{\alpha}}^{(1)} \in \overline{D}^1$, and $\overline{\boldsymbol{\alpha}}^{(2)} \in \overline{D}^2$ are vectors corresponding to $\overline{\boldsymbol{\alpha}}$. It is known that $\boldsymbol{\Phi}(\overline{\boldsymbol{\alpha}}^{(1)})$ and $\boldsymbol{\Phi}(\overline{\boldsymbol{\alpha}}^{(2)})$ can be approximated by the vectors $\boldsymbol{\Phi}(\hat{\boldsymbol{\alpha}}^{(1)})$ and $\Phi(\hat{\boldsymbol{\alpha}}^{(2)})$, respectively, with accuracy up to $\boldsymbol{\epsilon}^{(j)}$, where $\boldsymbol{\epsilon}^{(j)}$ are specified beforehand. Hence, $\boldsymbol{\Phi}(\overline{\boldsymbol{\alpha}})$ is approximated by the vector $\boldsymbol{\Phi}(\hat{\boldsymbol{\alpha}})=\boldsymbol{\Phi}(\hat{\boldsymbol{\alpha}}^{(1)}, \hat{\boldsymbol{\alpha}}^{(2)})$, with accuracy up to $\boldsymbol{\epsilon}$.

In case $m>2$ (m is the number of subsystems), we successively concatenate the pseudofeasible solutions set for the ith subsystem with the result of concatenation of corresponding sets for previous $i-1$ subsystems. When doing this, we take into account the design variables that are common for the ith subsystem and for at least one of the $i-1$ subsystems.

Corollary. The pseudofeasible solutions set $\overline{D}$ contains the approximation of the feasible solutions set, D, for the whole system.

Proof. Indeed, let us calculate the whole system at the points of the set $\overline{D}$. According to Chapter 1, we determine the constraints Φ_ν^{**}. Now it is easy to see that the points of $\overline{D}$ satisfying the constraints Φ_ν^{**} just form the desired approximation of the set of feasible design variables for the whole system. This completes the proof.

In connection with the aforementioned, it is interesting to discuss the possibility of reducing the number of vectors taking part in the concatenation of subsystems, and by this, reducing the very set $\tilde{D}$ on which criteria of the whole system are calculated.

In systems that we usually deal with, the monotonicity, as a rule, takes place, that is, performance criteria of the whole system, $\Phi_\nu(\boldsymbol{\alpha})=\tilde{\Phi}_\nu(\Phi_\nu(\boldsymbol{\alpha}^{(1)}), \ldots,\Phi_\nu(\boldsymbol{\alpha}^{(m)}))$, monotonically depend on $\Phi_\nu(\boldsymbol{\alpha}^{(i)})$.

Let us establish the condition ensuring the possibility of reducing the number of the design-variable vectors. Let the design-variable vector of the ith subsystem be $\boldsymbol{\alpha}^{(i)}=(\boldsymbol{\alpha}_{(1)}, \boldsymbol{\alpha}_{(2)})$, where $\boldsymbol{\alpha}_{(1)}$ and $\boldsymbol{\alpha}_{(2)}$ represent internal design variables of the subsystem and the external couplings design variables, respectively. Let us consider $\Phi_\nu(\boldsymbol{\alpha}^{(i)})$, $\boldsymbol{\alpha}^{(i)} \in \overline{D}^i$. Let for any $\boldsymbol{\beta}_{(2)}$, for which there exists the vector $\boldsymbol{\beta}^{(i)}=(\boldsymbol{\alpha}_{(1)}, \boldsymbol{\beta}_{(2)}) \in \overline{D}^i$, one can find the vector $\overline{\boldsymbol{\beta}}_{(1)}$ such that $\overline{\boldsymbol{\beta}}^{(i)}=(\overline{\boldsymbol{\beta}}_{(1)}, \boldsymbol{\beta}_{(2)}) \in \overline{D}^i$ and $\Phi_\nu(\boldsymbol{\alpha}^{(i)}) \approx \Phi_\nu(\overline{\boldsymbol{\beta}}^{(i)})$. In other words, the changes of criteria $\Phi_\nu(\boldsymbol{\alpha}^{(i)})$ caused by changing the design variables of external couplings can be compensated by changing internal design variables of the subsystem. Then the following statement is valid.

Let $\boldsymbol{\alpha} \in \overline{D}$ result from the concatenation of vectors $\boldsymbol{\alpha}^{(1)},\ldots,\boldsymbol{\alpha}^{(m)}$, where $\boldsymbol{\alpha}^{(i)} \in \overline{D}^i$ and $\boldsymbol{\alpha}^{(i)} \notin \overline{P}^i$($\overline{P}^i$ is the Pareto optimal set in $\overline{D}^i$). Then $\boldsymbol{\Phi}(\boldsymbol{\alpha}) \notin \Phi(P)$, where P is the Pareto optimal set of design-variable vectors of the whole system.

This formulation makes it possible not to concatenate vectors that do not belong to Pareto optimal sets of respective subsystems. This does not influence the Pareto optimal set of the whole system. As a result, we can construct the set reduced in comparison with $\tilde{D}$, and it will be sufficient to carry out calculations of the whole system at points reduced in number.

The main applicability condition for Scheme C requires the existence of a sufficient number of design variables that do not influence the ith subsystem $i=\overline{1,m}$. Let us illustrate this condition by an example of a slotting machine[10] (unlike Section 4-3, here we consider another dynamic model of the machine). We can indicate the subsystems design variables that are most essential when calculating performance criteria for these subsystems.

The whole system contains 25 design variables: α_1, α_2, α_3, α_4 are the stiffness and damping coefficients of the joint between the bed and column; α_5 is the ram mass; α_6, α_7 represent the stiffness and damping coefficients of the junction between the ram and guideways; α_8, α_9, α_{10}, α_{11} are the stiffness and damping coefficients of hydraulic drive units; etc. When considering performance criteria of a subsystem, one can conclude that it is impossible to calculate these criteria regarding only the design variables immediately related to this subsystem. For example, performance criteria of the column depend on the table design variables, α_1–α_4. Therefore, to calculate criteria for each of the subsystems, one should take into account all system design variables influencing these criteria.

It has been established that design variables of the table do not influence the hydraulic drive criteria. The column criteria depend on four design variables of the hydraulic drive, α_8, α_9, α_{10}, and α_{11}. Design variables of the column, in turn, do not influence the hydraulic drive, but criteria of the table depend on the column design variables (α_5, α_6, α_7). The hydraulic drive design variables do not influence performance criteria of the table. It has been established that the first subsystem (the table) contains 10 design variables, the second subsystem

[10]This material was kindly given us by E.V. Khlebalov, the researcher of Experimental Research and Development Institute for Metalcutting Machine Tools.

(the column) contains 16 design variables, and the third one (the hydraulic drive) contains seven design variables. Thus, each subsystem contains significantly fewer design variables than the whole system, and the condition in question is fulfilled. (In general, to investigate the influence of design variables on performance criteria of subsystems or the system as a whole, one can use the methods given in Sections 5-1 and 5-2.)

In our example, Scheme C is used as follows. After having optimized the subsystems of the slotting machine, the set $\overline{D}^1$ of design-variable vectors is concatenated with vectors of $\overline{D}^2$, taking into account their common design variables, and then, with vectors of $\overline{D}^3$. The first and second subsystems are concatenated by their common design variables, α_1–α_7. Vectors of the resultant superstructure, $\overline{D}_{1,2}$, are concatenated with vectors of the third subsystem by the design variables α_8–α_{11}. As a result, we obtain the set $\tilde{D}$ for the whole system, and then, the set D.

Thus, when describing Schemes A, B, and C, we have consecutively considered the basic ways of simplifying the original model, depending on relationships among design variables and criteria of subsystems and the whole system. Many other ways of optimizing large systems can be obtained by combining the schemes given previously. However, two important cases have not been considered: when the very subsystems are large-scale systems and cannot be effectively optimized, and when there is no mathematical model of the whole system. Both these cases are typical for such machines as airplanes, ships, spacecrafts, and motor cars.

In the first case, it is reasonable to optimize subsystems independently. For the optimization of each of the subsystems, it is advisable to use one of the three schemes described previously or their combinations. The resultant pseudofeasible solutions sets for subsystems are to be aggregated to form the feasible solutions set for the whole system.

In the second case, as has already been mentioned, it is impossible or difficult to create a mathematical model of the system. For example, it is very difficult to create the general model of an airplane that could be calculated in reasonable time and, at the same time, would take into account all basic criteria (aerodynamics, weight, dynamics, strength, economics, altitude, speed, different characteristics of the engine). As a rule, they choose another way. They create the bank of various mathematical models. In conformity with an airplane, these are aerodynamic, weight, economic, and other models. Taken together, these models describe all basic criteria. Many of the models have common design variables and criteria, these criteria often being contradictory.

Applying one of these aforementioned schemes (or their combination) and obtaining the set $\tilde{D}$ for the whole system as a result of concatenation are recommended. Although it is practically impossible to calculate criteria of the whole system exactly, very often one can evaluate the quality of the system using indirect approaches, for example, experiments. Estimates of the performance criteria obtained thus can be used for optimization of the system over the feasible solutions set $\bar{D}$. Some other approaches can be found in Statnikov and Matusov (1989). It should be noted, however, that by now, there are no satisfactory optimization methods for the overwhelming majority of large-scale complex systems. This circumstance stimulates the development of approaches similar to those mentioned here.

3-3. Example: The Construction of Consistent Solutions to the Problem of Calculation of a Car for Shock Protection

As has already been mentioned, optimization of large-scale systems envisages the substantiation of the decomposition of the system (including the generation of mathematical models for subsystems and finding out impacts influencing the subsystems), the determination of constraints and criteria vectors for the subsystems, finding the set of solutions consistent with all the subsystems, and the search for the optimal solution for the entire system. Because of great CPU consumption required for calculating the criteria vector of the whole system, the procedure of searching for the optimal solution must be organized so as to reduce the number of calls for the system as a whole when calculating its separate criteria and other characteristics as much as possible. This is demonstrated by the example in question (Bondarenko et al. 1994).

Cars of serial production must meet modern requirements concerning reliability, safety, noise level, etc. For example, there are different norms of testing cars for shock protection. These norms require that the car body acquires no damage after having been hit by a block head that has mass equal to that of the car and moves at a speed of 8.9 km/h on a horizontal plane, at an angle of 30 degree to the longitudinal axis of the car, and at a height of 500 mm above the ground surface. Figure 3-1*b* shows the tested unit of the car, which consists of a plastic bumper, an insert made of expanded polyurethane, and a rear panel of the car body.

The experiments carried out at the plant show that the structure in question is imperfect: In the case of lateral impact the bumper is damaged and dents are left on the car body. Therefore, it is necessary to try to improve the prototype of the structure, to give recommendations that would provide damage protection of the car body, and also to find the optimal solution.

In the problem of the car protection against a lateral impact, following factors must be taken into account:

1. Large deformations and the possibility of loss of the structure's stability.
2. Contact interaction of bodies with variable contact boundary in two pairs of touching surfaces (the block head with the bumper and the bumper with the car body, Fig. 3-1*b*).

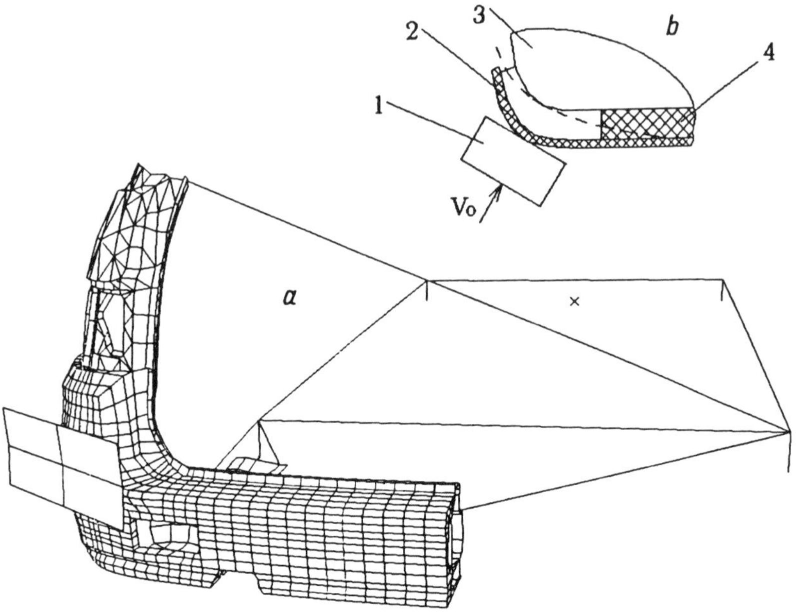

Figure 3–1 General view of the structure. (a) Finite element model; (b) the schematic of contact interaction of bodies under the impact: (1) Block head; (2) bumper; (3) rear panel; (4) insert. Dashed line shows the contact between the bumper and the rear panel.

3. The transition of loaded parts of the structure to the plastic state, the generation of cracks, and the material fracture.

When calculating, we have considered the left-hand half of the rear bumper and rear panel of the car (Fig. 3-1*a*). Figure 3-1*a* sketchily shows the contact between the block head and the bumper. The other part of the car has been modelled by beam elements. The structure is represented by the finite element model consisting of 2,016 elements and 1,986 nodes.

One calculation of the performance criteria vector for the aforementioned finite element model, provided the parameters values are fixed, requires more than 15 hours. Of course, with such a consumption of computer time, the optimization of design variables is hardly implementable.

Figure 3-2 presents the results of solving the dynamic contact problem in which the interaction forces and contact areas for block head - bumper and deformed bumper - rear panel pairs are to be determined. Time histories of the reaction force in the contact area between the block head and the bumper (curve 1), the reaction force in the contact area between the deformed bumper and the rear panel (curve 2), and the system energy (curve 3) are shown. Curve 3 takes into account the energy dissipation.

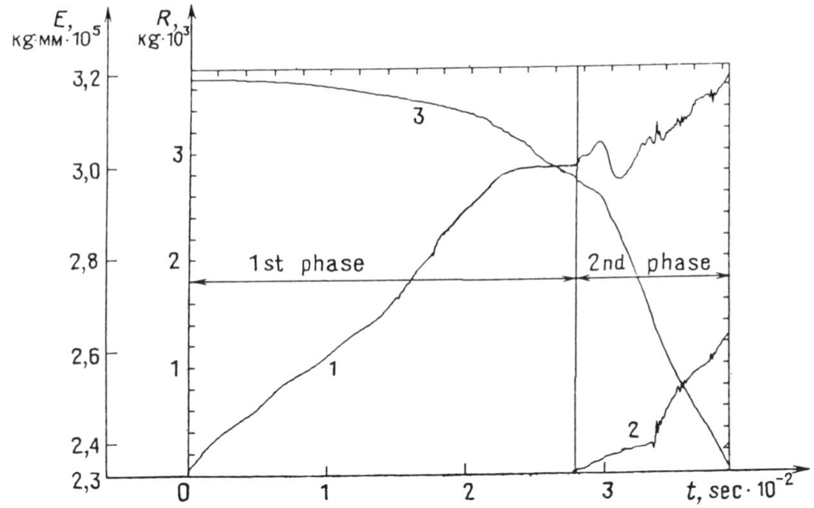

Figure 3–2 The histories of the reaction force in the area of contact between the block head and the bumper (curve 1); the reaction force in the area of contact between the bumper and the rear panel (curve 2); the energy of the system, with dissipation being taken into account (curve 3).

One can see from Figure 3-2 that the total time of impact, during which deformations increase, can be divided into two phases: before bumper comes into contact with the rear panel ($\sim 2.7 \cdot 10^{-2}$s), and after contact. During the first phase, only the bumper contacts with the block head, at this time the force of interaction between the bumper and rear panel is equal to zero (curve 2). Here, insignificant energy change occurs due to the weak strength of the plastic (low elasticity modulus). By the beginning of the second phase, the bumper fails to resist a load because of the damage. This fact is confirmed by some stabilization of the reaction force acting on the bumper at this time (curve 1). The fracture is taken into account in a nonlinear model of the bumper material. In the second phase, the rear panel makes contact with the bumper, the energy dissipation grows due to irreversible plastic deformations.

The investigations performed show the possibility of the decomposition of the finite element model into two subsystems. One of the subsystems describe the interaction of the block head with the bumper, while the other corresponds to the interaction of the deformed bumper with the rear panel. The second subsystem can be also defined as the subsystem describing the interaction of the block head with the rear panel after the destruction of the bumper.

Thus, the behavior of the whole system during impact can be represented by the first subsystem on the first phase, and by the second subsystem on the second phase.

Formulation and solution of the optimization problem.

Let us formulate performance criteria reflecting the requirements imposed on the system as a whole:

1. The mass of the structure must be minimal
$$\Phi_1(\boldsymbol{\alpha}) \rightarrow \min;$$
2. Residual deformations of the car body after the impact must be minimal, i.e.
$$\Phi_\nu(\boldsymbol{\alpha}) \rightarrow \min, \ \nu=\overline{2,10}.$$

   Here, $\boldsymbol{\alpha}$ is the vector of design variables of the bumper and the rear panel of the body; and $\Phi_\nu(\boldsymbol{\alpha})$ are residual deformations at certain monitored points. These deformations are calculated on the basis of the full model describing the whole system. We take the finite element grid nodes to which the external load is reduced as the monitored points.

There are nine of these points. Thus, the total number of performance criteria is 10.

Now, let us introduce the subsystem performance criteria and find out their relation to the performance criteria of the whole system. At the subsystem level, we specify performance criteria as follows. Criteria $\Phi_1^1(\boldsymbol{\alpha})$ and $\Phi_2^1(\boldsymbol{\alpha})$ are related to the bumper. $\Phi_1^1(\boldsymbol{\alpha})$ being the mass of the bumper while $\Phi_2^1(\boldsymbol{\alpha})$ characterizes the potential energy of the bumper deformation; $\boldsymbol{\alpha}$ is the vector of design variables of the bumper. The increase of the bumper deformation energy leads to the reduction of residual deformations of the rear panel.

Criteria $\Phi_1^2(\boldsymbol{\alpha}),\ldots,\Phi_{10}^2(\boldsymbol{\alpha})$ are related to the rear panel: $\Phi_1^2(\boldsymbol{\alpha})$ is the mass of the rear panel and $\Phi_2^2(\boldsymbol{\alpha}),\ldots,\Phi_{10}^2(\boldsymbol{\alpha})$ are its residual deformations at the monitored points, $\boldsymbol{\alpha}$ is the design-variable vector.

It is obvious that $\Phi_1=\Phi_1^1+\Phi_1^2$. Criterion Φ_2^1 is to be maximized while the others must be minimized.

We have analyzed the structure of the bumper prototype with 5 stiffening ribs and carried out relevant calculations of the criteria for the bumper and the rear panel. These calculations show that for the examined structure, it is impossible to find a feasible solution in which residual displacements of the body do not exceed the limiting admissible values:

$$\Phi_\nu^2(\boldsymbol{\alpha}) \leq 1\text{mm}, \ \nu=\overline{2,10}.$$

In this case, the structural optimization is necessary, that is the search for the structure configuration that would allow the solution of the posed problem.

Structural optimization of the bumper design variables based on the analysis of the first subsystem

As the reaction force grows on the first phase (Curve 1 in Fig. 3-2), the absorption of the block head kinetic energy by the bumper grows until the bumper starts interacting with the rear panel. To determine the deformation energy of the bumper we specify displacements of the nodal points at which the interaction occurs. Internal forces generated by this interaction produce work over the displacements. This work characterizes absorption properties of the bumper.

For strengthening the structure it is possible to either add stiffening ribs or use materials with better stiffness characteristics.

Let us consider the case of using additional stiffening ribs without changing the material. Three arrangements of the additional stiffening ribs, regarded by experts as being most promising, have been considered. For each of the arrangements, we have conducted the optimization of design variables. Later, the results are presented that relate to the best of the three configurations. In this configuration, the bumper contains 12 stiffening ribs, and the rear panel 6.

When optimizing design variables of the bumper, the thickness of the shell of the bumper, α_1^1, and the heights of twelve stiffening ribs, $\alpha_2^1,\ldots,\alpha_{13}^1$, were varied.

The feasible solutions set containing four solutions represented by vectors 126, 254, 257, and 494 i.e. $\boldsymbol{\alpha}^{126}$, $\boldsymbol{\alpha}^{254}$, $\boldsymbol{\alpha}^{257}$ and $\boldsymbol{\alpha}^{494}$ (see Table 3-1) was obtained.

The investigation was carried out on the finite element model containing 727 nodes and 950 elements. Figure 3-3 shows 12 stiffening ribs of the feasible structures of the bumper, as compared with five of the prototype. The ribs with numbers from 6 to 12 have been added.

Structural optimization of design variables of the body rear panel based on the analysis of the second subsystem. The search for consistent solutions

The second subsystem represents the structure part located in the zone of contact interaction between the block head and rear panel. The kinetic energy of the block head by the time it starts interacting with the rear panel is determined on

Table 3–1

N \ Φ_ν^1	Φ_1^1 (kg)	Φ_2^1 (N·m)
126	3.03	34.9
254	3.03	34.0
257	3.04	36.1
494	2.97	33.9

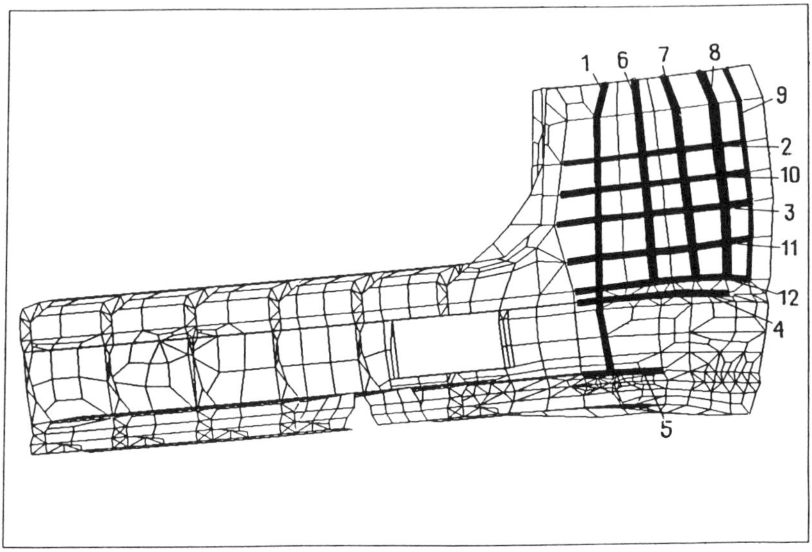

Figure 3–3 Finite element model of the optimal solution of the bumper.

the basis of the full model (block head, bumper, rear panel). Note here that different feasible design-variable vectors of the bumper correspond to different values of the initial kinetic energy. Therefore, for each of the design-variable vectors of the bumper, one has to consider the special problem of rear panel optimization. In doing this, we calculated the load on the rear panel for each of the feasible design-variable vectors of the bumper.

After having determined feasible design-variable vectors of the bumper (Table 3-1), one must find design-variable vectors of the rear panel consistent with the bumper design-variable vectors, taking into account the fact that the velocity of the block head strike against the rear panel is determined for each design-variable vector of the bumper. When optimizing the rear panel, for each of the four design-variable vectors of the bumper, the search for consistent design-variable vectors of the rear panel was conducted.

Besides the rear panel thickness (α_1^2), design variables of the stiffening ribs: thicknesses, $\alpha_2^2,\ldots,\alpha_7^2$, and heights of horizontal stiffening ribs, α_8^2, α_9^2, α_{10}^2, were also varied. Depending on the heights of horizontal ribs, design variables of vertical ribs are determined automatically.

In our example, the finite element model contained 305 nodes and 340 elements, see Fig. 3-4. For 257 design-variable vector of the bumper, six consistent and feasible design-variable vectors of the rear panel have been found that satisfy all the aforementioned criteria constraints on Φ_ν^2, $\nu=\overline{2,10}$, see Table 3-2. It turned out that for these design-variable vectors (just as for the prototype) $\Phi_1^2 \approx 2.72$ kg. As has been mentioned, this structure of the rear panel contains six stiffening ribs, three vertical and three horizontal. The ribs pass through

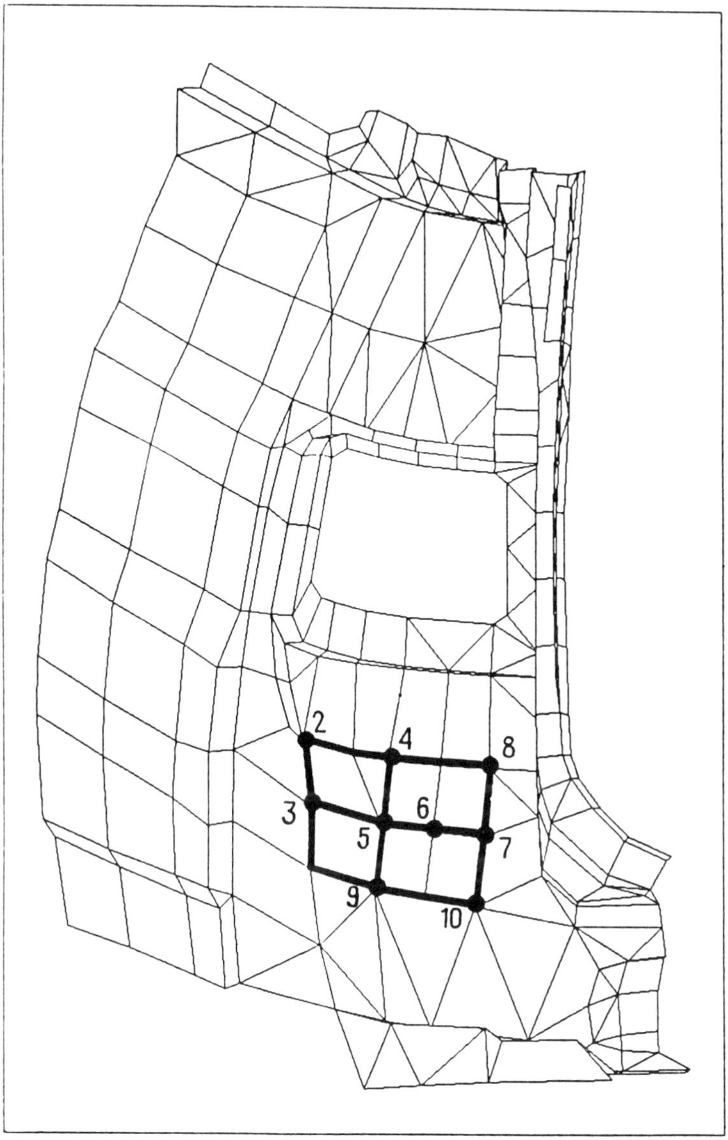

Figure 3–4 Finite element model of the optimal solution of the rear panel.

corresponding nodal points shown in Figure 3-4. The stiffening ribs are drawn by thick lines in the figure. The prototype contains no ribs.

The values of residual deformations at the nodal points 2,...,10 are presented in corresponding columns of Table 3-2. These values ($\Phi_2^2,\ldots,\Phi_{10}^2$) are presented for six consistent solutions.

The search for consistent design-variable vectors of the rear panel has also

Table 3–2

N \ Φ_ν^2	Φ_2^2 (mm)	Φ_3^2 (mm)	Φ_4^2 (mm)	Φ_5^2 (mm)	Φ_6^2 (mm)	Φ_7^2 (mm)	Φ_8^2 (mm)	Φ_9^2 (mm)	Φ_{10}^2 (mm)
67	0.22	0.17	0.46	0.39	0.59	0.85	0.43	0.14	0.14
90	0.27	0.21	0.40	0.28	0.37	0.49	0.47	0.14	0.11
98	0.23	0.19	0.44	0.33	0.48	0.66	0.44	0.15	0.12
125	0.23	0.31	0.45	0.37	0.59	0.94	0.50	0.13	0.14
148	0.24	0.19	0.42	0.32	0.45	0.61	0.39	0.14	0.12
181	0.24	0.22	0.39	0.29	0.35	0.43	0.47	0.15	0.11

been conducted for the three design-variable vectors of the bumper, 126, 254, and 494.

Aggregation of subsystems: Search for the optimal solution

In the aggregation procedure, design variables of each of the feasible design-variable vectors of the bumper are united (concatenated) with design variables of each of the corresponding consistent design-variable vectors of the rear panel. For instance, to the design-variable vector of the bumper 257, design variables of each of the six consistent design-variable vectors of rear panel have been added. As a result, the vectors $(\boldsymbol{\alpha}^{257}, \boldsymbol{\alpha}^{67})$, $(\boldsymbol{\alpha}^{257}, \boldsymbol{\alpha}^{90})$, $(\boldsymbol{\alpha}^{257}, \boldsymbol{\alpha}^{98})$, $(\boldsymbol{\alpha}^{257}, \boldsymbol{\alpha}^{125})$, $(\boldsymbol{\alpha}^{257}, \boldsymbol{\alpha}^{148})$, and $(\boldsymbol{\alpha}^{257}, \boldsymbol{\alpha}^{181})$ have been formed, and we have obtained six design-variable vectors of the whole system. An analogous procedure has also been performed with design-variable vectors of bumper 126, 254, and 494. As a result, 15 design-variable vectors of the whole system have been generated. The number of feasible solutions satisfying the criterion constraint on the total mass of the structure was nine, six of the solutions were just given. For all the feasible solutions, we have calculated criteria Φ_ν, $\nu = \overline{1,10}$ related to the whole system.

The installation of additional stiffening ribs leads to the reduction of residual deformations, their values not exceeding 1 mm.

After all the investigations, we preferred the structure using design-variable vector of bumper 257 and design-variable vector of rear panel 181. Characteristic of this structure is more uniform distribution of residual deformations (as compared with other versions) and an acceptable mass.

Conclusion

The proposed approach allows a designer to generate recommendations concerning the choice of design variables of the bumper and rear panel of a car, and also to reduce the time of operational development of the car structure. It should be added that the mass production of cars is in question. Therefore, the effect of the optimization is evident here.

4

Multicriteria Identification of Mathematical Models and Problems of Operational Development

Identification of parameters and structures of mathematical models is fairly well reflected in the literature (see, e.g., Red'ko, et al. (1985); and Ljung (1987). Multicriteria methods allow us to treat this important problem in a different way. Multicriteria identification is a new direction that is of great value in applications.

4-1. Problems of Multicriteria Identification and Their Features

So many times we were impressed by the results of optimization: The first criterion is improved twice as much, the second one, by 80%, etc. However, such advances always cause doubts. How trustworthy are those figures? How adequate is the mathematical model? Without having answered these questions, it is hardly possible to assert that the optimization is of some practical sense. To construct the model of a complex system so that all performance criteria (there may be many dozens of them) were determined with acceptable accuracy is unusual. As a rule, in practice some of the criteria are calculated with comparatively high accuracy, while others are determined with considerable errors.

This is the most typical situation when investigating complex mathematical models. Therefore, it is very important to have complete information about the mathematical model. In other words, we must be sure that our model is adequate for the system under study. The adequacy can be established by using different identification methods.

In the most common usage, the term identification means the construction of the mathematical model of a system and determining the parameters of the model by using the information about the system response to known external disturbances. In a sense, identification problems are inverse with respect to optimization problems.

By their nature, applied identification problems are multicriteria. However, as a rule, these problems have been treated as single-criterion problems (Red'ko et al. 1985; Ljung 1987). Let us briefly dwell on the most widespread identification methods and show the necessity of using multicriteria identification techniques.

When constructing a mathematical model, one first defines the class and structure of the model operator, that is, the law according to which disturbances (input variables) are transformed into the system response (output processes). This is called the structural identification. For mechanical systems, the structural identification means determining the type and number of equations constituting the mathematical model of the system. Parametric identification is reduced to finding numerical values of the equation coefficients, based on the realization of the input and output processes. In doing this, frequency responses, transfer functions, and unit step functions are often used (Graupe 1976). A number of problems require preliminary experimental determination of basic characteristics of a mechanical system (e.g., the frequencies, shapes, and decrements of natural oscillations). The structural identification is necessary if there is no preliminary information about the system structure or this information is not sufficient for compiling equations. In the general case, the structural identification problem is very difficult to solve. Apparently, this accounts for the absence of general methods for solving this problem.

The construction of a good (adequate) mathematical model of a complex system is a rare and great success for a researcher. Mostly, one has to represent the examined system as a set of mathematical models. For example, the planar rigid-body model of a truck (considered in Section 4-5) describes fairly well the behavior of the truck at low frequencies, whereas for high frequencies we have to use a three-dimensional nonlinear model.

Parametric identification (provided the structure of the mathematical model is known) is usually reduced to the minimization of a functional[11]

$$I=I[\boldsymbol{\epsilon}(\boldsymbol{\alpha})],$$

where $\boldsymbol{\alpha}=(\alpha_1,\ldots,\alpha_r)$ is a vector of variables (parameters) to be estimated; and $\boldsymbol{\epsilon}(\boldsymbol{\alpha})$ is a generalized error or the difference between the measured output processes of the system and respective responses of the mathematical model.

Let the system be linear and governed by

$$\mathbf{M}\ddot{\mathbf{g}}(t)+\mathbf{B}\dot{\mathbf{g}}(t)+\mathbf{C}\mathbf{g}(t)=\mathbf{x}(t),$$

where **M, B,** and **C** are $n\times n$-matrices of inertia, damping, and stiffness coefficients, respectively; $\mathbf{g}(t)$ and $\mathbf{x}(t)$ are the vectors of generalized coordinates and disturbances. For this system, the generalized error is expressed by

[11]For minimization of the functional I, single-criterion methods are mostly used, including gradient methods, stochastic search algorithms, and their numerous modifications (Bekey 1970).

$$\boldsymbol{\epsilon}(t,\boldsymbol{\alpha})=\mathbf{M}\hat{\ddot{\mathbf{g}}}(t)+\mathbf{B}\hat{\dot{\mathbf{g}}}(t)+\mathbf{C}\hat{\mathbf{g}}(t)-\hat{\mathbf{x}}(t),$$

where (ˆ) means the presence of errors in experimental data.

The most important stage of solving the identification problem is the choice of the criterion of agreement between the mathematical model and the real system, that is, the functional I. In publications, one can find several types of this criterion (Red'ko et al. 1985; Tsypkin 1982), the most frequently used being:

- The minimum of the mean square of the generalized error or difference between the responses of the model and system,
- The minimum of the weighted mean square of $\boldsymbol{\epsilon}$ (Markov's estimate),
- The maximal likelihood,
- The minimum of the average risk.

When determining the desired variables from the minimality of the generalized error mean square, the functional I has the form

$$I(\boldsymbol{\alpha})=\frac{1}{T}\int_0^T \boldsymbol{\epsilon}'(t,\boldsymbol{\alpha})\boldsymbol{\epsilon}(t,\boldsymbol{\alpha})\,dt$$

and, as a rule, is quadratic with respect to $\boldsymbol{\alpha}$ (here, the prime denotes the transposition operation). Therefore, the determination of the extremum of the functional I is reduced to solving the following system of algebraic equations

$$\boldsymbol{\Gamma}\boldsymbol{\alpha}=\mathbf{d}$$

This system is obtained by equating to zero partial derivatives of $I(\boldsymbol{\alpha})$ with respect to the components of the vector $\boldsymbol{\alpha}$, that is, $\partial I(\boldsymbol{\alpha})/\partial\alpha_j=0, j=\overline{1,r}$. Here, $\boldsymbol{\Gamma}$ is an $r\times r$-matrix, and $\mathbf{d}$ is the right-hand side vector. Very often, the solution of these problems reduces to the investigation of nonsingularity conditions for the matrix $\boldsymbol{\Gamma}$. For more complete information about other criteria, one can see, for example, (Tsypkin 1982).

Experimental data are known to be always determined with some errors. The nature of these errors determines the choice of a criterion (i.e., the functional I) for establishing the correspondence between the mathematical model and the real system. Therefore, it is very important to study the nature of the measurement errors, to analyze their influence on the results of identification, and to elaborate recommendations for obtaining the solution with prescribed accuracy.

In theory, identification methods for mathematical models of linear systems are the most developed. Red'ko et al. (1985) describe the identification methods using special signals (steplike, impulse, sinusoidal, etc.) applied to the system. These methods can serve for identification of steady-state processes with a single input or many inputs, provided only one of them is engaged at a time.

The aforementioned methods are based on the Fourier transformation. Note that the frequency method of identification of linear systems is based on works by Nyquist (1932) and Bode (1945) and uses amplitude-frequency characteristics (i.e. the dependence of amplitudes of the system oscillations upon the disturbance frequency). The frequency method implies that a sinusoidal signal whose frequency changes within a prescribed range is applied to the system input. This method uses the Laplace transformation for input-to-output ratio.

Some methods for the identification of mathematical models of linear systems are presented, for example, in Bode (1945); and Strobel (1968). According to these methods, the functional evaluating the discrepancy between the experimental and computed transfer functions is given by

$$I=\frac{1}{N}\sum_{j=1}^{N}\left|\epsilon(i\omega_j)\right|^2\Omega(\omega_j).$$

Here, $\epsilon(i\omega_j)$ is the generalized error, due to errors in determining the input data and the discrepancy between the structures of the system and its model; $i=\sqrt{-1}$; N is the number of the exciting force frequency values at which the experimental measurements be carried out; $\Omega(\omega_j)>0$ is the weighting function allowing for relative significance of the input data.

The apparent simplicity of formulas similar to those given here hides a complicated problem of determining the weighting coefficients. The weighting functions (coefficients) are used here in order to avoid the multicriteria consideration. We have already mentioned in Chapter 1 that such an approach is not effective. In some cases, more complex criteria are used as well. Note that the approach in question allows us not only to identify the variables of a linear system but also to determine its number of degrees of freedom (Red'ko et al. 1971; Woodside 1971).

One of the basic drawbacks of many identification methods for linear systems is the fact that these methods reduce to solving high-order systems of linear algebraic equations. The matrices of coefficients in such systems may appear to be ill-conditioned, and that leads to unstable solutions. The solution errors increase to unacceptable amounts, as the system order increases. In a number of cases, the way out consists in the application of so-called modal methods that do not require the solution of high-order systems with ill-conditioned matrices (Tsypkin 1982).

The identification techniques have also been developed for some classes of nonlinear systems such as chain systems, (Sprague and Kohr 1969; Tumanov et al. 1981). However, the general issues of identification of nonlinear systems have been studied poorly. One of the essential reasons for this lies in the impossibility (or great complexity) of constructing the functional I evaluating the adequacy of the model to a real system.

Thus, we can draw the following conclusion. The available identification methods are reliable enough in cases where the model structure is established exactly and one can construct the functional I. This is mostly related to linear systems (however, these systems are also associated with certain complications we have already mentioned). In case it is impossible to establish the model structure exactly or to construct the functional I, the identification methods are mostly ineffective, that takes place for the majority of nonlinear systems.

In this connection, a new approach is proposed in Genkin et al. (1987). In all basic units of the structure under study, we experimentally measure the values of characteristic quantities being of our interest (e.g., displacements, velocities, accelerations, etc.). Parallel with this, we calculate the corresponding quantities by using the mathematical model. As a result, particular adequacy (proximity) criteria are formed as functions of the difference between the experimental and computational data. Thus we arrive at a multicriteria problem. Its solution allows us to avoid the difficulties mentioned before. The multicriteria consideration makes it possible to extend essentially the application area of the identification theory.

Let us discuss some basic features of multicriteria (or vector) identification problems.

1. In the majority of conventional problems, the system is tacitly assumed to be in full agreement with its mathematical model. However, for complex engineering systems (e.g., machines), generally, we cannot assert a sufficient correspondence between the model and the object. This does not permit us to use a single criterion to evaluate the adequacy. In multicriteria identification problems, there is no necessity in artificially introducing a single criterion to the detriment of the physical essence of the problem.
2. Unlike conventional identification approaches, the adequacy of the mathematical model is evaluated by using a number of particular criteria of proximity, as already mentioned. For example, when identifying the parameters of the dynamic model of an automobile, it is necessary to take into account such important indexes (particular criteria) as vibration accelerations at all characteristic points of the driver's seat, driver's cab, frame, and engine; vertical dynamical reactions at contact areas between the wheels and the road; relative (with respect to the frame) displacements of the cab supports, springs, engine, etc (Perminov and Statnikov 1987). Such a multicriteria approach is very important for determining to what extent the mathematical model corresponds to the physical system. For complex systems, the number of particular proximity criteria used for the evaluation of the mathematical model adequacy can achieve many dozens (see Section 4-5).
3. Very often, when solving the problems of the class in question, the designer has no information about the limits α_j^* and α_j^{**} (see (1-1)) for

many of the variables to be identified. The improper specification of these limits can lead to a huge number of calculations, and nevertheless, the results of the identification will be unsatisfactory or incomplete.

Fig. 4-1 shows the initial parallelepiped Π^1 where the search for the variables to be identified has been started. Then the search continues in the parallelepiped Π^2 constructed after the correction of the boundaries of Π^1. The process goes on in a similar way, until the parallelepiped Π^k is found containing the set of feasible solutions $D_{\boldsymbol{\alpha}}$ ($D_{\boldsymbol{\alpha}} \subset \Pi^k$). If for some parallelepiped Π, the relation $\Pi^k \not\subseteq \Pi$ is valid, then $D_{\boldsymbol{\alpha}} \not\subseteq \Pi$. Note that the parallelepipeds $\Pi^1, \Pi^2, \ldots$ can contain some feasible vectors $\boldsymbol{\alpha}^i \in D_{\boldsymbol{\alpha}}$. To find Π^k it is necessary to use the recommendations given in Sections 1-3 and 1-4.

4. In structural identification, when investigating different mathematical models of the system, the number and limits of the variables to be identified, as well as the number of proximity criteria, can essentially change. In this connection, the problem arises on how to make the identification results for different structure agree.

4-2. Parameter Space Investigation Method in Problems of Multicriteria Identification

Denote by $\Phi_\nu^c(\boldsymbol{\alpha})$, $\nu=\overline{1,k}$, the indexes (criteria) resulting from the analysis of the mathematical model that describes a physical system, $\boldsymbol{\alpha}=(\alpha_1,\ldots,\alpha_r)$ being

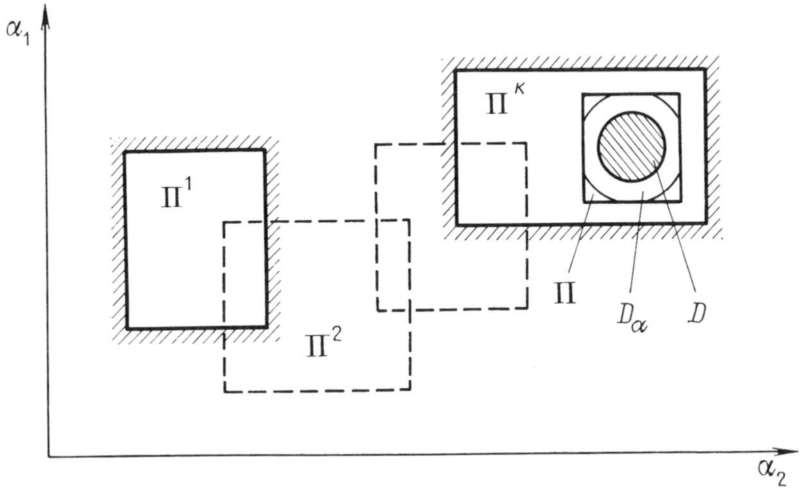

Figure 4–1 The search for domains of variables in multicriteria identification problems. Determination of the feasible solutions set for operational development.

the vector of parameters of the model. The criteria $\Phi_\nu^c(\boldsymbol{\alpha})$ can be functionals of integral curves of differential equations or functions of $\boldsymbol{\alpha}$ that are not associated with solutions of differential equations.

Let Φ_ν^{exp} be the experimental value of the νth criterion measured directly on the prototype. (The experiment is assumed to be sufficiently accurate and complete. By completeness we mean that criteria Φ_ν^{exp} are measured in all basic units, at most characteristic points of the structure. The amount of measurement data must be sufficient for correct formulation of the identification problem).

Suppose there exists a mathematical model or a hierarchical set of models describing the system behavior. Let $\boldsymbol{\Phi}=(||\Phi_1^c-\Phi_1^{exp}||, \ldots, ||\Phi_k^c-\Phi_k^{exp}||)$, where $||\cdot||$ is a particular adequacy (closeness, proximity) criterion. This criterion, as has already been mentioned, is a function of the difference (error) $\Phi_\nu^c-\Phi_\nu^{exp}$. Very often it is given by $(\Phi_\nu^c-\Phi_\nu^{exp})^2$ or $|\Phi_\nu^c-\Phi_\nu^{exp}|$. In the cases where the experimental values Φ_ν^{exp}, $\nu=\overline{1,k}$, are measured with considerable error, they can be regarded as a random quantity. If this random quantity is normally distributed, the corresponding adequacy criterion is expressed by $M\{||\Phi_\nu^c-\Phi_\nu^{exp}||\}$,where $M\{||\cdot||\}$ denotes the mathematical expectation of the random quantity $||\cdot||$, see (Red'ko et al. 1985; Raybman 1970). In cases of other distribution functions, more complicated methods of estimation are used, for example, the maximal likelihood method.

We formulate the following problem: by comparing the experimental and calculation data, determine to what extent the model corresponds to the physical system and find the parameters of the model. In other words, it is necessary to find variable vectors $\boldsymbol{\alpha}^i$ satisfying conditions (1-1) and (1-2) and, besides, the inequalities

$$\|\Phi_\nu^c(\boldsymbol{\alpha}^i)-\Phi_\nu^{exp}\|\leq\Phi_\nu^{**}. \tag{4-1}$$

Conditions (1-1), (1-2), and (4-1) define the feasible solutions set $D_{\boldsymbol{\alpha}}$ (Genkin et al. 1987). Here, Φ_ν^{**} are criteria constraints that are determined in the dialogue between the researcher and a computer. To a considerable extent, these constraints depend on the accuracy of the experiment and the physical sense of the criteria Φ_ν.

The Search for the Identified Solutions

The formulation and solution of the identification problem are based on the parameter space investigation method. In accordance with the algorithm given in Section 1-3, we specify the values Φ_ν^{**} and find vectors meeting conditions (1-1), (1-2), and (4-1). The vectors $\boldsymbol{\alpha}^i$ belonging to the feasible solutions set $D_{\boldsymbol{\alpha}}$ will be called adequate vectors.

The restoration of parameters of a concrete model on the basis of (1-1), (1-2), and (4-1) is the main purpose and essence of multicriteria parametric identification. Having performed this procedure for all structure (mathematical models), we carry out multicriteria structural identification as well.

The vectors $\boldsymbol{\alpha}_{id}^{i}$ that belong to the set of adequate vectors and have been chosen, by using a special decision-making rule, will be called identified vectors.

The role of the decision-making rule is often played by nonformal analysis of the set of adequate vectors. If this analysis separates several equally acceptable vectors $\boldsymbol{\alpha}_{id}^{i}$, the solution of the identification problem is nonunique.

The identified vectors $\boldsymbol{\alpha}_{id}^{i}$ form the identification domain $D_{id}=\bigcup_{i}\boldsymbol{\alpha}_{id}^{i}$. Sometimes by carrying out additional physical experiments, revising constraints Φ_{ν}^{**}, etc. one can reduce the domain D_{id} and even achieve that this domain contains only one vector. Unfortunately, this is far from being usual. Nonunique restoration of variables is a recompense for the discrepancy between the physical object and its mathematical model, incompleteness of physical experiments, etc.

If a mathematical model is sufficiently good (i.e. it rightly describes the behavior of the physical system), then multicriteria parametric identification leads to nonempty set $D_{\boldsymbol{\alpha}}$. The most important factors that can lead to empty $D_{\boldsymbol{\alpha}}$ are the imperfection of the mathematical model and lack of information about the domain in which the desired solutions should be searched for.

The search for the set $D_{\boldsymbol{\alpha}}$ is very important, even in case the results are not promising. It enables the researcher to judge the mathematical model objectively (not only intuitively), to analyze its advantages and drawbacks on the basis of all proximity criteria, and to correct the problem formulation.

Thus, multicriteria identification includes the determination and nonformal analysis of the feasible solutions set $D_{\boldsymbol{\alpha}}$ regarding all basic proximity criteria, as well as finding identified solutions $\boldsymbol{\alpha}_{id}^{i}$ belonging to this set.

Often multicriteria identification is the only way to evaluate the quality of the mathematical model and, hence, to optimize this model.

This algorithm is successfully used in practice. In Sections 4-3 and 4-5, we discuss some important problems solved by using this algorithm.

By analogy to the optimization problem, we can formulate and solve the problem of constructing the adequate solutions set with prescribed accuracy.

The Search for Identified Solutions with a Prescribed Accuracy

Let ϵ_{ν} $(\nu=\overline{1,k})$ characterize the desired accuracy of the correspondence between the physical system and its mathematical model with respect to the criterion Φ_{ν}^{c} (i.e., the inequality $\|\Phi_{\nu}^{c}-\Phi_{\nu}^{exp}\|<\epsilon_{\nu}$ must hold). Then the values of all criteria restoring the experimental characteristics with a prescribed accuracy can be found through the approximation of the adequacy criteria range.

In multicriteria identification, we are interested not only in values of adequacy criteria, but also in values of variables. For example, let $\boldsymbol{\alpha}$ and $\boldsymbol{\beta}$ be vectors giving "good" values to adequacy criteria, $\boldsymbol{\Phi}(\boldsymbol{\alpha})\approx\boldsymbol{\Phi}(\boldsymbol{\beta})$ while the vectors $\boldsymbol{\alpha}$ and

$\boldsymbol{\beta}$ being considerably different. In this case, if there is no additional information for making the choice between the vectors $\boldsymbol{\alpha}$ and $\boldsymbol{\beta}$, we can regard $\boldsymbol{\Phi}(\boldsymbol{\alpha})$ and $\boldsymbol{\Phi}(\boldsymbol{\beta})$ as being equally adequate to the physical experiment. However, the researcher must keep in mind all vectors corresponding to good values of adequacy criteria. This is explained by the following considerations. In practice it is usually impossible to formalize all requirements imposed on the physical or engineering system. If we take into account only one of two vectors corresponding to approximately the same values of adequacy criteria, we can possibly lose the better vector with respect to nonformalized criteria. Suppose we have succeeded in meeting all demands of the system. In this case, we should consider all the aforementioned vectors when working with the mathematical model after having completed the identification. Suppose we are to optimize the parameters of the model with respect to some criteria. If we have eliminated one of two equally adequate vectors, the dropped vector can turn out to be preferred with regard to the performance criteria. Taking into account these considerations, we can modify the definition of the solution of the multicriteria identification problem.

Denote by $V_{\boldsymbol{\epsilon}}(\Phi(P))$ an $\boldsymbol{\epsilon}$-neighborhood of the Pareto optimal set $\Phi(P)$ in the space of adequacy criteria. It is reasonable to define the solution of the multicriteria identification problem as a set $W_{\boldsymbol{\epsilon}}$ of all variable vectors $\boldsymbol{\alpha}$ belonging to the feasible solutions set $D_{\boldsymbol{\alpha}}$ and satisfying the inclusion $\boldsymbol{\Phi}(\boldsymbol{\alpha}) \in V_{\boldsymbol{\epsilon}}(\Phi(P))$.

As a result of nonformal analysis of the set $W_{\boldsymbol{\epsilon}}$, the researcher can choose the most preferred models.

Let us show how one can solve the problem by using the parameter space investigation method.

The solution algorithm is based not only on the approximation of the criteria space, but also on the approximation of the variable space. Let $\Phi_{k+j}(\boldsymbol{\alpha})=\alpha_j$ and δ_{k+j} be the admissible error for the variable α_j, where k is the number of adequacy criteria. By using the algorithm of Chapter 2, let us construct the approximation of the set $D_{\boldsymbol{\alpha}}$ to the accuracy $\boldsymbol{\delta}=\{\delta_{k+j}\}$, $j=\overline{1,r}$, and the approximation of its image, $\Phi(D_{\boldsymbol{\alpha}})$, to the accuracy $\boldsymbol{\epsilon}=\{\epsilon_\nu\}$, $\nu=\overline{1,k}$. The fact that we have declared the variables α_j as criteria Φ_{k+j}, enables us to approximate $\Phi(D_{\boldsymbol{\alpha}})$ and $D_{\boldsymbol{\alpha}}$ simultaneously. In this case, the set $V_{\boldsymbol{\epsilon}}(\Phi(P))$ can be approximated to the accuracy $\boldsymbol{\epsilon}$, and any vector of $D_{\boldsymbol{\alpha}}$ can be determined to the accuracy $\boldsymbol{\delta}$. Using the approximations of $D_{\boldsymbol{\alpha}}$ and $\Phi(D_{\boldsymbol{\alpha}})$ we can find the set $W_{\boldsymbol{\epsilon}}$, and thus obtain the solution of the multicriteria identification problem.

Let us call the set $W_{\boldsymbol{\epsilon}}$ the set of $\boldsymbol{\epsilon}$-adequate vectors. The vectors $\boldsymbol{\alpha}_{id}$ that belong to the set of $\boldsymbol{\epsilon}$-adequate vectors and are determined with the help of a decision-making rule will be called identified vectors. The set D_{id} of all identified vectors is called the identification set.

Thus, the described method makes it possible to find solutions of the multicriteria identification problem with any prescribed accuracy. Also, this method is universal and can be applied to linear and nonlinear systems, both with distributed and lumped parameters. However, the computer implementation of this algorithm can be time-consuming.

Let us draw some conclusions. The formulation and solution of the multicriteria identification problem combined with nonformal analysis of the obtained results make it possible to:

1. Determine the sets of adequate and identified solutions and thus judge about the agreement between the physical system and its mathematical model
2. Assess relative merits of models (in case there is a hierarchical set of models), decide about the expedience of complication of the model, and establish the accuracy, completeness, and reliability of the obtained results.
3. Correctly specify the limits of the variables range and justify the list of performance criteria (performance indexes) for subsequent multicriteria optimization having established the adequacy of the mathematical object to the physical system (e.g., in the course of operational development of a machine).

4-3. Example 1: Multicriteria Identification of the Parameters of a Slotter

The methodology of multicriteria identification will be considered in the example of a slotter whose thrust is 30 kN. Slotters are widely used for machining irregular-shaped internal surfaces. Figure 4-2 shows a slotter consisting of column I and bed II joined by the bolted joint A. Slotting ram III is mounted on the column. It holds a slotting cutter and reciprocates in the vertical plane. On the bed, table IV is mounted and a workpiece is clamped on it. Machining is implemented by reciprocating the slotting cutter with respect to the workpiece.

Unfortunately, slotters are prone to intense vibration within the most important range of cutting speeds from 6 m/min to 12 m/min. The vibrations during the cutting process limit the productivity of the machine, reduce its reliability, shorten the service life, and deteriorate both the accuracy and the quality of the processed surfaces.

Experimental Study of Vibration Stability of the Slotter's Hydromechanical Systems

The system's response to dynamic disturbances was measured by transducers mounted at eight points located on different levels over the bed guideways (Fig. 4-2). This allowed for measuring vibration amplitudes as functions of the disturbing-force frequency.

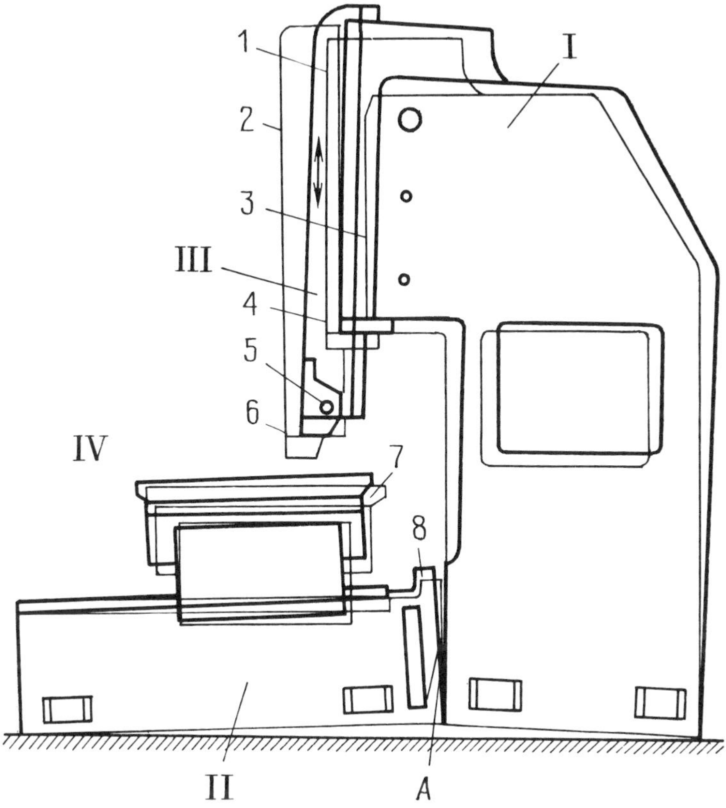

Figure 4–2 The layout of transducers on the machine tool. The "opening" of joint *A*.

The study of the dynamic stiffness of the mechanical system and the modes of vibration of the machine tool has shown that the stiffness of the joint between the column and the bed was insufficient. Excited by a periodic disturbing force, the column starts vibrating, and the amplitudes of vibration of individual points vary linearly depending on their vertical position. The experiments have shown that the maximum-amplitude resonance vibration takes place at the natural frequency 15.9 Hz and causes "opening" of the joint, upon which the column and the bed start displacing with respect to each other (Fig. 4-2).

Construction of a Dynamic Model of the Machine Tool

The analysis of experimental results shows that the most reliable data have been obtained for the horizontal-torsional and vertical modes of vibration. The former is more intense: it causes "opening" of the joint A between the column and the bed, large amplitudes of the slotting tool vibration with respect to a workpiece, and deterioration of the surface finish.

Since the two basic modes of vibration correspond to the frontal plane of the machine tool (which passes through the symmetry axes of the column and the bed), it was decided to solve the problem of identification using the plane model shown in Figure 4-3. The model comprises a column and a bed, a table, a ram, a slotting tool, joints between the table and the bed and between the machine's supports and the foundation, the joint between the column and the bed, the joint between the ram and the column guideways, fastening units of the hydraulic cylinder, and ram slide.

Parameters of the dynamic model include the masses and the moments of inertia, the coordinates of fastening units of vibroisolating and elastoinertial elements, the angles of rotation of local coordinate systems, stiffness characteristics, the damping factors of structural elements, and mechanical characteristics of elastoinertial elements.

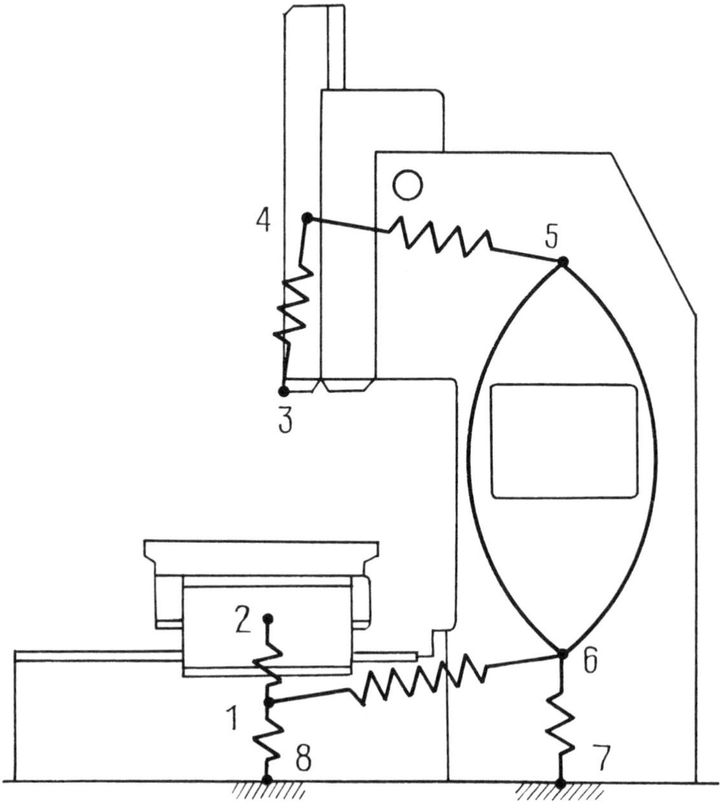

Figure 4–3 Dynamic model of the machine tool.

The Horizontal, Vertical, and Torsional Stiffnesses of the Column and the Bed Supports on the Foundation (Links 1-8 and 6-7 in Fig. 4-3).

During the slotter operation the concrete foundation is subjected to periodic loads and is gradually destroyed. The support stiffnesses are responsible, to a considerable extent, for the main drawback of the machine tool, the "opening" of the joint between the column and bed. Since the stiffness coefficients of the supports may hardly be accurately determined experimentally or specified a priori by some alternative means, they were included in the list of the model parameters to be identified.

The Horizontal, Vertical, and Torsional Stiffnesses of the Joint Between the Column and the Bed (Link 1-6 in Fig. 4-3).

The column and the bed of the machine tool under consideration are connected with bolts distributed over the surface of the joint. Since the ram mass and its distance from the joint are large, slotting is accompanied by the appearance of a large bending moment that loosens the bolted joints in a nonuniform manner. Due to this, it is rather difficult to measure the joint stiffnesses precisely.

It was decided to identify the following nine variables determining the modes of the machine tool vibration: $\boldsymbol{\alpha}=(c_y^{1\text{-}8}, c_z^{1\text{-}8}, p_x^{1\text{-}8}, c_y^{1\text{-}6}, c_z^{1\text{-}6}, p_x^{1\text{-}6}, c_y^{6\text{-}7}, c_z^{6\text{-}7}, p_x^{6\text{-}7})$ where c_y, c_z, and p_x are the horizontal, vertical, and torsional stiffnesses respectively, and 1-8, 6-7, and 1-6 are the superscripts of the links.

Adequacy criteria

The mathematical model of the machine tool is used for the determination of the set of criteria characterizing the most vibroactive modes of oscillations in the low-frequency range, because the full-scale test data are sufficiently complete for this range only. Therefore, the adequacy criteria characterize the degree of correspondence of the model to the real object mainly within the given range. Table 4-1 presents experimental characteristics Φ_ν^{exp} defining the list of the adequacy criteria. The latter were calculated using the formula

$$\Phi_\nu = \left| \frac{\Phi_\nu^{exp} - \Phi_\nu^c}{\Phi_\nu^{exp}} \right| \cdot 100\%, \; \nu=\overline{1,8}.$$

Functional constraints

In solving the problem of identification of the dynamic model parameters, a list of eight constraints taking into account the accuracy of manufacture and assemblage of the machine tool units as well as the accuracy of mounting the machine tool on the supports, has been generated.

Thus, a full-scale experiment has been conducted, and the mathematical model, proximity criteria, and functional and variable constraints were generated. Solution of the identification problem, and subsequently, the optimization problem,

Table 4–1

Φ_ν^{exp}	Physical meaning	Experimental value
Φ_1^{exp}	Basic frequency of natural oscillations, Hz	16
Φ_2^{exp}	The second natural frequency, Hz	30
Φ_3^{exp}	Coefficient of the oscillation shape in node 3, relative units	0.738
Φ_4^{exp}	Coefficient of the oscillation shape in node 5, relative units	0.934
Φ_5^{exp}	Coefficient of the oscillation shape in node 6, relative units	0.283
Φ_6^{exp}	Vibratory compliance value in the cutting zone at the first resonance frequency, mm/N	$3.2 \cdot 10^{-4}$
Φ_7^{exp}	Static displacement value in node 3, mm, under cutting force of 1 kN	$0.492 \cdot 10^{-2}$
Φ_8^{exp}	Static displacements in the cutting zone, mm, under cutting force of 1 kN	$0.288 \cdot 10^{-2}$

was aimed at improving the basic characteristics of the machine tool, namely its reliability, service life, machining accuracy, and vibration stability.

Solution of the problem of multicriteria identification

In the initial parallelepiped Π^1, N=2,048 trials were conducted, and it was found that $D_{\boldsymbol{\alpha}}=\varnothing$. A similar result has been obtained by analyzing new parallelepipeds Π^2 and Π^3 obtained by correcting the Π^1 boundaries (see Fig. 4-1).

Upon analyzing the test tables and histograms of corrected variables it was decided to change the variation boundaries again, changing at the same time the number of variables. This led to parallelepiped Π^4. Variables α_{10}–α_{13} described later, have a substantial effect on the character of vibration in the low-frequency range; these variables cannot be accurately measured in an experiment. Here α_{10} is the stiffness of the screw of the table feed drive (modeled by the horizontal stiffness of the joint between the table and the bed, link 1-2); α_{11} is the torsional stiffness of the ram guideways on the column (link 3-4); and α_{12} and α_{13} are the corrections for taking into account shear deformations in the column (link 5-6).

These variables permitted taking into account the possible effect of the column vibrational compliance on the nature of vibroactive modes of oscillations.

Thus, the subsequent search was carried out within a 13-dimensional variable space.

Let us analyze the results obtained in parallelepiped Π^4. The calculations yielded acceptable values of discrepancies in all the static and dynamic criteria, except for Φ_6. The latter criterion is the only one depending on both dissipation and stiffness parameters of the model. All the criteria pertaining to free vibration of the machine tool mechanical system and its static rigidity, are fully defined by its stiffness and inertia variables. Therefore, large initial discrepancies in criterion Φ_6 against the background of satisfactory results for the majority of the rest criteria, invited doubt in the correctness of the damping coefficients specification.

In line with what was said before we have analyzed and corrected the damping coefficients of the table and the ram guideways, the column and bed supports, and elastoinertial elements of the fixed joint between the column and the bed.

Correctness of this decision was confirmed by obtaining satisfactory values of discrepancies in criterion Φ_6 in subsequent calculations in parallelepiped Π^5.

From the viewpoint of further correction of the parallelepiped Π^4 boundaries, of special importance is analysis of the correlation matrix constructed using the results of trials implemented in Π^4. The analysis of the coefficients of pair correlation between criteria and variables has revealed a strong influence of the stiffness of the joint between the column and the bed on criteria Φ_4 and Φ_7. This was conclusively confirmed by the results of a full-scale experiment that has shown that the low vibration stability of the machine tool under consideration is caused by the "opening" of the joint between the column and the bed. Also analyzed were the dependences of closeness criteria on variables (see Chapter 5). A strong dependence of the criterion of vibration compliance within the cutting zone on the stiffness of the joint has been demonstrated. Also, the experimental study has shown that the total vibration displacements of the tool and the workpiece are mostly determined by the column vibration. The effect of horizontal stiffness of the bed support on displacements at node 6 is obviously caused by the specific features of the machine tool design.

Thus, this analysis has allowed determination of the boundaries of the new parallelepiped Π^5.

Search in Parallelepiped Π^5: Analysis of the Results

Upon introducing the improved values of the damping factors, the discrepancies in the vibration displacement amplitude in the cutting zone, criterion Φ_6, proved to make up to 8%. The character of the modes of vibration remained the same. In order to construct the feasible solutions set of adequate models, eight designer-computer dialogues were conducted. The interactive mode was used for determining the feasible solutions set of models $\boldsymbol{\alpha}^i$ depending on the values of criteria constraints Φ_ν^{**}, $\nu=1,\ldots,8$. Next, the set was subjected to nonformal analysis.

Models $\boldsymbol{\alpha}^{59}$ and $\boldsymbol{\alpha}^{395}$ should be considered the best ones, because (see Table 4-2):

1. The discrepancies in the frequency criteria are quite acceptable.
2. The discrepancies in criterion Φ_8 correspond to the accuracy of the experiment (taking into account displacements of a statically loaded

Table 4–2

Models	Adequacy criteria values, %							
	Φ_1	Φ_2	Φ_3	Φ_4	Φ_5	Φ_6	Φ_7	Φ_8
$\boldsymbol{\alpha}^{59}$	3.9	18.6	4.19	8.2	35.6	4.95	6.72	2.08
$\boldsymbol{\alpha}^{395}$	2.56	16.5	5.1	8.32	31.7	5.72	5.8	0.48
$\boldsymbol{\alpha}^{185}$	3.51	14.1	4.35	8.16	36.1	4.54	5.38	1.61

cutter and a workpiece in the cutting zone, which were measured in the experiment with the greatest accuracy).

3. The discrepancies in the criteria Φ_3, Φ_6, and Φ_7 are approximately equal (especially so for model $\boldsymbol{\alpha}^{395}$), since the three criteria characterize vibration in the cutting zone.

The variables of a set of adequate models have been analyzed, and it was found out that for the best solutions the tendency to large values of horizontal stiffness of the table-bed joint persists.

These studies have allowed formation of parallelepiped Π^6 within which five adequate vectors were determined. Model $\boldsymbol{\alpha}^{185}$ characterized by comparatively small discrepancies in the basic closeness criteria, proved to be most preferred (see Table 4-2).

This model was preferred for the following reasons:

1. The variables of the model take into account the low torsional stiffness of the column, characteristic of the present machine tool design, correctly.
2. The model correctly takes into account the high compliance of the longitudinal feed drive and the contacting surfaces of the table (the compliance leads to large vibrational displacements in the cutting zone).
3. All the closeness criteria, taking into account the oscillatory nature of slotting, are characterized by the values of Φ_3, Φ_6, and Φ_7 being approximately equal (see Table 4-2). This means that alongside with a good agreement of the criteria subjected to analysis, the general character of vibration in the zone is also reproduced correctly.

Stability of this solution with respect to the variable changes was analyzed in the vicinity of model $\boldsymbol{\alpha}^{185}$. The analysis has shown that the model is stable. (A parallelepiped centered at $\boldsymbol{\alpha}^{185}$ was constructed in accordance with the variables tolerances, and $N=256$ trials were conducted. This is quite sufficient for the parallelepiped of such a small volume. All the models proved to be feasible, and the values of the criteria changed insignificantly.)

In order to determine the domain of admissible variations of design variables (see Section 4-4), the variables of adequate models were analyzed in parallelepipeds Π^5 and Π^6, and a parallelepiped for solving the problem of optimization was constructed on $D_{\boldsymbol{\alpha}}$.

Let us draw some conclusions.

1. The problem of multicriteria identification of the parameters (variables) of a slotter whose thrust is 30 kN was formulated and solved using eight adequacy criteria that take into account the oscillation frequencies and shapes, vibrational displacement in the cutting zone, and static characteristics.

   In solving the problem, 13 stiffness variables of the machine tool's joints and connections were identified.
2. The boundaries of the variables were found. Although prior to solving the problem the boundaries were rather indefinite, the five-fold correction has allowed finding such values of the boundaries, which ensure solution of the problem of multicriteria identification.
3. The set of adequate models was found.
4. Taking into account all the adequacy criteria, model $\boldsymbol{\alpha}^{185}$ was identified, characterized by low stiffness of the joint between the column and the bed (this determines the loss of vibration stability of the machine tool in the low-frequency range), and by the predominant effect of the column vibration on the vibrational displacements of the tool with respect to the workpiece.

   In designers' opinions, this model adequately describes the character of vibration in the cutting zone and agrees well with experimental results.
5. The solution of the problem of multicriteria identification has allowed objective estimation of the mathematical model of the slotter. In turn, this permitted correct formulation and solution of the problem of improving the machine tool operation as concerns the criteria associated with stability, machining accuracy, the hydraulic drive, ram, and slotting tool service lives, the consumption of metal, and reduced costs. The ways of improving the machine tool design are indicated.

4-4. Operational Development of Prototypes

In this section, we will discuss the problems of perfecting engineering systems (machines). Mainly, these problems are related to the operational development of a prototype of a machine designed for serial and mass production. First, the machine is tested. the structure of the test is determined by the type of machine (an airplane, car, ship, machine tool, etc.). For example, cars are subjected to laboratory (bench) tests, including strength, fatigue, and vibration investigations of both individual units and the car as a whole.

Great attention is paid to road tests. These are mostly carried out on proving grounds where the car is tested on properly profiled road sections, in different conditions depending on the carried load and velocity of the car. Apart from this, cars are tested on regular roads under conditions close to operational ones. Thus, cars are subjected to bench and road tests. These tests are aimed at the detection of imperfections with subsequent operational development of the prototype so as to satisfy the customer's demands. The operational development is aimed at increasing the durability and reliability, reducing vibrations and noise, etc.

It is of essential importance to make the process of operational development as short as possible. Perhaps this is the main problem faced by designers of cars and other machines. As a rule, the decision about termination of the operational development is made after a number of successive improvements of the prototype.

We propose a new technique for the operational development of mechanical structures. The technique is shown in Figure 4-4. The operational development starts with testing the prototype. Then two options are available. In the first approach, based on the results of the tests only, we improve the prototype, and then repeat the tests. If a series of successive improvements of the prototype gives acceptable results, the decision is made to terminate the operational development. However, if the designer considers the results of the procedure insufficient, the second approach is advisable.

This approach envisages the construction of a mathematical model of the system on the basis of the tests conducted. The subsequent investigation is carried out in two stages. In the *first stage*, we perform the multicriteria identification of parameters of the mathematical model. If, after the identification has been

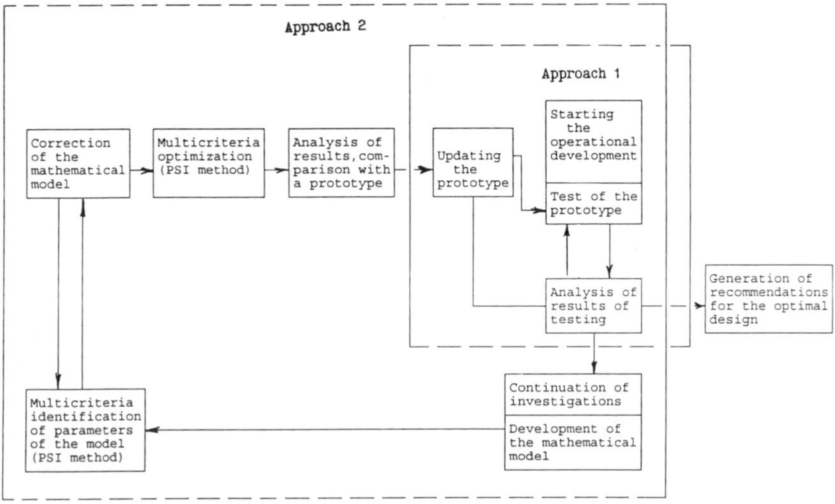

Figure 4–4 The block diagram of operational development of engineering systems.

completed, a significant disagreement takes place between the results of tests and computations, one should correct the mathematical model and repeat the identification procedure. This continues until the discrepancies between the experimental and computational data are within tolerable limits.

In the *second stage,* after the multicriteria identification, the designer formulates and solves the multicriteria optimization problem. In doing this, he uses the mathematical model whose adequacy was established in the first stage. Based on the results of the optimization, the improvement of the prototype is done, and then the tests are reproduced. This cycle is repeated until the designer decides about the termination of the operational development.

Thus, in the first stage, the set $D_{\boldsymbol{\alpha}}$ is found as a result of the multicriteria identification. In the second stage, the optimization problem is solved: we construct the parallelepiped Π in $D_{\boldsymbol{\alpha}}$, determine the vector of performance criteria, and find the feasible solutions set D (see Fig. 4-1).

We already mentioned that a weak point of optimization when used in design is a significant discrepancy between the mathematical model and the physical system, as well as improperly specified constraints. Therefore, very often the results of optimization were of no practical value. In our approach, we obtain a confirmed model and the set $D_{\boldsymbol{\alpha}}$ resulting from the multicriteria identification. This, to a sufficient extent, justifies the optimization performed at the second stage, and substantiates the recommendations for improving the prototype of a machine. In addition, this approach is expected to significantly reduce the amount of expensive and durable tests in the course of operational development of machines.

Note that having the approximation of the set $D_{\boldsymbol{\alpha}}$, we can construct the set D with required accuracy. First, we construct the domain of admissible variations of design variables, $D_v = \bigcup_i \Pi_i$. Here, the parallelepiped Π_i must satisfy the following conditions: (1) The inclusions $\boldsymbol{\beta} \in \Pi_i$ implies $\boldsymbol{\beta} \in D_{\boldsymbol{\alpha}}$ for any vector $\boldsymbol{\beta}$; and (2) Π_i is the maximal parallelepiped satisfying condition 1. In other words, there is no parallelepiped in $D_{\boldsymbol{\alpha}}$ that contains Π_i.

The boundaries of parallelepipeds Π_i can be constructed when having analyzed test tables compiled by using the results of approximation of the set $D_{\boldsymbol{\alpha}}$.

Let us make some comments as to the necessity of determining D_v. In the set $D_{\boldsymbol{\alpha}}$ we are to determine the sets where it is possible to vary the design variables continuously when searching for the optimal solution. It seems to be inexpedient to search for the optimal solution over all the parallelepiped Π in the case of the rigid constraints, since the volume of the set $D_{\boldsymbol{\alpha}}$ can be considerably less than the volume of Π. Therefore, when searching over the set $D_{\boldsymbol{\alpha}}$, we increase the percentage of found feasible models, as compared with the search over the entire Π. Obviously, the probability of obtaining better results also increases in this case.

After having obtained D_v, we construct the feasible solutions set $D (D \subset D_v)$ as described in Section 1-3.

The optimization in Π can fail to give desired results, for instance, if the volume of Π is comparatively small. In this case, the designer, having analyzed the results of identification and determined significant design variables (see Sections 1-3 and 1-4, and Chapter 5), and also using his experience, can find it possible and advisable to vary some of the design variables within essentially wider ranges.

Then a new parallelepiped is to be constructed in which the optimal solution $\boldsymbol{\alpha}^0$, with respect to criteria $\Phi_1,\ldots,\Phi_k$, is searched for according to the method of parameter space investigation. Usually, after having manufactured the optimal prototype, it is advisable to investigate it and confirm that the design variables have been found correctly, and the mathematical model is adequate to the physical system.

4-5. Example 2: Operational Development of a Vehicle

The problem of operational development of a prototype was formulated and solved on the example of a truck.

The solution was obtained in two stages. In the first stage, the mathematical model of the truck was identified on the basis of experimental data obtained in road tests (the problem of multicriteria identification).

In the second stage, the results of solving the problem were used for developing the optimal recommendations for improving the vibroprotective properties of the suspension system (the problem of multicriteria optimization).

Mathematical Model of a Truck

Vibrations of a truck were studied and calculated by analyzing its simplified scheme.

Both the analysis of experimental data and numerical studies of the truck vibration have shown that for the case under consideration it suffices to consider vertical and longitudinal-angular vibrations over the frequency range from 0.5 Hz to 16 Hz, using the model shown in Figure 4-5. The latter is composed of concentrated masses connected by inertialess elastic and damping elements, and has been developed subject to the following assumptions (Khachaturov 1976):

1. The platform, engine, frame, and cab are absolutely rigid bodies.
2. The moments of inertia of axles with respect to the wheel rotation axes are zero.
3. The inertia forces due to unbalanced rotating masses are equal to zero.
4. The vibrations of the truck masses are small.
5. Each tire contacts the road at a point.
6. The center of mass of the platform stays in the longitudinal symmetry plane and moves in such a way that the projection of its velocity onto the horizontal plane remains constant.
7. Both the elastic and damping elements have linear characteristics.

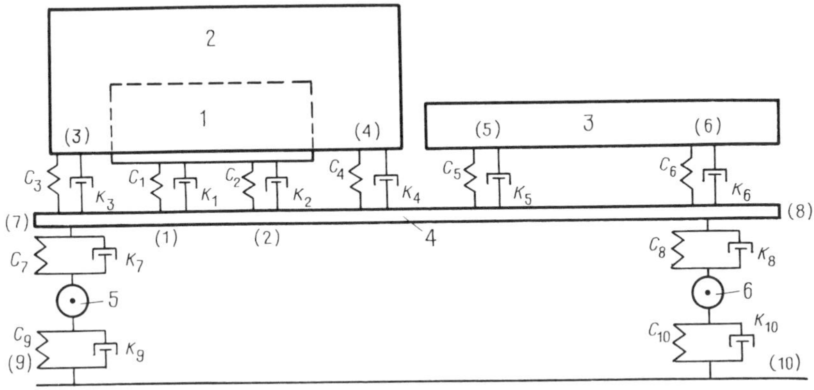

Figure 4–5 Automobile schematic. 1 is the engine; 2 the cab; 3 the platform; 4 the frame; 5 and 6 the axles; (1) and (2) the front and rear engine supports; (3) and (4) the front and rear cab supports; (5) and (6) the front and rear platform supports; (7) and (8) the front and rear suspensions; and (9) and (10) the front and rear tires.

The external excitation of the system is defined by functions $q_1(t)$ and $q_2(t)$, which take into account the road microprofile under the front and rear axles, respectively, and whose spectral density is given by the expression

$$K_{q_i}(W)=\frac{A\cdot V_t^{n-1}}{W^n}\cdot\frac{W^2+W_1^2}{W^2+W_2^2} \qquad (4\text{-}2)$$

where A is a coefficient characterizing the road roughness (measured in l/m); n is the exponent corresponding to the type of road under consideration; V_t is the truck's velocity (m/s); W_1 and W_2 are the coefficients characterizing the power spectral density of excitation (1/s); $i=1, 2$; and W is frequency (1/s).

Functions $q_1(t)$ and $q_2(t)$ differ only in the time lag T due to the distance between the front and rear axles and are related by $q_2(t)=q_1(t-T)$. The time lag T is given by

$$T=\frac{B}{V_t} \qquad (4\text{-}3)$$

where B is the truck wheel base (m).

The vibrations of the structure were estimated using the power spectral densities (PSD) of accelerations at the characteristic points of the units linked to each other by connecting elements, the PSD of relative displacements, and the angles through which the bodies rotate. The spectra were determined over a frequency range five octaves wide, starting from 0.5 Hz with the resolution of 1/12 octave.

Small vertical oscillations of the center of mass of a rigid body are described by the following linear differential equation:

$$MZ'' = \sum F_i$$

where M is the body mass; and F_i is the force exerted by the ith connecting element.

Force F_i is given by the equation

$$F_i = C_i \Delta_i + K_i \Delta_i'$$

where Δ_i is the ith element deformation; Δ_i' is the rate of deformation of the ith element; C_i is the equivalent stiffness of the ith element; and K_i is the equivalent damping factor of the ith element.

Small angular oscillations of a rigid body are described by the following linear differential equation:

$$JU'' = \sum M_i$$

where J is the moment of inertia of a body about its center of mass; $M_i = F_i X_i$ is the moment of the force exerted by the ith connecting element; and X_i is the arm of the force.

Thus, having written down the equations for each body, we get a full system of linear differential equations. Then, using the Laplace transform, we arrive at a system of linear algebraic equations in the transfer functions describing the effect of an input disturbance on the points of the system (see Fig. 4-5) where C_i and K_i, $i=1,\dots,10$, are the values of equivalent stiffnesses and damping factors of the ith element of the truck suspension.

Accelerations and deflections were calculated for the points of the truck (shown in Fig. 4-6).

For each of the three types of road (asphalt, smooth cobblestone, and rough cobblestone) the equivalent stiffnesses and damping factors of elements 7 and 8 (see Fig. 4-5) were taken from the results of dynamic bench tests. Accordingly, in calculating the vibrations, the nonlinear properties of springs were taken into account by choosing the corresponding equivalent stiffnesses and damping factors for different levels of excitation. This was done in line with the method of linearization discussed in detail in (Voevodenko and Pevzner (1985)). The method implies that for a given disturbance (determined by the road roughness), a leaf spring may be adequately simulated by a linear element with the corresponding values of equivalent stiffnesses and damping factors within the said frequency range.

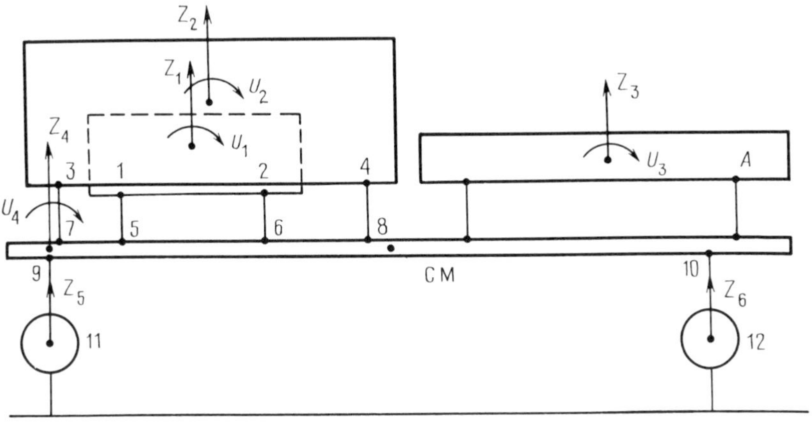

Figure 4–6 Z_1–Z_6 are linear displacements; U_1–U_4 are rotations; *CM* is the frame center of mass; 1–12 are the positions of the road tests measurements points.

Truck Tests

The tests have been carried out in order to determine the spectra of vibration accelerations of the structural elements of a truck subjected to a nominal loading. Measurements were implemented in moving over various roads at testing grounds.

The accelerations induced by vibration at 12 points of the structure were measured with accelerometers and recorded using a tape recorder (Fig. 4-6).

Note that the PSD of accelerations practically coincide for the points positioned symmetrically with respect to the longitudinal symmetry axis. Therefore, the data obtained for only one of a pair of symmetric points were used.

The experimental data were processed using an FFT-analyzer allowing plotting the PSD of accelerations versus the frequency of excitation. The statistical error of the experiment lay within $\pm$ 15%.

Formulation and Solution of the Problem of Multicriteria Identification of the Parameters of a Truck

The mathematical model is characterized by the following variables: the masses and moments of inertia of units, the coordinates of the structural elements, and the stiffnesses and damping factors of the connecting elements.

In carrying out laboratory tests the following variables of the truck under study were determined: the masses and moments of inertia of the engine, cab, platform, and frame, the coordinates of their supports, and the coordinates of the points at which the engine, cab, and platform are attached to the frame. The dynamic characteristics of the leaf springs, C_7, C_8, K_7, K_8, were also determined experimentally.

Table 4–3

Ordinal numbers of variables	Denotation	Dimension	Values of the prototype variables	Boundaries of variables: lower α_j^*	Boundaries of variables: upper α_j^{**}	Optimal model
1	C_1	N/m	$7.7\cdot10^5$	$6.6\cdot10^5$	$8.8\cdot10^5$	$7.08\cdot10^5$
2	C_2	N/m	$3.6\cdot10^6$	$3.05\cdot10^6$	$4.14\cdot10^6$	$3.29\cdot10^6$
3	C_3	N/m	$1.14\cdot10^6$	$9.6\cdot10^5$	$1.32\cdot10^6$	$1.06\cdot10^6$
4	C_4	N/m	$6.5\cdot10^5$	$6.1\cdot10^5$	$8.8\cdot10^5$	$6.83\cdot10^5$
5	C_5	N/m	$2.0\cdot10^8$	$1.6\cdot10^8$	$2.4\cdot10^8$	$1.8\cdot10^8$
6	C_6	N/m	$2.0\cdot10^8$	$1.6\cdot10^8$	$2.4\cdot10^8$	$1.92\cdot10^8$
7	C_7	N/m	$4.2\cdot10^5$	$4.2\cdot10^5$	$4.2\cdot10^5$	$2.99\cdot10^5$
8	C_8	N/m	$1.05\cdot10^6$	$1.05\cdot10^6$	$1.05\cdot10^6$	$7.21\cdot10^5$
9	C_9	N/m	$2.0\cdot10^6$	$1.6\cdot10^6$	$2.4\cdot10^6$	$1.63\cdot10^6$
10	C_{10}	N/m	$3.6\cdot10^6$	$2.88\cdot10^6$	$4.32\cdot10^6$	$3.01\cdot10^6$
11	K_1	N·s/m	2,000	1,200	2,800	3,400
12	K_2	N·s/m	7,800	4,700	11,000	9,210
13	K_3	N·s/m	5,000	4,000	6,000	4,920
14	K_4	N·s/m	4,000	3,200	4,800	4,376
15	K_5	N·s/m	25,000	20,000	30,000	25,700
16	K_6	N·s/m	25,000	20,000	30,000	24,467
17	K_7	N·s/m	14,000	14,000	14,000	14,665
18	K_8	N·s/m	36,000	36,000	36,000	42,588
19	K_9	N·s/m	4,000	3,200	4,800	3,908
20	K_{10}	N·s/m	8,000	6,000	10,400	5,902

The following variables were to be identified: (1) The stiffnesses of connecting elements, C_1–C_6, C_9, and C_{10}; and (2) the damping factors of the connecting elements, K_1–K_6, K_9, and K_{10} (see Fig. 4-5). These variables could hardly be determined by either bench or road tests.

The boundaries of the variables form a 16-dimensional parallelepiped Π^1, see Table 4-3.

Figure 4-7 shows the curves of the RMS spectra of accelerations: curve 1 has been calculated using the mathematical model, and curve 2 was obtained by road tests. (RMS is a square root of the PSD, $G=\sqrt{g(f)}$).

In order to estimate the closeness of the calculated and experimental curves one has to introduce adequacy criteria[12].

The degree to which curves match each other is estimated using the following three groups of criteria:

[12]In practice, a researcher must often construct a multitude of curves by varying the parameters of a mathematical model, and select the curve that approximates a given one in the best possible way. This is a typical multicriteria problem in which the closeness of curves is estimated using various adequacy criteria.

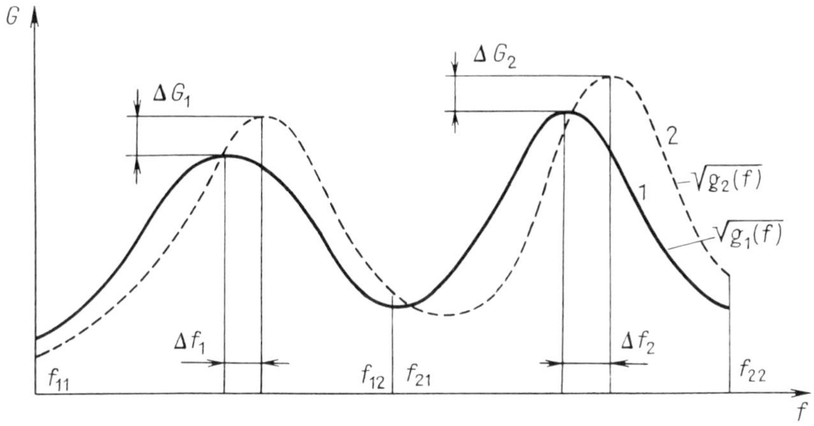

Figure 4–7 1 is the calculated RMS spectrum of accelerations; 2 is an analogous plot drawn using the road tests results; f_{1j} and f_{2j}, $j = 1, 2$, are the frequency range boundaries.

1. Mismatch in the frequencies at which the local RMS maximum is observed:

$$\Phi_i = |\Delta f_{i\max}|, \quad i=1,\ldots,n;$$

2. The difference in the values corresponding to the local RMS maximum:

$$\Phi_i = |\Delta G_{i\max}|, \quad i=n+1,\ldots,2n$$

   where n is the number of local maxima at the RMS plot.
3. The difference in the root-mean-square of accelerations within a given range:

$$\Phi_i = \left|\left(\int_{f_{1j}}^{f_{2j}} g_1(f)df\right)^{1/2} - \left(\int_{f_{1j}}^{f_{2j}} g_2(f)df\right)^{1/2}\right|, \quad i=2n+1,\ldots,m$$

   where j is either 1 or 2 depending on the measurement point under consideration; $f_{11}=0.5$ Hz; $f_{12}=f_{21}=6$ Hz; and $f_{22}=16$ Hz.

Next, it is shown that there are 65 such criteria. Therefore, it was decided to ignore for the time being the values of the spectra of relative displacements and the angles of the structural elements rotations. The latter were taken into account after the construction of the feasible solutions set using the aforementioned three groups of criteria.

The mathematical model was calculated for 60 frequencies f_i belonging to the range [0.5 Hz–16 Hz].

Let us denote by $f_{i\max}^{exp}$ the frequencies at which the experimental RMS spectrum of accelerations attains its local maximum, and by $G_{i\max}^{exp}$ the values of the maxima. Depending on the measurement point position, n varied from 1 to 3.

For all 12 measurement points 65 closeness criteria were needed: The first group consisted of 22 criteria determining the values of frequencies (Hz), the second group incorporated the next 22 criteria representing RMS ($m/s^2 \cdot \sqrt{Hz}$), finally, the third group included 21 criteria representing the RMS accelerations (m/s^2). First, an interval $[f_{i\max}^{exp} \pm 0.1 f_{i\max}^{exp}]$ Hz was chosen. Then, the local maxima $G_{i\max}$ and the corresponding values of $f_{i\max}$ belonging to the interval were determined, and criteria $\Phi_i = |f_{i\max}^{exp} - f_{i\max}|$, $i=1,\ldots,n$, and $\Phi_i = |G_{i\max}^{exp} - G_{i\max}|$, $i=n+1,\ldots, 2n$, were calculated.

It is known that the differences between experimental and calculated frequencies corresponding to local maxima of RMS spectra of accelerations may hardly be compared for different frequency ranges. In this case, the estimated mismatch between experimental and calculated values depends on the frequency range.

The experience of comparing experimental and calculated characteristics allows making the following statement. Let the difference between experimental and calculated frequencies Δf_i be measured not in Hertz but in one-twelfths of the octave intervals between $f_{i\max}^{exp}$ and $f_{i\max}$. Then these values characterize the criteria Φ_i irrespective of the frequency range the values belong to.

Table 4-4 presents experimental[13] values of Φ_ν^{exp} $\nu=1,\ldots,65$. In line with Section 4-2, Φ_ν^{exp} is included into the expression for respective closeness criterion.

The set of variables and the corresponding criteria vector determined by the designers of the truck manufacturer will be called a prototype, see Table 4-3. The studies were aimed at finding out how well the prototype was selected, whether it can be perfected in at least the basic variables, and whether there exist alternative solutions of interest to the designers.

Analysis in Π^1

In all, N=4,096 trials were carried out, and constraints for the first 22 criteria were determined. For the rest of the criteria no constraints were introduced, since they could not be determined with sufficient accuracy. Only seven models (the prototype excluded) proved to meet the aforementioned criteria constraints. Since the latter were not determined for all the criteria, the seven models were subjected to further analysis.

For each calculated solution, the RMS spectra of acceleration plots were considered (12 such plots were drawn for each solution, corresponding to the aforementioned points of the truck). Then, these plots were compared with the ones obtained in the road tests.

Figure 4-8 shows the plots obtained for point 2 (see also Fig. 4-6). The solid line corresponds to one of the feasible models, namely model 552, obtained by

---

[13]The experiment was carried out for a smooth cobblestone road and the truck speed 60 km/h.

Table 4–4

Ordinal numbers of criteria	Measurement point number (Fig. 4-6)	The value of Φ_ν^{exp}	Ordinal numbers of criteria	Measurement point number (Fig. 4-6)	The value of Φ_ν^{exp}	Ordinal numbers of criteria	Measurement point number (Fig. 4-6)	The value of Φ_ν^{exp}
1	1	2.12	23	1	0.712	45	1	2.04
2	1	6.5	24	1	0.9	46	1	5.46
3	1	9.875	25	1	0.836	47	2	1.28
4	2	10.0	26	2	1.3	48	2	4.06
5	3	1.75	27	3	0.546	49	3	0.69
6	3	10.125	28	3	0.787	50	3	1.67
7	4	1.25	29	4	0.537	51	4	0.507
8	4	10.75	30	4	1.07	52	4	1.125
9	5	2.125	31	5	0.771	53	5	2.07
10	5	6.5	32	5	0.769	54	5	3.4
11	6	9.5	33	6	0.794	55	6	0.833
12	7	1.25	34	7	0.527	56	6	1.98
13	7	9.75	35	7	0.937	57	7	1.01
14	8	1.25	36	8	0.530	58	7	2.71
15	8	4.5	37	8	0.595	59	8	0.537
16	8	9.5	38	8	0.409	60	8	0.570
17	9	1.75	39	9	0.433	61	9	2.5
18	9	9.5	40	9	1.61	62	10	1.26
19	10	1.62	41	10	0.865	63	11	2.0
20	11	9.75	42	10	0.225	64	11	12.4
21	11	9.87	43	11	4.05	65	12	6.85
22	12	12.0	44	12	2.26	—	—	—

solving the multicriteria identification problem. The dashed line corresponds to road tests.

The characteristics of the model proved to be rather close to the road-test results for all the other measurement points too.

Similar plots were drawn for the rest of the feasible solutions. No considerable departures of the curves from the results of road tests were revealed. This allowed concluding that all the seven calculated solutions have entered the feasible solutions set D_α in all 65 criteria.

The correction of the boundaries of the variables was of special importance in solving the multicriteria identification problem. Since in problems of multicriteria identification one cannot define a priori constraints on the variables, the latter had to be corrected in order to be able to construct a parallelepiped Π^k, such that the feasible solutions set $D_\alpha \subset \Pi^k$, and if for a parallelepiped Π, $\Pi^k \not\subset \Pi$ holds then $D_\alpha \not\subset \Pi$. Figure 4-9 shows the histograms constructed for three, α_3, α_{10}, and α_{11}, of the 10 variables being corrected, for which the acceptable solutions concentrate near the ends of segments $[\alpha_j^*, \alpha_j^{**}]$. The designers have altered the corresponding boundaries of the variables. The other six variables in

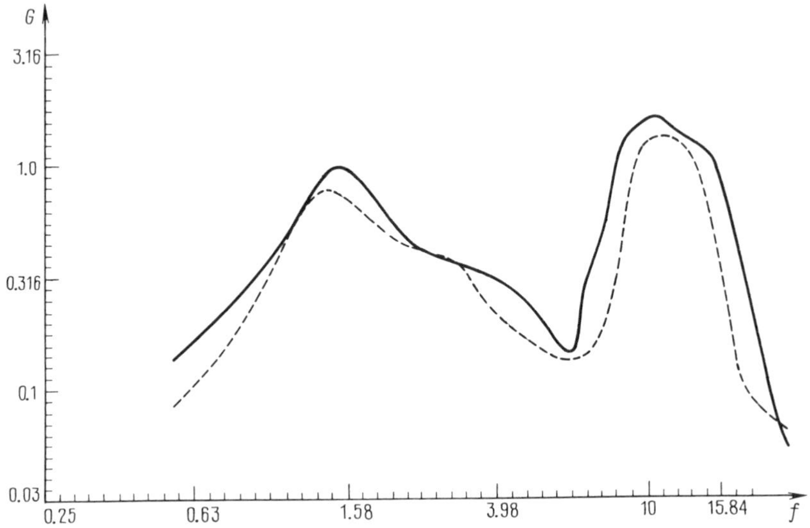

Figure 4–8 RMS spectra for the point 2 acceleration.

Π^2 were the same as in Π^1. (The construction of histograms and analysis are discussed in Sections 1-3 and 1-4.)

Analysis in Π^2

The same number of trials, N=4,096, were carried out in the 16-dimensional parallelepiped Π^2. The same constraints on the first 22 criteria were preserved. As a result, 11 models were obtained, for which the constraints are met.

The models were analyzed taking the remaining criteria, Φ_{23}–Φ_{65}, into account. All of them have entered the feasible solutions set $D_{\boldsymbol{\alpha}}$. Hence, our assumption about the presence of acceptable solutions outside Π^1 proved to be true.

In analogy to the seven models found earlier in Π^1, these models proved to be acceptable as regards the closeness criteria taking into account the values of the spectra of the relative displacements and rotation angles of the structural elements.

In line with the design and technological requirements no further corrections of constraints on the variables were carried out.

Thus, parallelepiped Π^2 proved to be the final one in the present analysis.

Similar analyses were conducted taking into account the tests carried out on the other proving-ground roads. As a result, feasible solutions set $D_{\boldsymbol{\alpha}}$ being the intersection of the feasible solutions sets corresponding to each of the three roads, was constructed.

Three of the models obtained by solving the problem under consideration, $\boldsymbol{\alpha}^{552}$ among them, have entered the set. These acceptable models were used for

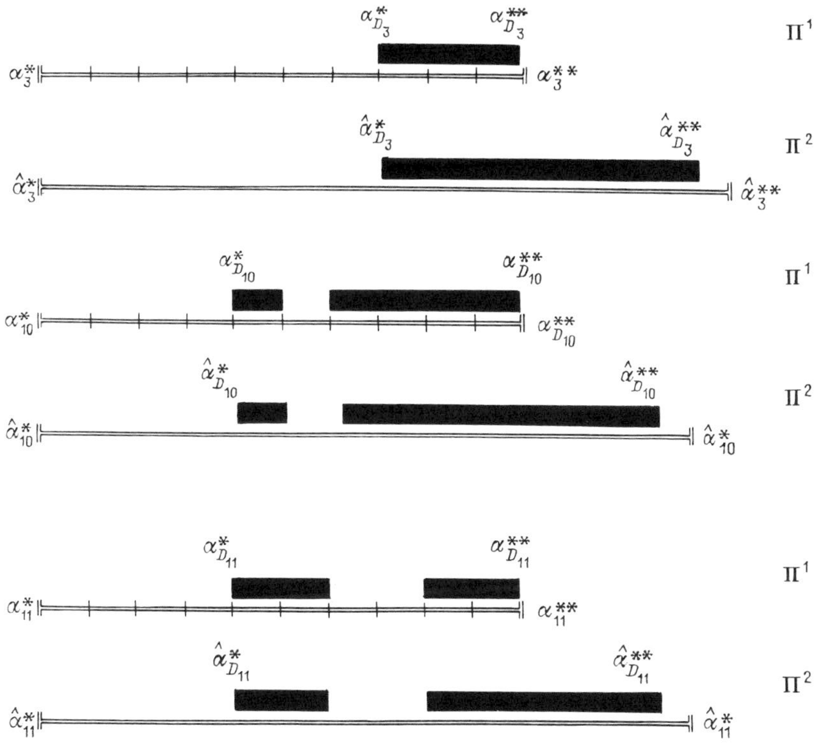

Figure 4–9 Initial, $\alpha_j^{*(*)}$, and new $\hat{\alpha}^{*(*)}$, boundaries of the corrected variables 3, 10, and 11. $\alpha_{Dj}^{*(*)}$ and $\hat{\alpha}_{Dj}^{*(*)}$ are the feasible solutions set boundaries in Π^1 and Π^2 respectively.

constructing parallelepiped Π within which the problem of optimization was being solved.

Formulation and Solution of the Optimization Problem

Let us consider the choice of criteria and design variables. As the design variables, we have taken the stiffnesses and damping factors presented in Table 4-3, and also the stiffnesses and damping factors of the leaf springs. Together they form a 20-dimensional parallelepiped Π. The boundaries of design variables were chosen by the designers who took into account both the results of solving the problem of the set $D_{\boldsymbol{\alpha}}$ determination and the technological potential of the plant.

The criteria may be conditionally divided into the groups of: (1) Comfort; (2) durability; (3) load preservation; and (4) safety (see Table 4-5).

In line with ISO 2631/1-1985(E)[14] the comfort criterion Φ_1 was set equal to

[14]The International Organization for Standardization, 1985. Evaluation of human exposure to whole-body vibration-Part 1: General requirements, pp. 1–17.

Table 4–5

Ordinal numbers of criteria	Range	Dimension	Measurement point number (Fig. 4-6)	Performance criteria values of the prototype	Performance criteria values for the optimal design (model 1,820)	Meanings of performance criteria
1	W	m/s^2	4	2.149	1.55	comfort
2	L	m	1	$1.09 \cdot 10^{-3}$	$9.84 \cdot 10^{-4}$	durability
3	H	m	1	$3.8 \cdot 10^{-4}$	$2.8 \cdot 10^{-4}$	durability
4	L	m	2	$1.75 \cdot 10^{-4}$	$1.50 \cdot 10^{-4}$	durability
5	H	m	2	$6.01 \cdot 10^{-4}$	$4.8 \cdot 10^{-4}$	durability
6	L	m	3	$7.85 \cdot 10^{-4}$	$7.38 \cdot 10^{-4}$	durability
7	H	m	3	$6.94 \cdot 10^{-4}$	$6.71 \cdot 10^{-4}$	durability
8	L	m	4	$5.52 \cdot 10^{-4}$	$4.42 \cdot 10^{-4}$	durability
9	H	m	4	$5.33 \cdot 10^{-4}$	$3.57 \cdot 10^{-4}$	durability
10	W	m/s^2	A	10.66	8.3	load preservation
11	L	m	9	$1.03 \cdot 10^{-2}$	$1.037 \cdot 10^{-2}$	durability
12	H	m	9	$7.78 \cdot 10^{-4}$	$6.012 \cdot 10^{-4}$	durability
13	L	m	10	$1.02 \cdot 10^{-2}$	$9.065 \cdot 10^{-3}$	durability
14	H	m	10	$5.62 \cdot 10^{-4}$	$4.11 \cdot 10^{-4}$	durability
15	L	m	11	$2.75 \cdot 10^{-3}$	$2.879 \cdot 10^{-3}$	durability
16	H	m	11	$3.46 \cdot 10^{-3}$	$3.36 \cdot 10^{-3}$	durability
17	L	m	12	$3.72 \cdot 10^{-3}$	$3.528 \cdot 10^{-3}$	durability
18	H	m	12	$2.58 \cdot 10^{-3}$	$2.53 \cdot 10^{-3}$	durability
19	W	—	11	$6.26 \cdot 10^{-1}$	$5.12 \cdot 10^{-1}$	safety
20	W	—	12	$6.59 \cdot 10^{-1}$	$5.296 \cdot 10^{-1}$	safety

the frequencies-weighted RMS acceleration. Since in the case under consideration the driver's seat was unsprung, the accelerations at its surface and at the cab floor under the seat (see Fig. 4-6) are approximately equal over the frequency range of 0.5 Hz–16 Hz. Hence, criterion Φ_1 may be set equal to the ISO weighted acceleration at the surface of the cab floor (see point 4 in Fig. 4-6). The vibration of a driver was ignored.

The second group of criteria, Φ_2–Φ_9 and Φ_{11}–Φ_{18}, incorporates the RMS relative displacements.

As the load-preservation criterion, Φ_{10}, the trebled RMS acceleration of point *A* of the truck platform over the entire frequency range has been taken (see Fig. 4-6). The value of this criterion must not exceed g=9.8 m/s^2, since otherwise a load may lose contact with the platform.

The final group was composed of safety criteria Φ_{19} and Φ_{20}, which characterize the probability of a tire losing contact with the road surface:

$$\Phi_i = 3 \cdot \frac{\text{RMS}d_j}{R_j} c_j, \quad i=19, 20, \quad j=11, 12$$

where RMSd_j is the root-mean-square tire deformation at a point j (m); R_j is the static load within the tire-road contact zone (measured in Newtons) (see Fig. 4-6); and c_j is the tire stiffness (measured in N/m).

Each of the criteria has been calculated for the front and rear truck wheels.

If at least one of these criteria is equal to unity, then a tire may lose contact with the road surface. Therefore, the constraints are imposed on these criteria, and they must not exceed unity. Besides, Table 4-5 presents the values of the performance criteria for the prototype.

The whole of the frequency range 0.5 Hz–16 Hz, denoted in the table by W, was divided into two intervals, 0.5 Hz–6 Hz and 6 Hz–16 Hz, denoted by L and H, respectively. The performance criteria were calculated either for each interval taken alone or for the entire range.

The numbers of the points where the performance criteria values were determined (see Fig. 4-6) are presented in the fourth column of the table. In the designers' opinion, the prototype should have been substantially improved as regards the comfort and durability criteria. Optimization was aimed at reaching this goal.

In all, N=4,096 trials were conducted in parallelepiped Π, 21 of which have entered set D. Of the latter, 20 solutions were Pareto optimal.

Upon analyzing set D the designers preferred solution 1,820.

Table 4-5 presents the values of the performance criteria for the prototype and the optimal solution. The design variables of the optimal solution are presented in Table 4-3. By comparing them with the prototype design variables presented in the same table we see that the optimal solution surpasses the prototype in 18 criteria, which include the most important ones. In fact, by comparing, for example, the values of the first and the 10th criteria for the optimal solution and the prototype, we see that they were improved up to 30%. However, solution 1,820 lags behind the prototype in the 11th and 15th criteria. It should be noted that the values of the criteria have decreased insignificantly, by less than 5%, and the criteria themselves do not belong to the basic ones. Thus, the solution of the optimization problem has resulted in finding a solution surpassing the prototype considerably.

Conclusions

1. By solving the problem of multicriteria identification, the values of the stiffness and damping factors ensuring the adequacy of the truck model under consideration were found. Feasible boundaries of the design variables for the subsequent solution of the optimization problem were also determined.
2. The set of feasible models of interest to designers has been found. Some of the models surpass the prototype in the basic performance criteria.

Recommendations for improving the truck suspension and, as a result, the prototype basic performance criteria (comfort, safety, durability, and load preservation) have been formulated.

Thus, to improve the design, the reduction of the stiffnesses of two engine supports as well as of the front cab support and tires, was recommended. At the same time, the stiffnesses of the suspension and of the rear cab support should be somewhat increased. The values of the damping factors should be altered as shown in Table 4-3 for the optimal model 1,820.

5

Determination of Significant Design Variables

Many real objects are described by means of high-order systems of equations with a large number of coefficients for unknowns. These coefficients play the role of design variables to be varied when optimizing an object.

For the purpose of multicriteria optimization, especially if the dimension r of the design-variable space is large enough, it is necessary to carry out a sufficiently great number of trials for the construction of a feasible solutions set. This number increases considerably with the r growth. This requires so great an amount of time for seeking the optimum solution that the optimization cannot be carried out in many cases.

To solve multicriteria optimization problems, it is advisable to use methods allowing the reduction of the dimensions of the design-variable space by eliminating the insignificant design variables—those that do not perceptibly influence the values of the criteria Φ_ν, $\nu=\overline{1,k}$. In other words, q of the significant design variables, $q<r$, to which the criteria are sensible, are determined as a result of the evaluation of the criteria sensibility to the design variables change. And further, in solving the optimization problem, these q of the design variables are varied. Here we can single out two approaches in solving this problem. The first approach is universal. The regression analysis technique may be referred to this approach. The second approach comprises methods allowing us to tackle problems of some particular classes taking into account their inherent features. The energy balance principle for investigating dynamic problems is an example of the approach. To this approach we can refer methods taking into account the influence of the system design variables that determine interactions between the subsystems. If these interactions happen to be weak, the influence of the aforementioned design variables may be neglected under certain conditions.

5-1. Evaluating Performance Criteria Sensitivity Through Regression Analysis Technique

We now introduce certain definitions and notations needed in the sequel.

The mean of a function $f(\boldsymbol{\alpha})$ in a domain D ($D \subset \Pi$) is the integral:

$$E_D(f) = \int_D f(\boldsymbol{\alpha}) d\boldsymbol{\alpha} \tag{5-1}$$

The variance of a function $f(\boldsymbol{\alpha})$ in a domain D is defined by:

$$\mathrm{var}_D(f) = [\int_D (f(\boldsymbol{\alpha}) - E(f))^2 \, d\boldsymbol{\alpha}]^{1/2} \tag{5-2}$$

The L_2-norm of a function $f(\boldsymbol{\alpha})$ in a domain D is defined by:

$$\|f\|_D = [\int_D (f(\boldsymbol{\alpha}))^2 d\boldsymbol{\alpha}]^{1/2}$$

The L_2-distance between two functions $f(\boldsymbol{\alpha})$ and $g(\boldsymbol{\alpha})$ in a domain D is defined by:

$$d(f,g) = \|f - g\|_D = [\int_D (f(\boldsymbol{\alpha}) - g(\boldsymbol{\alpha}))^2 d\boldsymbol{\alpha}]^{1/2}.$$

The L_2-distance satisfies the triangle inequality:

$$d(f,g) \leq d(f,z) + d(g,z). \tag{5-3}$$

The scalar product of two functions $f(\boldsymbol{\alpha})$ and $g(\boldsymbol{\alpha})$ in a domain D is defined to be the integral:

$$(f,g)_D = E(f,g) = \int_D f(\boldsymbol{\alpha}) g(\boldsymbol{\alpha}) d\boldsymbol{\alpha}$$

The coefficient of covariance between two functions $f(\boldsymbol{\alpha})$ and $g(\boldsymbol{\alpha})$ is defined to be the scalar product of the centered functions $f(\boldsymbol{\alpha}) - Ef(\boldsymbol{\alpha})$ and $g(\boldsymbol{\alpha}) - Eg(\boldsymbol{\alpha})$:

$$c_{fg} = E(f - Ef, g - Eg) = \int_D (f(\boldsymbol{\alpha}) - Ef(\boldsymbol{\alpha}))\,(g(\boldsymbol{\alpha}) - Eg(\boldsymbol{\alpha})) d\boldsymbol{\alpha}\ ,$$

and the coefficient of correlation between two functions $f(\boldsymbol{\alpha})$ and $g(\boldsymbol{\alpha})$ is defined as follows:

$$\rho_{fg} = \frac{c_{fg}}{(\mathrm{var}_D(f)\mathrm{var}_D(d))^{1/2}}\ .$$

If integration is taken over the entire parallelepiped Π, we will not indicate the domain of integration.

Standardization of Variables

Here let us make a remark. Irrespective of what method we will apply to evaluate the significance of design variables, the first step always consists in standardizing the variables, that is, the design variables and the criteria $\Phi_\nu(\boldsymbol{\alpha})$ $(\nu=\overline{1,k})$. Let z be one of these variables. Standardization of z lies in changing over to a new variable:

$$z_{st}=\frac{z-E(z)}{(\mathrm{var}_D(z))^{1/2}}$$

After this transformation, we have $E(z_{st})=0$ and $\mathrm{var}_D(z_{st})=1$. In the sequel these equalities are always assumed to hold for all design variables and criteria.

Evaluation of the Significance of Design Variables

How can the significance of a design variable α_i be evaluated for some particular criterion? There are many ways of doing this.

Significance measures based on the norms of partial derivatives

The first of these methods is based on the use of the mean of the square or the absolute value of the partial derivative $\partial\Phi/\partial\alpha_i$ in the parallelepiped Π. In what follows we consider the mean squares of derivatives only. Accordingly, the proposed criterion can be expressed as

$$I_{1,\alpha_i}=\left\|\frac{\partial\Phi(\boldsymbol{\alpha})}{\partial\alpha_t}\right\|^2=\int_\Pi\left(\frac{\partial\Phi(\boldsymbol{\alpha})}{\partial\alpha_i}\right)^2 d\boldsymbol{\alpha} \qquad (5\text{-}4)$$

Indeed, if the design variable α_i has in general no influence on the functional $\Phi(\boldsymbol{\alpha})$, then the derivative $\partial\Phi/\partial\alpha_i\equiv 0$ in the parallelepiped Π and the mean of the square of the derivative should also be zero ($I_{1,\alpha_i}=0$). At the same time, it is reasonable to assume that the criterion $\Phi(\boldsymbol{\alpha})$ is more sensitive (on the average) to the variations in the parameter (the design variable) α_i than to the variations in the parameter α_j, provided $I_{1,\alpha_i}>I_{1,\alpha_j}$.

In the expression (5-4) we first determine the derivative, then square it, and finally integrate it to calculate the norm. Or we may proceed differently. We can first average (integrate) the criterion $\Phi(\boldsymbol{\alpha})$ with respect to a set of design variables $\boldsymbol{\beta}_i$, i.e., $\boldsymbol{\beta}_i=(\alpha_1,\ldots,\alpha_{i-1}, \alpha_{i+1},\ldots,\alpha_r)$ derived from $\boldsymbol{\alpha}$ after eliminating the design variable α_i, in other words, we pass on to the function:

$$\overline{\Phi}_i(\alpha_i)=\int_{\Pi_{\boldsymbol{\beta}_i}} \Phi(\alpha_i,\boldsymbol{\beta}_i)d\boldsymbol{\beta}_i, \tag{5-5}$$

where $\Pi_{\boldsymbol{\beta}_i}$ is a parallelepiped in the space of the design variables $\boldsymbol{\beta}_i$. Now we differentiate the function thus obtained with respect to α_i and then calculate its norm. As a result, we arrive at the significance measure

$$I_{2,\alpha_i}=\left\|\frac{d\overline{\Phi}_i(\alpha_i)}{d\alpha_i}\right\|^2=\int_{\Pi_{\alpha_i}}\left(\frac{d\overline{\Phi}_i(\alpha_i)}{d\alpha_i}\right)^2 d\alpha_i, \tag{5-6}$$

However, in averaging a function of the type (5-5), we may obtain $\overline{\Phi}_i(\alpha_i)\equiv$ const. For example, consider a function of two variables, say $\Phi(\boldsymbol{\alpha})=\sin\alpha_1 \sin\alpha_2$ $(0\leq\alpha_i\leq 2\pi,\ i=1,2)$. Now on averaging with respect to α_2, we obtain $\overline{\Phi}_1(\alpha_1)\equiv 0$. Therefore, in general, we have to use modified criteria, which leads us to the function

$$\overline{\Phi}_i(\alpha_i)=\int_{\Pi_{\boldsymbol{\beta}_i}} \Phi^2(\alpha_i,\boldsymbol{\beta}_i)\, d\boldsymbol{\beta}_i, \tag{5-7}$$

Other modifications of criteria helpful in increasing the computational efficiency are considered in describing the respective algorithms (see Estimates of the significance measures, I_1 and I_2).

Significance estimate based on averaging with respect to a design variable

In this case, to estimate the sensitivity of a criterion $\Phi(\boldsymbol{\alpha})$ to a design variable α_i, we first eliminate the influence of this design variable on the criterion $\Phi(\boldsymbol{\alpha})$ by averaging the criterion with respect to α_i and thus obtain a function of design variables $\boldsymbol{\beta}_i$:

$$\psi_i(\boldsymbol{\beta}_i)=\frac{1}{\alpha_i^{**}-\alpha_i^{*}}\int_{\Pi_{\alpha_i}}\Phi(\boldsymbol{\alpha})d\alpha_i$$

Then we compute the sensitivity measure

$$I_{3,\alpha_i}^2=\frac{\int_{\Pi}(\psi_i(\boldsymbol{\beta}_i)-\Phi(\boldsymbol{\alpha}))^2\, d\boldsymbol{\alpha}}{\mathrm{var}_D(\Phi)}, \tag{5-8}$$

If $\Phi(\boldsymbol{\alpha})$ does not depend on α_i, then obviously $\psi_i(\boldsymbol{\beta}_i)\equiv\Phi(\boldsymbol{\alpha})$ and $I_{3,\alpha_i}=0$. It is also reasonable to assume that, if $I_{3,\alpha_i}>I_{3,\alpha_j}$, then the design variable α_i exerts greater influence on $\Phi(\boldsymbol{\alpha})$ than the design variable α_j.

Use of approximations of dependencies for estimating the significance of design variables

In order to estimate the degree of the significance of the design variables we may from the very beginning make an attempt to use, instead of the criterion $\Phi(\alpha)$ itself, some or other its approximation that has a sufficiently simple analytical expression

$$\Phi(\boldsymbol{\alpha})=\hat{\Phi}(\boldsymbol{\alpha})+d(\boldsymbol{\alpha})+\epsilon, \tag{5-9}$$

where $\hat{\Phi}(\boldsymbol{\alpha})$ is the approximating function from some class of functions $\mathbb{F}$.

In (5-9) $d(\boldsymbol{\alpha})$ is the approximation error due to the improper choice of the approximating function $\hat{\Phi}(\boldsymbol{\alpha})$. This error must satisfy the condition $\int d(\boldsymbol{\alpha})d\boldsymbol{\alpha}=0$. While "fitting" the approximation, an attempt is made to make "on the average" the quantity $d(\boldsymbol{\alpha})$ minimal, say by minimizing $\|d\|^2$ or, in other words, to minimize the L_2-distance between the initial function $\Phi(\boldsymbol{\alpha})$ and its approximation.

Also, ϵ is the random error due to the inaccuracy in the measurement of the values of $\Phi(\boldsymbol{\alpha})$, to the presence of design variables which have not been taken into account, etc. Its mathematical expectation should be zero, that is, $\mathbf{E}(\boldsymbol{\epsilon})=0$.

In our case (active computational experiment) measurement error may arise solely due to the insufficient accuracy in computing the values of $\Phi(\boldsymbol{\alpha})$, for example, in solving differential equations, and/or due to round-off errors. In what follows we neglect the error ϵ.

Suppose that we have derived the approximation $\hat{\Phi}(\boldsymbol{\alpha})$ of the criterion $\Phi(\boldsymbol{\alpha})$ and let $\delta^2=d^2(\Phi,\hat{\Phi})/\mathrm{var}_D(\Phi)$. Now eliminating α_i and using the triangle inequality, we obtain

$$d(\psi_i,\Phi)\leq d(\Phi,\hat{\Phi})+d(\psi_i,\hat{\Phi})\leq d(\Phi,\hat{\Phi})+d(\psi_i,\hat{\psi}_i)+d(\hat{\Phi},\hat{\psi}_i).$$

Hence, we have

$$I_{3,\alpha_i}\leq\hat{I}_{3,\alpha_i}+2\delta,$$

where

$$\hat{I}^2_{3,\alpha_i}=\frac{\int\limits_{\Pi}(\hat{\psi}_i(\boldsymbol{\beta}_i)-\hat{\Phi}(\boldsymbol{\alpha}))^2d\boldsymbol{\alpha}}{\mathrm{var}_D(\Phi)}$$

Therefore, if the approximation is of good quality (δ is small), the estimate of the measure of significance I_{3,α_i} will be close to its true value.

Use of regression analysis for approximating the performance criteria

We will apply the regression approach to derive the approximation $\hat{\Phi}(\boldsymbol{\alpha})$.

From the viewpoint of regression analysis the design variables $\alpha_1,\ldots,\alpha_r$ are input variables governing the conditions for the functioning of the system. We will also call these design variables *predictors* or *explaining variables*.

The criteria Φ_i are output variables that we may also call dependent or resulting variables. Suppose, as a result of modeling, we have obtained N vectors of dimension $(r+k)$ of the type:

$$\Phi_1^{(j)},\ldots,\Phi_k^{(j)},\ \alpha_1^{(j)},\ldots,\alpha_r^{(j)} \quad j=\overline{1,N}.$$

The vectors form a data matrix $\mathbf{Z}$ of dimension $N\times(r+k)$ with its rows consisting of components of these vectors. The matrix $\mathbf{A}$ composed of the values of the design variables, with its rows being formed by components of the vectors $\boldsymbol{\alpha}^{(j)}$ $(j=\overline{1,N})$, is, from the viewpoint of regression analysis, a design matrix. Therefore a data matrix is obtained in the course of an active experiment.

Basic regression models used in estimating the significance measures

Now, we will approximate the dependence $\Phi(\boldsymbol{\alpha})$ with the help of linear or generalized linear model:

$$\Phi(\boldsymbol{\alpha})=b_0+\sum_{i=1}^{r} b_i\alpha_i+d_1(\boldsymbol{\alpha}) \tag{5-10}$$

$$g(\Phi(\boldsymbol{\alpha}))=b_0+\sum_{i=1}^{r} b_i\overline{\Phi}_i(\alpha_i)+d_2(\boldsymbol{\alpha}) \tag{5-11}$$

$$\Phi(\boldsymbol{\alpha})=\sum_{j=1}^{q} g_j\left(\sum_{i=1}^{r} b_i\alpha_i\right)+d_3(\boldsymbol{\alpha}) \tag{5-12}$$

In all the three cases, parameters b_i are to be estimated. Moreover, functions $g(\,)$ and $\overline{\Phi}_i(\,)$ in (5-11) and $g_j(\,)$ in (5-12) are to be estimated too. The functions and parameters are evaluated from the minimality condition of the integrals $\delta_j^2=\int_{\Pi} d_j^2(\boldsymbol{\alpha})d\boldsymbol{\alpha}$ $(j=1, 2, 3)$.

Linear multiparametric regression (LMR)

Linear multiparametric regression (Draper and Smit 1966) gives the approximation of the function to be evaluated as follows:

$$y=(\mathbf{B}'\mathbf{X})+b_0+\delta, \tag{5-13}$$

where $\mathbf{B}=(b_1,\dots,b_r)'$ is a vector of unknown coefficients of the regression equation (5-13) and $(\mathbf{B}'\mathbf{X})$ is the scalar product of the vector $\mathbf{B}$ and vector of the values of variables $\mathbf{X}$. Here, any criteria $\Phi_i(\boldsymbol{\alpha})$ $(i=\overline{1,k})$ can be taken as the variable y, and the vector $\mathbf{X}$ represents some subset of the variables from $\boldsymbol{\alpha}$ or predefined functions of these variables. For example, if there is only one variable x, then by introducing the variables $x_i=x^i$, $i=\overline{1,q}$, we obtain a polynomial model. Of importance here is the fact that the model should be linear with respect to the unknown coefficients of the regression equation (5-13).

The parameters of $\mathbf{B}$ and b_0 are estimated by the least squares method (LSM) (Seber 1977; Aivazyan et al. 1986), i.e., from the condition of minimum of the sum of the squares of residuals or mismatches:

$$\delta^2=\sum_{i=1}^{N}\frac{(y_i-(\mathbf{B}'\mathbf{X}_i)-b_0)^2}{N}\to\min_{\mathbf{B},b_0},$$

where y_i is the criterion value at the ith point; and X_i is an appropriate set of design variables values or functions of them.

This result in a so-called normal system of equations (Seber 1977; Aivazyan et al. 1986), whose solution yields the desired estimates of the parameters:

$$\begin{cases}\mathbf{SB}=\hat{\mathbf{C}}_{y\mathbf{X}}\\ b_0\ =E(y)-(\mathbf{B}'\mathbf{E}(\mathbf{X}))\end{cases}\tag{5-14}$$

where the matrix $\mathbf{S}=\mathbf{C}'\mathbf{C}$ is called the matrix of normal system of equations; $\mathbf{C}$ being a design matrix (in case the vector $\mathbf{X}$ coincides with the vector of the design variables $\boldsymbol{\alpha}$, i.e., when we approximate the criterion $\Phi(\boldsymbol{\alpha})$ in the space of the initial variables without using any additional functions of the initial variables, we find $\mathbf{C}=\mathbf{A}$).

$\mathbf{C}_{y\mathbf{X}}=\boldsymbol{E}(y\mathbf{X})$ is the vector of covariance between the variable y and the variables from $\boldsymbol{X}$, and $\hat{\boldsymbol{C}}_{yX}$ is its estimate obtained by averaging over the elements of the design matrix of the experiment.

Quality of linear regression equation we will measure with the help of the *determination coefficient*

$$R^2=\frac{1-\delta^2}{\mathrm{var}_D(y)}.$$

Let us consider in more detail two regression models linear in parameters of the regression equation. The first model uses only the α_i as the variables, whereas the other involves the squares of α_i (i.e. α_i^2, $i=\overline{1,r}$) as additional variables.

Regression linear both in parameters of the regression equation and in $\boldsymbol{\alpha}$

The analytical expression of the model is given by (5-10). Its design matrix **C** is simply the matrix **A** consisting of the design variables $\boldsymbol{\alpha}$ found in an experiment. Since, by conditions of generation, the variables $\alpha_1,\ldots,\alpha_r$ are independent, in this case the matrix of the set of normal equations is also given by $\mathbf{S}=\mathbf{AA}'$. Normalization conditions imply that the diagonal elements of the matrix **S** are equal to unity ($E(\alpha_i^2)=1$), while off-diagonal elements are close to zero. (Since $\rho_{ij}=E(\alpha_i\alpha_j)=0$ for $i\neq j$, the difference from zero is only due to the error in evaluating the integral where ρ_{ij} is approximated by a sum over a finite set of points.)

Hence, it follows that the matrix **S** is well-conditioned (Aivazyan et al. 1986; Seber 1977; Belsley et al. 1980) (correlations between the explaining variables are small) and the solution of the system of normal equations computationally does not cause any difficulty.

Since the variables Φ and α_i ($i=\overline{1,r}$) are normalized, the regression coefficients b_i coincide with the correlation coefficients between Φ ($\Phi\equiv y$) and α_i.

Regression linear in parameters of the regression equation with the addition of squares of explaining variables $\boldsymbol{\alpha}$

The regression equation in this case is of the form:

$$y=b_0+b_1\alpha_1+,\ldots,+b_{r+1}\alpha_1^2+,\ldots,+b_{r+r}\alpha_r^2. \qquad (5\text{-}15)$$

Here, we have a $2r+1$-dimensional vector of coefficients (the regression equation parameters) b_0, $b_1,\ldots,b_{r+1},\ldots,b_{r+r}$.

The rows of the design matrix **C** contain $2r$ elements (in each row). The first r elements of the ith row of the matrix **C** coincide with the elements of the corresponding row in the design matrix **A.** The other r elements are the squares of the values of the design variables $\alpha_i^{(j)}$ corresponding to the point $\boldsymbol{\alpha}^{(j)}$. Thus the jth row (the row corresponding to the jth experiment) of the matrix **C** is

$$\alpha_1^{(j)},\ldots,\alpha_r^{(j)},\ (\alpha_1^{(j)})^2,\ldots,(\alpha_r^{(j)})^2.$$

The off-diagonal elements of the matrix of normal elements are again close to zero here, because

$E(\alpha_i\alpha_j^2)=0$ ($i\neq j$) (by virtue of independence) and
$E(\alpha_i^3)=0$ (by virtue of symmetry).

(Recall that the variables are normalized and centered.) Off-diagonal elements are, as before, different from zero only due to the replacement of integrals by the sums over finite sets of points.

Thus, in this case the matrix of a system of normal equations is also well-conditioned.

Estimation of significance measures on the basis of linear regression

In the case of regression linear both in parameters and in variables $\boldsymbol{\alpha}$, for the three significance measures we have

$$I_{1,\alpha_i}=I_{2,\alpha_i}=I_{3,\alpha_i}=b_i^2,$$

and for regression involving the second powers of the variables, we have

$$I_{1,\alpha_i}=I_{2,\alpha_i}=b_i^2+4b_{i+r}^2$$
$$I_{3,\alpha_i}^2=b_i^2+2b_ib_{i+r}+\frac{9}{5}b_{i+r}^2$$

These formulas can be easily derived by direct calculating expressions (5-4), (5-6), (5-8), after substitution of y from (5-13) or (5-15) for Φ.

Drawbacks of estimating the design variables significance by linear regression. Suppose we have obtained the coefficients b_i of the variable α_i to be close to zero, that is, $b_i \approx 0$. Does it imply that the design variable is of small significance? Let us examine in detail how to compute the coefficient b_i. Since in this case the matrix of the system of normal equations is almost a diagonal matrix with diagonal elements being equal to unity, we have ($y=\Phi(\boldsymbol{\alpha})$)

$$b_i \approx \rho_{\Phi(\boldsymbol{\alpha})\alpha_i}=\int_{\Pi}\alpha_i\Phi(\boldsymbol{\alpha})d\boldsymbol{\alpha}.$$

Integrating by parts, we obtain

$$b_i \approx \int_{\Pi}\alpha_i^2\left(\frac{\partial\Phi}{\partial a_i}\right)d\boldsymbol{\alpha},$$

that is, b_i is equal to the weighted mean (with weight $w(\boldsymbol{\alpha})=\alpha_i^2$) of the partial derivative of $\Phi(\boldsymbol{\alpha})$ with respect to α_i. Its value may be close to zero even if the value of the integral I_{1,α_i} is considerable.

In practice, we may use the following rule: If the determination coefficient is not very close to 1, smallness of the coefficient b_i does not imply that the significance of the design variable α_i is small. If, however, the coefficient b_i is considerably different from zero, the design variable should be regarded as significant.

Analogous reasoning can also be applied to more complicated parametric regression models. Therefore a need arises for direct estimating the measures, I_1 and I_2, of significance of the design variables. These estimates are obtained with the help of modified local parametric regression analysis techniques.

Projection pursuit regression

This approach is used for estimating parameters and functions in regression models of the kind (5-12). The technique using the projection pursuit regression for approximating regression functions was suggested by Friedman and Stuetzle (1981).

Suppose, we have a data matrix consisting of N $(p+1)$-vectors $(y,\mathbf{X})$ (p is the number of the components of the vector $\mathbf{X}$, i.e., the number of explaining variables) and our aim is to restore the function of regression of the variable y using the components of the vector $\mathbf{X}$. Assume now that the regression function can be represented as follows

$$y=\sum_{j=1}^{q} g_j(\mathbf{U}_j'\mathbf{X})+\epsilon, \tag{5-16}$$

where $g_j(\ )$ are unknown functions; $\mathbf{U}_j$ are unknown vectors, and q is the number of projections, which may also be unknown.

Let us describe the computation procedure. First, we seek a function $g_1(\ )$ and a vector $\mathbf{U}_1$ such that

$$\delta_1^2=\sum_{j=1}^{N}(y_j-g_1(\mathbf{U}_1'\mathbf{X}_j))^2\rightarrow\min.$$

Since the variable y is normalized, we find the quantity $1-\delta_1^2$ to coincide with the determination coefficient.

Nonparametric estimates of local regression (knn-smoothing)

One of the possible and sufficiently effective methods of nonparametric estimation of a one-dimensional regression function (and thereby, the conditional mean) lies in evaluating the local polynomial regression in the neighborhood of the point under consideration. We will study polynomials of orders 0, 1, and 2 that correspond to local mean, local linear regression, and local second-order polynomial regression, respectively.

For this purpose we apply order statistics. Let $z_1,\ldots,z_n$ be a sample of the explaining variable values, and $z_{(1)},\ldots,z_{(n)}$ be the corresponding order statistics (Aivazyan et al. 1983), that is, the values $z_1,\ldots,z_n$ arranged in the increasing order.

Let t be a positive integer such that $t<n/2$. Now we search for the predicted

value of the dependent variable $\hat{y}_{(j)}$ at the point $z_{(j)}$, using a polynomial regression of orders 0,1, and 2 constructed on the basis of observation results

$$z_{(j-t)_+},\ z_{(j-t)_+ +1},\ z_{(j)},\ldots,\ z_{(j+t)_- -1},\ z_{(j+t)_-}, \tag{5-17}$$

where $(j+t)_- = \min(n,j+t)$, $(j-t)_+ = \max(1,j-t)$.

The number of observations may vary from t to $2t$, depending on the position of the point $z_{(j)}$. Linear regression and polynomial regression of the second order are evaluated by the least squares method.

This procedure can be modified in different ways, for example, the very point $z_{(j)}$ can be excluded from "training."

Now let $y_{(k)}$ and $\hat{y}_{(k)}$ denote, respectively, the observed and predicted values of the dependent variable at the point $z_{(k)}$.

Let us now introduce a mean normalized square deviation:

$$SD=\delta^2=\frac{1}{N}\sum_{k=1}^{N}\frac{(y_{(k)}-\hat{y}_{(k)})^2}{\mathrm{var}_D(y)}. \tag{5-18}$$

The quantity $(1-\delta^2)$ can be regarded as a nonparametric estimate of the determination coefficient for the dependence between one-dimensional y and z.

If the points z_i are the projections of multidimensional points, that is $z_i=(\mathbf{U}'\mathbf{X}_i)$, then we can use the derivative of δ^2 with respect to $\mathbf{U}$ and thereby obtain an estimate of the gradient needed for implementing the effective optimization procedure.

Estimates of the significance measures I_1 and I_2

We will evaluate the criteria $I_{1,\alpha_i}, I_{2,\alpha_i}$ using the local-parametric regression (see, for example, Aivazyan et al. (1983) and Aivazyan et al. (1986)) and modified ACE-regression (see Breiman and Friedman (1985)), respectively.

Local-parametric approach to estimating the criterion I_1

The basic idea here lies in approximating the functional dependence $\Phi(\boldsymbol{\alpha})$ not over the entire experiment domain Π, but in the neighborhoods of a certain subset $\mathbb{S}$ of randomly chosen elements of the design matrix $\mathbf{A}$. In the limiting case the subset $\mathbb{S}$ may contain all the points (rows) of the matrix $\mathbf{A}$; but usually this subset is formed by choosing $n\leq N$ points from $\mathbf{A}$.

Without loss of generality, we assume that the variables $\boldsymbol{\alpha}$ are preliminarily normalized so that the mean of each variable is zero and the variance is equal to unity. As the neighborhoods, we will take spheres.

If the function $\Phi(\boldsymbol{\alpha})$ is twice differentiable with respect to $\boldsymbol{\alpha}$, we can write its Taylor expansion in the neighborhood of a sphere $\Omega_\rho(\boldsymbol{\alpha}_0)$ of radius ρ with the center at $\boldsymbol{\alpha}_0 \in \mathbb{S}$, as

$$\Phi(\boldsymbol{\alpha})=\Phi(\boldsymbol{\alpha}_0)+\left(\mathbf{B}'(\boldsymbol{\alpha}-\boldsymbol{\alpha}_0)\right)+\left((\boldsymbol{\alpha}-\boldsymbol{\alpha}_0)'\mathbf{S}(\boldsymbol{\alpha}-\boldsymbol{\alpha}_0)\right)+\Theta(\rho^3). \tag{5-19}$$

Here, $\mathbf{B}=\text{grad}(\Phi(\boldsymbol{\alpha}))|_{\boldsymbol{\alpha}=\boldsymbol{\alpha}_0}$ stands for the vector of the first derivatives of the function Φ and $\mathbf{S}$ represents the matrix of the second derivatives.

Now applying the least squares method to the elements of the design matrix that fall inside the sphere $\Omega_\rho(\boldsymbol{\alpha}_0)$ to evaluate the linear regression (see Linear multiparametric regression (LMR)), we obtain the estimates of the first derivatives as the coefficients in the linear regression equation (5-13).

In practice, we find for the point $\boldsymbol{\alpha}_0$ the sphere of minimal radius that contains exactly K neighboring points, rather than specify the radius ρ of the sphere beforehand. Thus, the coefficients of the linear regression equation are evaluated from the data matrix corresponding to K points. Evidently, the inequality $K<r$ must hold in this case.

Estimate of the significance measure I_2: Specific features of the alternate condition expectation (ACE) algorithm

Let us fix one of the design variables, say, α_i. The significance measure I_2 can be evaluated with the help of a modified ACE-regression algorithm (Breiman and Friedman 1985).

Estimating the averaged criterion $\overline{\Phi}_i(\alpha_i)$ (see (5-7))

In order to apply the ACE-regression for approximating the unknown function $\overline{\Phi}_i(\alpha_i)$ and its derivative, first we have to derive the values of this function at least with some error, i.e., to average the criterion $\Phi(\alpha_i)$, at $\alpha_i=z$, over the design variables contained in the set $\boldsymbol{\beta}_i$.

Several approaches can be utilized for this purpose. We will apply the technique based on the use of zero-order knn-smoothing (for details, see Nonparametric estimates of local regression (knn-smoothing)).

In our case this approach is applied as follows. First, the values of $\Phi(\boldsymbol{\alpha})$, where $\boldsymbol{\alpha} \in \mathbf{A}$ are arranged in the increasing order of the values of α_i (α_i is the ith coordinate). Then we fix a certain number K_0 of "neighbors". Let $\alpha_{i(1)}<\ldots<\alpha_{i(N)}$ be the variational sequence (order statistics) derived for the values of α_i. For the estimate of the function $\overline{\Phi}_i(\alpha_i)$ at the point $\alpha_i=\alpha_{i(j)}$, let us take its mean over the smoothing interval $L_{(j)}$ (K_0), i.e., over the points with the numbers $(j-K_0)_+,\ldots,(j+K_0)_-$

$$\overline{\Phi}_{i(j)}=\overline{\Phi}_i(\alpha_{i(j)})=\frac{1}{n_{(j)}}\sum_l \Phi(\alpha_{i(l)}, \boldsymbol{\beta}_{i(l)}),$$

where $n_{(j)}$ is the number of points within the interval $L_{(j)}$ (K_0) and summation is taken over the points of this interval. The quantity $\overline{\Phi}_{i(j)}$ is exactly the estimate of the value of the function $\overline{\Phi}_i(\alpha_i)\,|_{\alpha_i=\alpha_{i(j)}}$. This estimate can be expressed as

$$\overline{\Phi}_{i(j)}=\overline{\Phi}_i\,(\alpha_{i(j)})+d_{(j)}+\tau_{(j)}, \tag{5-20}$$

where $d_{(j)}$ is the regular error due to the averaging of the variable α_j over an interval $L_{(j)}$ (K_0) of a finite length h_j; $\tau_{(j)}$ is the random error due to the estimate of the integral (5-5) made using a finite number $n_{(j)}$ of randomly distributed points in the parallelepiped $\Pi_{\boldsymbol{\beta}_i}$. Since the experimental points are uniformly distributed, $h_j \approx n_{(j)}/N$; $d_i \sim 0(h_j^3)$. The mean square of $\tau_{(j)}$ is related to the variability of the criterion $\Phi(\boldsymbol{\alpha})\Big|_{\alpha_i=\alpha_{i(j)}}$ in the parallelepiped $\Pi_{\boldsymbol{\beta}_i}$: the greater the scatter of the values, the greater the mean square of $\tau_{(j)}$. Here, as everywhere in nonparametric estimation, we find ourselves in a situation where the regular and random errors exist in a balance: On reducing the regular error by decreasing the interval length $h_{(j)}$ (in our case, the number of neighbors K_0), we increase the random error $\tau_{(j)}$ and vice versa. Both the errors can be decreased simultaneously only by increasing N—the number of experiments.

Removal of the criterion component linear in design variables

The random component $\tau_{(j)}$ can however be decreased by modifying the criterion $\Phi(\boldsymbol{\alpha})$ so that the modified criterion would have less scatter in the parallelepiped $\Pi_{\boldsymbol{\beta}_i}$. But here, care should be taken so that calculation of the modified criterion would not become essentially more complicated as compared with the calculation of the initial criterion.

If the criterion $\Phi(\boldsymbol{\alpha})$, linear in design variables, $\boldsymbol{\alpha}$ is approximated by the least squares method, the determination coefficient quite often lies in the range ≈ 0.5–≈ 0.8. Though such an approximation cannot serve as a basis for choosing the insignificant design variables relying on the values of the regression coefficients, it can nevertheless be successfully applied to modify the criterion $\Phi(\boldsymbol{\alpha})$. It is just in evaluating the significance of the design variable α_i that we use the modified criterion

$$\tilde{\Phi}(\boldsymbol{\alpha})=\Phi(\boldsymbol{\alpha})-(\mathbf{G}_i\boldsymbol{\beta}_i)-g_{0i}, \tag{5-21}$$

where $\mathbf{G}_i$ is an $(r-1)$-dimensional vector of the coefficients of the linear regression of the criterion $\Phi(\boldsymbol{\alpha})$ using the design-variable vector $\boldsymbol{\beta}_i$ reduced by eliminating the design variable α_i and g_{0i} is the free term in the regression equation. Thereafter the value of the criterion $\tilde{\Phi}(\boldsymbol{\alpha})$ is squared and we obtain the modified criterion $\Phi_{\text{mod}}(\boldsymbol{\alpha})=\tilde{\Phi}^2(\boldsymbol{\alpha})$, which we apply in the sequel. In order to modify ("clean") the criterion it is more effective to use linear regression containing second-order terms or projection pursuit regression with a small number of term functions ($q=1,2$).

From the foregoing it is clear that when we change over to a different design variable (estimate of its significance), the vector $\mathbf{G}_i$ is to be recalculated.

Estimating the derivative

After determining the values of the function $\overline{\Phi}_i(\alpha_i)$ (or the function corresponding to the modified criterion) the derivative $\frac{d\overline{\Phi}_i}{d\alpha_i}$ is evaluated as follows: first a new number K_1 of nearest neighbors is fixed and then the first- or second-order knn-smoothing is applied to the function $\overline{\Phi}_i(\alpha_i)$ (in other words, the estimates of local linear or quadratic regression are constructed).

Suppose in the neighborhood of the point $\alpha_{i(j)}$ we obtain the following regression equation:

$$\overline{\Phi}_i(\alpha_i)=a\alpha_i+b \qquad \text{(first-degree equation)},$$
$$\overline{\Phi}_i(\alpha_i)=c\alpha_i^2+d\alpha_i+e \quad \text{(second-degree equation)}.$$

Accordingly, the local estimates of the derivatives are

$$\overline{\Phi}'_{i(j)}=\frac{d\overline{\Phi}_i}{d\alpha_i}=\begin{cases} a \text{ for linear regression,} \\ 2c\alpha_i+d \text{ for polynomial regression} \end{cases}$$

The estimate for the measure of informativeness of α_i is

$$I_{2,\alpha_i}=\frac{1}{N}\sum_{j=1}^{N}(\overline{\Phi}'_{i(j)})^2.$$

Algorithm

Finally, the algorithm for determining the significance of design variables for the criterion I_{2,α_i} can be written as follows:

Step 1. The design variables are normalized so that they all have unit variance and zero mean.

Step 2. Cycle with respect to $i=1,\ldots,r$.

1. Estimation of some approximation $\overline{\overline{\Phi}}(\boldsymbol{\beta}_i)$ of the criterion for the reduced design-variable vector $\boldsymbol{\beta}_i$, say, linear regression coefficients, $\overline{\overline{\Phi}}(\boldsymbol{\beta}_i)=(\mathbf{G}_i\boldsymbol{\beta}_i)+g_{0i}$.
2. Change over to the modified criterion.

$$\Phi_{\text{mod}}(\boldsymbol{\alpha})=(\Phi(\boldsymbol{\alpha})-\Phi(\boldsymbol{\beta}_i))^2.$$

3. Ordering the values of the design variable α_i. Zero-order smoothing of function $\Phi_{\text{mod}}(\boldsymbol{\alpha})$.
4. Estimating the derivatives at the points of the variational series $\alpha_{i(j)}$ and evaluating the significance measure for the design variable α_i.

End of the cycle with respect to i.

Step 3. Arranging the design variables of $\boldsymbol{\alpha}$ in the order of their significance increase.

Example

Determination of significant design variables in operational development of a truck (see Section 4-5).

Estimation of the closeness criteria, Φ_ν, sensitivity to the change of the mathematical model parameters, α_j, (the multicriteria identification problem)

As the variables, we consider here the equivalent stiffness coefficients and damping factors (see Table 4-3 and Fig. 4-5). The criteria of the aforementioned three groups, Φ_i^1, Φ_i^2, and Φ_i^3, are regarded as proximity indexes.

To investigate the sensitivity of the criteria with respect to the variables we use the significance measure I_2. A sample of 512 trials within parallelepiped Π^1 has been considered.

Table 5-1 gives the values of the significance measure I_2 for each reduced variable and each criterion. The closer the values in the last column to unity, the higher is the accuracy of approximating the respective criteria by using formula (5-11).

The dimension of the thereby obtained matrix of values of the measure I_2 values is 65×17 (65 proximity criteria and 16 variables have been considered, the last, 17th, column characterizes the accuracy of approximating each of the criteria by using formula (5-11). This column gives the values of determination coefficients). Because of the large size of the matrix, Table 5-1 gives only a fragment for three proximity criteria Φ_2, Φ_{23}, and Φ_{34}, and 11 variables.

Table 5–1

Φ_ν \ α_i	α_1	α_2	α_5	α_6	α_9	α_{10}	α_{14}	α_{15}	α_{16}	α_{19}	α_{20}	Determination coefficients
Φ_2	0.674	0.017	0.0	0.0	0.053	0.003	0.007	0.0	0.0	0.001	0.0	0.9
Φ_{23}	0.003	0.0	0.0	0.002	0.006	0.013	0.605	0.0	0.0	0.012	0.03	0.97
Φ_{34}	0.0	0.0	0.0	0.002	0.01	0.700	0.005	0.0	0.0	0.012	0.03	0.97

Analyzing the matrix, we select the variables whose influence on all criteria is sufficiently small. We consider the influence of a variable on a criterion as small if the corresponding value of I_2 is less than 1% of the sum of elements in the row corresponding to the criterion. In Table 5-1, such variables turn out to be α_5, α_6, α_{15}, and α_{16}. In terms of the truck characteristics, α_5 and α_{15} are the values of the equivalent stiffness coefficient and damping factor, respectively, for the front support of the platform; α_6 and α_{16} are analogous variables for the rear support of the platform.

Table 5-1 shows that the influence of these variables does not exceed 1% (for criteria that are not presented in Table 5-1, the results are similar).

We can try to reduce the number of variables to be varied when optimizing the system by eliminating insignificant variables. In this case, the insignificant variables are kept constant when calculating the characteristics of the mathematical model. For example, they can be chosen to be equal to the values of the respective parameters of the prototype. Reducing the variable vector dimension, we thereby reduce the time required for carrying out the computational experiment.

To confirm the applicability of such an approach, we randomly selected 100 models (solutions) within 16-dimensional parallelepiped Π^1 and calculated the values of all closeness criteria for each of the solutions. Then the closeness criteria were calculated for the 100 models with the values of the insignificant variables, α_5, α_6, α_{15}, and α_{16}, being fixed and equal to those of the prototype. Finally, the corresponding criteria values calculated for 16- and 12-dimensional variables vectors were compared with each other.

The analysis shows that for the majority of criteria, the discrepancies are zero or rather insignificant. It allows making a conclusion that the values of the closeness criteria changed insignificantly after having reduced the dimension of the parallelepiped. This confirms the expedience of reducing the dimension of the variables vector.

Determination of the performance criteria sensitivity to design variables in the problem of improving the automobile suspension system

When solving this problem we considered 20 performance criteria reflecting the requirements of comfort, durability, load preservation, and safety. Twenty design variables were varied.

We have taken a sample of 512 trials and determined the influence of the design variables on each of the criteria. Like in the identification problem, insignificant design variables have been determined. These turn out to be α_5, α_6, α_{15}, and α_{16}. Thereafter, these design variables were no longer varied. The optimization results in this case coincide with those obtained in Chapter 4 (see Table 4-5). In some cases, the reduction of time necessary for the optimization/identification of the variables can be achieved by using regression methods. The presence of insignificant variables can be of practical use. Without breaking the optimality, a designer can adjust these variables so as to meet some additional requirements, for instance, technological conditions of the manufacture process.

5-2. The Energy Balance Principle for Determining the Dependence of Criteria on Design Variables

The proposed technique is based on the parameter space investigation method and the energy balance principle (Masataka 1977; Gurychev et al. 1985). The technique can be used for optimization of a wide class of mechanical structures in which:

- The potential and kinetic energies are distributed between subsystems unevenly.
- The dissipation is small, and natural modes of oscillations differ insignificantly from natural modes of oscillations in the conservative system.
- performance criteria reflect the static and dynamic compliance of the system, resonant frequencies, vibration resistance, metal consumption, etc.

It is known that the kinetic (T) and potential (Π) energies of the system can be represented as follows

$$T=\frac{1}{2}\sum_{i=1}^{n} m_i \dot{x}_i^2, \quad \Pi=\frac{1}{2}\sum_{i=1}^{n} c_i x_i^2,$$

where x_i are physical coordinates of the system, while m_i and c_i are its inertia and stiffness coefficients, $1 \le i \le n$.

Following the energy balance principle, for each of the n natural frequencies, we determine the contribution of individual elements of the structure to the kinetic (T) and potential (Π) energies. The elements are characterized by different design variables (geometrical dimensions, stiffnesses in different directions, masses, damping factors, etc.).

To determine T_m and Π_m corresponding to the mth natural frequency ω_m ($1 \le m \le n$) we use the following relationships (Masataka 1977; Kaminskaya and Gringlaz 1989)

$$\begin{aligned} T_m &= \frac{\omega_m^2}{2} q_m \sum_{i=1}^{n}\sum_{j=1}^{n} M_{ij}\gamma_{im}\gamma_{jm}, \\ \Pi_m &= \frac{q_m}{2}\sum_{i=1}^{n}\sum_{j=1}^{n} K_{ij}\gamma_{im}\gamma_{jm} \end{aligned} \qquad (5\text{-}22)$$

Here, M_{ij} and K_{ij} are the elements of inertia and stiffness matrices; γ_{im} and γ_{jm} are coefficients characterizing the shape of the system oscillations corresponding to the natural frequency ω_m; i and j are the numbers of the generalized coordinates; n is the number of degrees of freedom; and q_m is a normalizing factor.

In many cases, for each of the natural frequencies, one can indicate some elements whose potential and kinetic energies many times exceed the energies of the other elements. To determine such significant elements it is convenient to normalize the coefficients γ_{im} and γ_{jm} so that both T_m and Π_m corresponding to the natural frequency ω_m are equal to unity. Then we select the significant elements whose contribution to the kinetic or potential energy is large. When doing this, it is necessary that the sum of energies of the other elements not exceed a prescribed level, such as 10% or 20% of the total energy of the system. Based on this, we determine insignificant elements. By analogy, we will consider the design variables describing the aforementioned elements as significant or insignificant.

However, we still have to check whether the insignificant design variables essentially influence the performance criteria. Depending on this, we can conclude about the expedience of varying the insignificant design variables when solving the optimization problem.

Let us describe the main stages of problem solving.

1. The designer determines an r-dimensional vector of the design variables and specifies the ranges of their variations: $\alpha_i^* \leq \alpha_i \leq \alpha_i^{**}$, $i = \overline{1,r}$.
2. According to the parameter space investigation method, n tests are carried out, with n being comparatively small. In each ith test, we determine significant design variables l_i.
3. Having completed all n tests, we find the number of significant design variables, l:

$$l = l'_1 + l'_2 + \ldots + l'_n,$$

   where l'_i is the number of the significant design variables in the ith test that were not significant in all previous tests.
4. A new, l-dimensional, vector of the design variables is formed, $l \leq r$.
5. The errors are determined:

$$\Delta\Phi_{\nu,i}^{r,l} = \left| \frac{\Phi_{\nu,i}^r - \Phi_{\nu,i}^l}{\Phi_{\nu,i}^r} \right| 100\%, \quad i = \overline{1,n} \tag{5-23}$$

   where $\Phi_{\nu,i}^r$ and $\Phi_{\nu,i}^l$ are the values of the νth criterion in the ith test; r and l being the dimensions of the respective vectors of design variables.
6. The conditions $\Delta\Phi_{\nu,i}^{r,l} \leq \Delta\Phi_\nu^{**}$ are checked, where $\Delta\Phi_\nu^{**}$ are the admissible errors. In case these conditions are satisfied, we solve the multicriteria

optimization problem, with the dimension of the design-variable vector being l, and the number of tests being N, $N>n$.

7. For all obtained feasible solutions, the conditions $\Delta\Phi_{\nu,i}^{r,l} \leq \Delta\Phi_{\nu}^{**}$ are verified. If these conditions are satisfied, we consider the feasible solutions as having been obtained with the prescribed accuracy.

In the next section, we will consider the determination of significant design variables as applied to the example of designing the grinding machine structure.

5-3. Example: Determination of Sensitivity of the Cylindrical Grinding Machine Structure Criteria

The Basic Elements of Calculation Schemes and Formation of the Equations of Motion

Numerous experimental data indicate that the most intense vibrations of tools and workpieces, caused by vibration of the machine-tool structure, occur, as a rule, within the frequency range up to 150 Hz. Therefore, in calculation, the majority of the structure elements may be modeled by beams and rigid bodies. The former simulate slides, columns, beds, and traverses.

Characteristically, the proper deformations of rigid bodies are small compared to contact strains in joints (e.g., spindle and wheel heads). Rigid bodies and beams are connected by weightless elastodissipative elements whose characteristics are determined by the parameters of joints, guideways, etc. A structure interacts with the foundation via the support elements of a machine tool.

In dynamic calculations structures are usually considered linear oscillatory systems described by the equation of motion

$$\mathbf{M}\ddot{\mathbf{X}}+\mathbf{D}\dot{\mathbf{X}}+\mathbf{K}\mathbf{X}=\mathbf{F}(t)$$

$\mathbf{F}(t)$ is an n-dimensional column vector of external forces; $\mathbf{M}$ is the matrix of inertia; $\mathbf{K}$ is the stiffness matrix; and $\mathbf{D}$ is the damping matrix.

In considering structures the model of viscous friction is used.

Construction of the Mathematical Model and Determination of the Static-Dynamic Characteristics of a Structure

Cylindrical grinding machines are intended for machining cylindrical, tapered, and end-face surfaces of rotational parts.

A workpiece is held between the centers of the workhead and tailstock, which are mounted on the rotatory table. The main motion is the grinding wheel rotation. A workpiece is rotated in the centers, and the cross and longitudinal feeds are implemented by displacing the wheelhead and the work table, respectively.

Experiments have shown that the cylindrical grinding machine under consideration loses stability at a frequency of 74 Hz.

The analysis of experimental modes of vibration has allowed construction of the calculation scheme of the machine tool's structure (see Fig. 5-1). The mathematical model shown in Figure 5-2 incorporates concentrated masses and elastodissipative links. The number of degrees of freedom is 48.

In Figures 5-1 and 5-2 K_{1-2} is the reduced stiffness of the grinding wheel-wheelhead joint; $\mathbf{K}_{2-3}$ is the vector of stiffnesses of the spindle head guideways and the feed drive, $\mathbf{K}_{2-3} = (K^x_{2-3}, K^y_{2-3}, K^z_{2-3}, K^{\varphi_x}_{2-3}, K^{\varphi_y}_{2-3}, K^{\varphi_z}_{2-3})$; $\mathbf{K}_{3-4}$ is the vector of stiffnesses of the joint between the front and rear portions of the machine tool bed; $\mathbf{K}_{4-5}$ is the vector of stiffnesses of the table guideways and drive; $\mathbf{K}_{5-6}$ and $\mathbf{K}_{5-7}$ are the vectors of stiffnesses of the joints connecting the workhead and the tailstock with the table; $\mathbf{K}_{6-8}$ and $\mathbf{K}_{7-8}$ are the vectors of stiffnesses of the workhead and tailstock centers; K_3 is the axial stiffness of the supports under the wheelhead bed; K_4 is the axial stiffness of the support under the table bed; and M_i, $i=1,\ldots,8$ are the masses of the basic units of the cylindrical grinding machine under consideration.

The frequency and amplitude/phase responses within the cutting zone were calculated using the finite element model of the structure and compared with the corresponding experimental characteristics. It was found that within the range of the most vibroactive frequencies, 50–70 Hz, the experimental and calculated characteristics compare favorably. Hence, the dynamic model ensures an adequate description of the machine tool prototype under study.

The Problem of the Machine Tool Structure Optimization: Performance Criteria

The performance criteria characterize the consumption of metal, the static and dynamic compliances of the structure within the cutting zone, and vibration

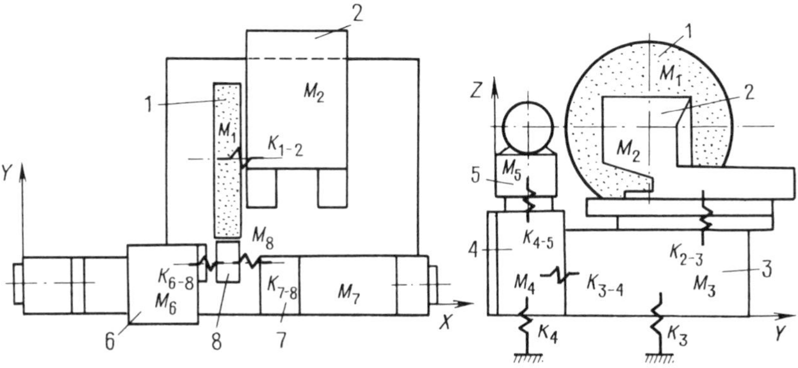

Figure 5–1 Schematic of a cylindrical grinding machine. 1 Grinding wheel, 2 wheelhead, 3 rear portion of the bed (under the wheelhead), 4 front portion of the bed (under the table), 5 table (together with the rotary portion), 6 headstock, 7 tailstock, and 8 workpiece.

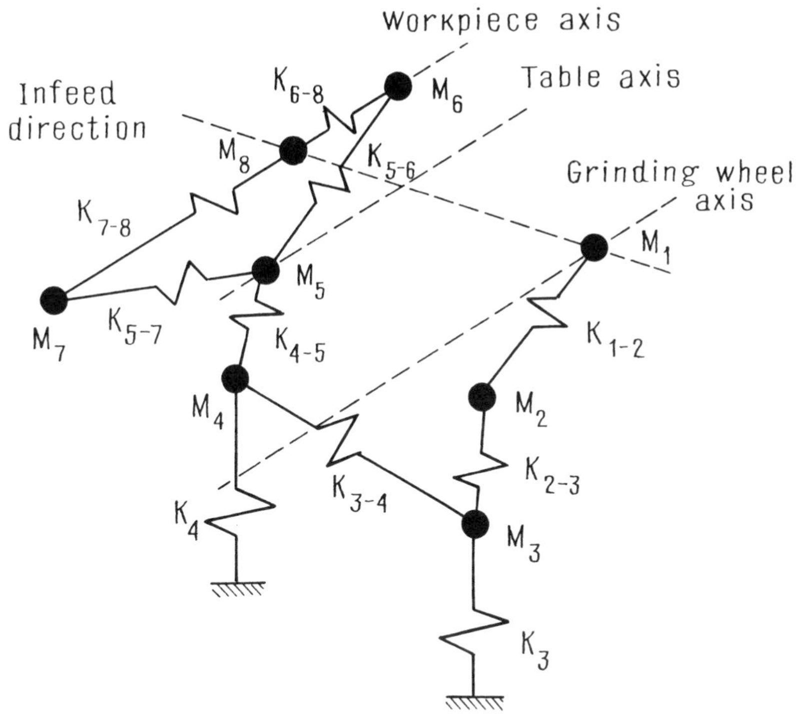

Figure 5–2 Dynamic model of the machine tool structure.

stability of the structure (using the Nyquist criterion). The expressions for the criteria under consideration (with the exception of the metal consumption) are based on the dependence of the structure dynamic compliance $w(i\omega)$ within the cutting zone.

$$w(i\omega)=\frac{\Delta y_{1-8}}{P}$$

where P is the variable component of the cutting force; Δy_{1-8} is the relative displacement of workpiece 8 and grinding wheel 1 along the Y-axis (see Fig. 5-1). Let us compile the performance criteria vector $\mathbf{\Phi}=(\Phi_1, \Phi_2, \Phi_3, \Phi_4)$.

1. The metal consumption per machine tool

$$\Phi_1=\sum_{i=1}^{8} M_i \rightarrow \min.$$

2. Static compliance

$$\Phi_2=w_{\omega=0} \rightarrow \min.$$

3. Vibration stability of the structure (the Nyquist criterion),

$$\Phi_3 = \max(-\text{Re}\, w) \rightarrow \min.$$

4. Vibration activity within the cutting zone.

$$\Phi_4 = \max|w| \rightarrow \min$$

where $|w| = \sqrt{(\text{Re}\, w)^2 + (\text{Im}\, w)^2}$.

Criteria Φ_2 and Φ_3 are related to the machine tool productivity, and Φ_4 to the machining accuracy.

Selection of Significant Design Variables

The following design variables were varied: masses, M_3 and M_4, of the two portions of the machine tool bed (these design variables are denoted by α_1 and α_2); the axial stiffness of the supports under the wheelhead bed, K_3 (or α_3); the axial stiffness of the supports under the table bed, K_4 (or α_4); the stiffness of the lead screw of the wheelhead feed, $K^y_{2\text{-}3}$ (or α_5); and the interference, (α_6). The latter two design variables determine the stiffness characteristics of the 2-3 joint. A specified value of interference was used for determining the wheelhead guideways stiffnesses $\mathbf{K}_{2\text{-}3}$. Six geometrical design variables of the joint 3-4 between the two portions of the bed were also varied. These design variables were used for calculating linear stiffnesses, $K^x_{3\text{-}4}$, $K^y_{3\text{-}4}$, and $K^z_{3\text{-}4}$, and angular stiffnesses $K^{\varphi_x}_{3\text{-}4}$, $K^{\varphi_y}_{3\text{-}4}$, and $K^{\varphi_z}_{3\text{-}4}$. Thus, at the start of the study the design-variable vector was $\boldsymbol{\alpha} = (\alpha_1, \ldots, \alpha_{12})$.

Using the calculated and experimental values of static and dynamic characteristics of the structure, it was found that the most intensive dynamic processes occur for the fifth, sixth, and seventh modes of vibration within the frequency range 50–70 Hz.

In line with the technique described in the preceding section, $n=32$ trials were conducted in the 12-dimensional parallelepiped. The energy balance was analyzed for the fifth, sixth, and seventh modes of vibrations. Table 5-2 presents the results of the analysis for model $\boldsymbol{\alpha}^1$. As we see, the major contributions into the kinetic energy are due to the masses of the wheelhead M_2, the rear portion of the bed (under the wheelhead) M_3, and the front portion of the bed (under the work table) M_4.

The potential energy of vibrations is mainly determined by stiffnesses of joint 2-3, as well as K_3 and K_4. Similar results were obtained for all the 32 trials.

The results have shown that the set of significant (substantial) design variables incorporates the first six design variables, α_1–α_6, of the total number of 12 design variables. In all trials the contribution of the linear and angular stiffnesses of joint 3-4 proved to be less than 5%. (In line with technological conditions, the wheelhead mass could not be varied.) These results imply that it is advisable

Table 5–2

Kind of Energy	Design Variables: Name	Design Variables: Designation	Distribution of Vibration energy (%), Natural frequencies (Hz): 59.8	64	68.2
Kinetic	Wheelhead mass	M_2	31	30	22
	Mass of the bed under the wheelhead	M_3	38	26	22
	Mass of the bed under the table	M_4	22	28	36
Potential	Stiffness of the lead screw for wheelhead feed	$K^y_{2\text{-}3}$	25	26	16
	Stiffness of the wheelhead guideways	$K^x_{2\text{-}3}, K^z_{2\text{-}3}, K^{\varphi_x}_{2\text{-}3}, K^{\varphi_y}_{2\text{-}3}, K^{\varphi_z}_{2\text{-}3}$	8	11	6
	Axial stiffness of the supports under the wheelhead bed	K_3	44	23	40
	Axial stiffness of the supports under the table bed	K_4	14	25	32

to optimize the structure by the first six design variables only. All other design variables were assumed to be constant and equal to those of the prototype.

This conclusion concerning the advisability of varying the first six design variables when optimizing the structure was confirmed by comparing the performance criteria calculated for 32 trials in the six- and 12-dimensional parallelepipeds.

Table 5-3 compares the values of $\Phi^{12}_{v,i}$ and $\Phi^{6}_{v,i}$ for the five trials, $\boldsymbol{\alpha}^2$, $\boldsymbol{\alpha}^7$, $\boldsymbol{\alpha}^{18}$, $\boldsymbol{\alpha}^{27}$, and $\boldsymbol{\alpha}^{32}$. We see that the errors $\Delta\Phi^{12,6}_{v,i}$ are insignificant.

The latter were calculated using the formula

$$\Delta\Phi^{12,6}_{v,i}=\frac{|\Phi^{12}_{v,i}-\Phi^{6}_{v,i}|}{\Phi^{12}_{v,i}}\cdot 100\%,\ \ v=1,\dots,4,\ \ i=1,\dots,32.$$

The boundaries of design variables are presented in Table 5-4. One had to find the feasible solutions set D and the Pareto optimal set P of the solutions, and choose the most preferred solution on this set. In all, $N=512$ trials were conducted in the six-dimensional parallelepiped. The criteria constraints corresponded to the prototype performance criteria.

Eighteen models, all of them Pareto optimal, have entered the feasible solutions set. Table 5-5 presents five of the models.

Table 5–3

Design variables	Performance criteria	Numbers of calculation tests (selected)				
		2	7	18	27	32
α_1–α_6	Φ_1	5,971	6,346	5,751	6.001	5.876
	Φ_2	0.000220	0.000228	0.000240	0.000196	0.000220
	Φ_3	0.000450	0.000530	0.000540	0.000418	0.000345
	Φ_4	0.001060	0.000970	0.001210	0.000731	0.000656
α_1–α_{12}	Φ_1	5,971	6,346	5,751	6,001	5,876
	Φ_2	0.000215	0.000230	0.000240	0.000196	0.000220
	Φ_3	0.000459	0.000545	0.000580	0.000416	0.000345
	Φ_4	0.001072	0.000982	0.001230	0.000730	0.000650
Error values $\Delta\Phi_{\nu,i}^{12,6}$	$\Delta\Phi_{1,i}^{12,6}$	0	0	0	0	0
	$\Delta\Phi_{2,i}^{12,6}$	2.2	0.8	0	0	0
	$\Delta\Phi_{3,i}^{12,6}$	1.9	2.7	6.8	0.5	0
	$\Delta\Phi_{4,i}^{12,6}$	1.1	1.2	1.6	0.1	1.0

Table 5–4

Design variables	Designation α_i	Lower boundary	Upper boundary
Masses of two bed portions	M_3(t)	1.5	2.0
	M_4(t)	1.6	2.1
Supports stiffness	K_3(kgf/mm)	20,000	40,000
	K_4(kgf/mm)	30,000	60,000
Stiffness characteristics of the joint 2-3	$K_{2\text{-}3}^{y}$ (kgf/mm)	15,000	40,000
	interference (mm)	0.003	0.005

We see that all the errors proved to be below the 10%-level specified by the designer, $\Delta\Phi_{\nu,i}^{12,6} \leq 10\%$. Hence, the feasible solutions found in the six-dimensional space of design variables have been obtained with a specified accuracy.

Solution $\boldsymbol{\alpha}^{17}$ has been preferred to all the rest. Tables 5-6 and 5-7 present the values of the criteria and design variables for both the optimal solution and the prototype.

For the optimal solution, the performance criteria of the machine tool structure exceeded the corresponding prototype criteria considerably: by 422 kg in the consumption of metal, by 18.46% in the static compliance, by a factor of 2.96 in vibration stability, and by a factor of 1.72 in the dynamic compliance of the structure within the cutting zone.

The analysis of the dynamic characteristics of the structure with the design variables corresponding to the optimal solution $\boldsymbol{\alpha}^{17}$, has shown that the most vibroactive is the frequency range 75–85 Hz.

Table 5–5

Design variables	Performance criteria	Numbers of feasible models (selected)				
		17	20	66	260	422
α_1–α_6	Φ_1	5,752	5,627	5,807	5,598	6,082
	Φ_2	0.000190	0.000207	0.000199	0.000206	0.000193
	Φ_3	0.000150	0.000150	0.000216	0.000240	0.000228
	Φ_4	0.000695	0.000877	0.000691	0.000766	0.000733
α_1–α_{12}	Φ_1	5,752	5,627	5,807	5,598	6,082
	Φ_2	0.000202	0.000210	0.000198	0.000198	0.000184
	Φ_3	0.000170	0.000163	0.000230	0.000259	0.000239
	Φ_4	0.000642	0.000910	0.000668	0.000801	0.000749
Error values $\Delta\Phi_{v,i}^{12,6}$	$\Delta\Phi_{1,i}^{12,6}$	0	0	0	0	0
	$\Delta\Phi_{2,i}^{12,6}$	5.4	1.4	0.5	4.0	4.8
	$\Delta\Phi_{3,i}^{12,6}$	8.0	7.9	6.0	7.3	4.2
	$\Delta\Phi_{4,i}^{12,6}$	8.2	3.6	3.4	4.3	2.1

Table 5–6

Models	Criteria			
	Φ_1(t)	Φ_2(mm/kgf)	Φ_3(mm/kgf)	Φ_4(mm/kgf)
Model 17	5.752	0.000190	0.000153	0.000695
Intial solution (a prototype)	6.174	0.000233	0.000450	0.001200

Table 5–7

Models	Design variables					
	α_1(t)	α_2(t)	α_3(kgf/mm)	α_4(kgf/mm)	α_5(kgf/mm)	α_6(mm)
Model 17	1.766	1.616	384400	439400	39220	0.0045
Intial solution (a prototype)	1.800	2.000	20000	30000	16000	0.0035

Conclusions

The most vibroactive elements of the dynamic model of the cylindrical grinding machine under consideration have been revealed. The significant design variables were found, thus allowing reduction in the design-variable space dimensionality for solving the optimization problem. In turn, this permitted both finding the optimal design variables, which ensure the best values of the structure performance criteria as compared with the prototype, and reduction in the time needed to solve the optimization problem.

5-4. Weakly Coupled Oscillatory Systems

In this section, we present an approach that can be used both for determining the sensitivity of the criteria with respect to parameters of systems and for decoupling the systems (Banach 1988). When using this approach, we seek the parameters responsible for weak interaction between subsystems. Such parameters can be omitted, and then we obtain the system whose order is less than the order of the original system. Note that, provided certain conditions are fulfilled, the difference between the solution of the reduced-order system and the original system does not exceed a prescribed quantity ϵ, and in many cases we can guarantee sufficient proximity between the criteria values of the original and modified systems.

The Method of Finding Weak Couplings

The equations of a complex system that consists of a number of subsystems can be written in the form $\mathbf{D}\boldsymbol{x}=0$, where $\mathbf{D}$ is a matrix of symmetric block-type structure whose blocks have the form $\mathbf{K}_{ij}-\lambda\mathbf{M}_{ij}$:

$$\mathbf{D}=\mathbf{K}-\lambda\mathbf{M}=[\mathbf{K}_{ij}-\lambda\mathbf{M}_{ij}],\ \ i,j=\overline{1,\mathrm{m}}. \tag{5-24}$$

Here, $\mathbf{K}$ is the system stiffness matrix whose elements are k_{ij}; $\mathbf{M}$ is the inertia matrix; diagonal blocks $\mathbf{K}_{ii}$ and $\mathbf{M}_{ii}$ are the stiffness and inertia matrices of the ith subsystem; off-diagonal blocks $\mathbf{K}_{ij}$ and $\mathbf{M}_{ij}$ describe the stiffness and inertia coupling between the ith and jth subsystems; m is the number of the subsystems.

Let us call a subsystem, which is obtained after rigidly fixing the remaining $m-1$ subsystems, a partial subsystem. Then it is evident that each diagonal block of the matrix $\mathbf{D}$ describes a certain partial subsystem, whereas off-diagonal blocks reflect the interaction between subsystems.

Suppose now that we know the natural frequency spectrum and natural oscillation shapes for each of the partial subsystems. Then for each of the subsystems, the following matrices can be formed: the diagonal matrix $\boldsymbol{\Lambda}_i=\mathrm{diag}\,[\lambda_i^p]$, $p=\overline{1,n_i}$, consisting of the natural frequencies of the ith subsystem, and the matrix $\boldsymbol{\Phi}_i$ whose columns $\boldsymbol{\Phi}_i^p$ represent the shapes of oscillations of the ith subsystem (n_i is the order of the ith subsystem.)

Let us find the conditions under which system (5-24) is weakly coupled. Let us form the block-diagonal matrix $\boldsymbol{\Phi}_0$ consisting of blocks $\boldsymbol{\Phi}_i$, and diagonal matrix $\boldsymbol{\Lambda}_0$ consisting of the blocks $\boldsymbol{\Lambda}_i$. It is evident, that these matrices describe the subsystems that are not coupled with each other.

Let us premultiply the matrix $\mathbf{D}$ (see 5-24)) by $\boldsymbol{\Phi}_0^T$ and postmultiply by $\boldsymbol{\Phi}_0$ (the superscript T marks the transposed matrix). Then we obtain the symmetric block matrix

$$\mathbf{D}^*=\boldsymbol{\Phi}_0^T\mathbf{D}\boldsymbol{\Phi}_0=[\boldsymbol{\Phi}_i^T\,(\mathbf{K}_{ij}-\lambda\mathbf{M}_{ij})\boldsymbol{\Phi}_j] \tag{5-25}$$

Since $\mathbf{\Phi}_i$ describes the natural oscillation shapes of the ith subsystem, the blocks of matrix (5-25) arranged on the main diagonal have the diagonal form. Let us introduce the following notation:

$$\mathbf{\Phi}_i^T \mathbf{M}_{ii} \mathbf{\Phi}_i = \text{diag}\,[\mu_i^p] = \boldsymbol{\mu}_i \tag{5-26}$$

$$\mathbf{\Phi}_i^T \mathbf{K}_{ii} \mathbf{\Phi}_i = \text{diag}\,[æ_i^p] = \mathbf{æ}_i \tag{5-27}$$

$$æ_i^p = \lambda_i^p \, \mu_i^p \tag{5-28}$$

Taking into account (5-26)–(5-28) we can represent the diagonal blocks of (5-25) as follows:

$$\mathbf{\Phi}_i^T (\mathbf{K}_{ii} - \lambda \mathbf{M}_{ii}) \mathbf{\Phi}_i = \mathbf{æ}_i - \lambda \boldsymbol{\mu}_i \tag{5-29}$$

In order for matrix $\mathbf{D}^*$ to describe weakly coupled subsystems, it is necessary to represent it as a matrix containing a small parameter $\epsilon \ll 1$ at off-diagonal blocks, that is (taking into account (5-29))

$$\mathbf{D}^* = \text{diag}\,[\mathbf{æ}_i - \lambda \boldsymbol{\mu}_i] + \epsilon \mathbf{B},$$

$$\mathbf{B} = \begin{bmatrix} 0 \; \mathbf{B}_{12} \ldots \mathbf{B}_{1m} \\ \cdots\cdots\cdots\cdots\cdots \\ \mathbf{B}_{1m}^T \; \mathbf{B}_{2m}^T \ldots 0 \end{bmatrix} \tag{5-30}$$

Then, following the perturbation theory (Kato 1966) we can represent the solution of the eigenvalue problem for matrix (5-30) as series expansions in powers of ϵ:

$$\begin{aligned} \mathbf{\Lambda} &= \mathbf{\Lambda}_0 + \epsilon \mathbf{\Lambda}_1 \mathbf{\Lambda} + \epsilon^2 \mathbf{\Lambda}_2 \mathbf{\Lambda} + \ldots, \\ \mathbf{\Phi} &= \mathbf{\Phi}_0 + \epsilon \mathbf{\Phi}_0 \mathbf{S} + \epsilon^2 \mathbf{\Phi}_0 \mathbf{T} + \ldots \end{aligned} \tag{5-31}$$

Here $\mathbf{\Lambda}_0$ and $\mathbf{\Phi}_0$ are the previously defined matrices representing the natural frequencies and shapes of natural oscillations of partial subsystems, and the matrix coefficients of ϵ describe correcting terms of the first and higher approximations; $\mathbf{S}$ is the matrix of spectral coupling coefficients S_{ij}^{ps} (see (5-37) below); and $\mathbf{T}$ characterizes second-order corrections to the eigenvectors. Some issues regarding the convergence of series like (5-31) are considered in (Kato (1966) and Dol'berg and Jasnitskaya (1973).

Matrix (5-25) can be reduced to the form (5-30) by different ways, and depending on this, we can obtain different types of weak couplings. In Banach and Perminov (1972), it is shown that the reduction can be done by premultiplying and postmultiplying matrix (5-25) by the diagonal matrix $\mathbf{N} = \text{diag}\,[(æ_i^p)^{1/2}]$. This results in the matrix $\mathbf{D}^{**} = \mathbf{N}\mathbf{D}^*\mathbf{N}$ whose diagonal elements are given by $1 - \lambda \mu_i^p / æ_i^p$, and off-diagonal blocks $\boldsymbol{\alpha}_{ij}$ consist of the elements

$$\alpha_{ij}^{ps}=[\mathbf{\Phi}_i^{pT}(\mathbf{K}_{ij}-\lambda\mathbf{M}_{ij})\,\mathbf{\Phi}_j^s]\,(æ_i^p æ_j^s)^{-1/2} \tag{5-32}$$

Taking into account (5-28) we can rewrite (5-32) as follows:

$$\alpha_{ij}^{ps}=[\mathbf{\Phi}_i^{pT}(\mathbf{K}_{ij}-\lambda\mathbf{M}_{ij})\mathbf{\Phi}_j^s](\lambda_i^p\,\lambda_j^s)^{-1/2}\,(\mu_i^p\mu_j^s)^{-1/2} \tag{5-33}$$

Weak Energy Coupling

Let us estimate the values of α_{ij}^{ps}. Note first, that $|\alpha_{ij}^{ps}|<1$, by virtue of positive definiteness of the matrices **K** and **M.**

The necessary condition of the system being weakly coupled is given by

$$|\alpha_{ij}^{ps}|\leq\epsilon\ll 1,\ \ p=\overline{1,n_i},\ \ s=\overline{1,n_j} \tag{5-34}$$

Indeed, if (5-34) holds, the matrix **D**** can be represented in the form (5-30), $\mathbf{D}^{**}=\mathbf{D}_0+\epsilon\mathbf{D}_1$, which describes weakly coupled systems.

Inequality (5-34) means that for *p*th and *s*th oscillations modes, the work of the elastic forces acting between the *i*th and *j*th subsystems is much less than the geometric mean of the potential energies of the *i*th ($\mathbf{V}_{ii}$) and *j*th ($\mathbf{V}_{jj}$) subsystems:

$$V_{ij}^{ps}\ll(V_{ii}^{p}V_{jj}^{s})^{1/2} \tag{5-35}$$

In case there exist inertia elements $\mathbf{M}_{ij}$ in the coupling matrix, we can obtain a similar condition for the kinetic energy:

$$W_{ij}^{ps}\ll(W_{ii}^{p}W_{jj}^{s})^{1/2} \tag{5-36}$$

Therefore, conditions (5-35) and (5-36) can be called conditions of weak energy coupling. Superscripts *p* and *s* indicate the respective numbers of oscillation modes.

As mentioned before, if (5-34) holds, the matrix **D*** has the form (5-30) that is, $\mathbf{K}=\mathbf{K}_0+\epsilon\mathbf{B}$, $\mathbf{M}=\mathbf{M}_0+\epsilon\mathbf{L}$, where blocks of the matrix are given by $\mathbf{B}_{ii}=0$, $\mathbf{B}_{ij}=\mathbf{\Phi}_i^T\mathbf{K}_{ij}\mathbf{\Phi}_j$. Analogously, we can obtain blocks of the matrix **L.**

One can seek the solution of the eigenvalue problem for matrix (5-25) in the form of series (5-31). Using the procedure given in Banach and Perminov (1972) we obtain first-order corrections (in the case of simple roots) for natural frequencies and oscillation shapes of the *i*th subsystem taking into account its coupling with the *j*th subsystem:

$$(\mathbf{\Delta\Lambda}_i)_j=0,\ \ S_{ij}^{ps}=\alpha_{ij}^{ps}\lambda_i^p(\lambda_j^s-\lambda_i^p)^{-1} \tag{5-37}$$

If the *i*th subsystem is coupled with several subsystems, the resultant correction is found by adding corrections (5-37).

Weak Spectral Coupling

Analyzing the first-order and higher-order corrections we see that these corrections contain the matrix $\mathbf{S}=[S_{ij}^{ps}]$, given by (5-37), as a multiplier. Consequently, the values of S_{ij}^{ps} determine the radius and rate of convergence of series (5-31). If $|S_{ij}^{ps}|>1$, then the series diverge.

Suppose, condition (5-34) of weak energy coupling is satisfied. However, as it follows from (5-34) and (5-37), in case $|1-(\lambda_j^s/\lambda_i^p)|<\epsilon$ the inequality $|S_{ij}^{ps}|>1$ implying the divergence of series in (5-31) can hold. This inequality means that different subsystems have close natural frequencies. In this case, as in Banakh and Perminov (1972), the system can turn out to be strongly coupled. It is evident that the satisfaction of the condition

$$|S_{ij}^{ps}|=|\alpha_{ij}^{ps}\,\lambda_i^p\,(\lambda_j^s-\lambda_i^p)^{-1}|<\epsilon \tag{5-38}$$

is sufficient for the convergence of series (5-31). If the condition (5-38) is satisfied, we say that there is weak spectral coupling between the *i*th and *j*th subsystems.

The condition of weak spectral coupling can be fulfilled in the following two cases: (1) When there is weak energy coupling between subsystems ($|\alpha_{ij}^{ps}|<\epsilon$), and there are no close frequencies in the subsystems, that is, $|1-(\lambda_j^s/\lambda_i^p)|\sim 1$; and (2) when the amount off resonance between subsystems is large, that is, $|1-(\lambda_j^s/\lambda_i^p)|\geq 1/\epsilon$. In the latter case, the weak energy coupling between subsystems is not necessary. This means weak coupling between the subsystems operating in different frequency ranges. Note that the condition of weak energy coupling alone is applicable only if the subsystems do not contain close frequencies. Otherwise, one can also check the convergence condition, $|S_{ij}^{ps}|<1$.

Separation of Weakly Coupled Subsystems in the General System

The decomposition of the system taking account of weak energy and spectral coupling allows us to reduce considerably the order of the examined system. If the solutions of eigenfrequency problems for the subsystems are known, one can construct the matrix $\boldsymbol{\alpha}_{ij}$ and estimate the degree of coupling between oscillation modes.

For the majority of practical problems, the order of the subsystems is very large (10^3–10^4), and it is difficult to obtain all necessary information about frequencies and shapes of oscillations. Therefore, to use effectively the weakness of couplings, we are to be able to estimate the quantities α_{ij}^{ps} approximately, without having a complete solution of the eigenvalue problems. Let us prove an important property of the energy couplings, namely that maximal matrix elements α_{ij}^{ps} decrease with mode numbers (p and s) increasing and, under certain conditions, remain less than a prescribed number ϵ, i.e.,

$$\max \alpha_{ij}^{11} \geq \ldots \geq \max \alpha_{ij}^{n_i n_j},\ p=\overline{1,n_i},\ s=\overline{1,n_j}\ . \tag{5-39}$$

For the proof, we simplify expressions (5-32) and (5-33) by using vector and matrix norms (Parlett 1980). Then the existence condition for weak energy coupling takes the form

$$\begin{aligned} |\alpha_{ij}^{ps}| &\leq \frac{\| \mathbf{\Phi}_i^{pT} \mathbf{K}_{ij} \mathbf{\Phi}_j^s \|}{(\mathbf{\Phi}_i^{pT} \mathbf{K}_{ii} \mathbf{\Phi}_i^p)^{1/2} (\mathbf{\Phi}_j^{sT} \mathbf{K}_{jj} \mathbf{\Phi}_j^s)^{1/2}} \\ &\leq \frac{\| \mathbf{Q}_i^{-1/2} \mathbf{K}_{ij} \mathbf{Q}_j^{-1/2} \|}{(æ_i^p\ æ_j^s)^{1/2}} \end{aligned} \tag{5-40}$$

where $\mathbf{Q}$ is determined from the orthogonality condition for the oscillation shapes: $\mathbf{\Phi}_i^T \mathbf{Q} \mathbf{\Phi}_\mathrm{i} = \mathbf{E}$. In particular, for rather widespread orthogonality conditions, with respect to energy ($1/2\ \lambda_i \mathbf{\Phi}_i^T \mathbf{M}_{ii} \mathbf{\Phi}_i = \mathbf{E}$) and with respect to the inertia matrix ($\mathbf{\Phi}_i^T \mathbf{M}_{ii} \mathbf{\Phi}_i = \mathbf{E}$), expression (5-40) takes the form

$$|\alpha_{ij}^{ps}| \leq \| \mathbf{M}_{ii}^{(-1/2)T} \mathbf{K}_{ij} \mathbf{M}_{jj}^{-1/2} \|\ (\lambda_i^p\ \lambda_j^s)^{-1/2} \tag{5-41}$$

Numerators in (5-40) and (5-41) do not depend on the solution and are determined by the elements of original matrices **K** and **M.** Hence, the change of α_{ij}^{ps} with numbers p and s increasing is determined by the values of λ_i^p and λ_j^p, from where relationships (5-39) follow immediately.

Now, we prove the second part of the statement. Let us choose the partition of the system into subsystems connected by weak energy couplings so that the inequalities

$$\| \mathbf{Q}_i^{-1/2} \mathbf{K}_{ij} \mathbf{Q}_\mathrm{j}^{-1/2} \|\ (\mu_i^{p1} \mu_j^{s1})^{-1/2} < \epsilon (\lambda_i^{p1} \lambda_j^{s1})^{-1/2} \tag{5-42}$$

hold. Then $|\alpha_{ij}^{ps}| < \epsilon$ for $p > p_1$, $s > s_1$. This completes the proof.

Condition (5-42) used for estimating $\max \alpha_{ij}^{ps}$ does not require the knowledge of higher frequencies and shapes of oscillations. Besides, from (5-42), we can determine frequencies λ_i^{p1} in order to provide the calculation accuracy equal to ϵ.

Thus, we can propose the following way for seeking weak energy couplings in a complex system. First, the system is partitioned into subsystems. Then, after having obtained natural frequencies and oscillation shapes for decoupled subsystems, we find from (5-42) the numbers p_1 and s_1 of the oscillation modes in the ith and jth subsystems connected by the weak energy coupling.

Hence, relationship (5-42) enables us to estimate the strength of energy couplings and is the existence criterion for weak energy couplings. In addition, this relationship determines the way of partitioning the system into weakly coupled subsystems.

Example (Banach 1988)

Consider the use of concepts of energy and spectral couplings for the analysis of free oscillations of a rotor mounted on an elastic foundation (see Fig. 5-3).

The foundation is considered to be consisting of beams with constant square-shaped cross section, the length of the square side being equal to 0.43 m. The lengths of longitudinal and transverse beams of the foundation are equal to $L_1=5$ m and $L_2=3$ m, respectively. The rotor is modeled as a beam of circular cross section. The length of the rotor is $L_r=5$ m, and the radius of its cross section is $r_r=0.6$ m. The finite element model of the system has been studied within the frequency range 0–100 Hz. The model has 12 elements. Each of the elements has six degrees of freedom, and hence, the total number of degrees of freedom is 72. The finite elements are numbered as shown in Figure 5-3.

The stiffness matrix of the system is given by

$$\mathbf{K} = \begin{bmatrix} \mathbf{K}_f \, \mathbf{K}_{fr} \\ \mathbf{K}_{fr}^T \, \mathbf{K}_r \end{bmatrix},$$

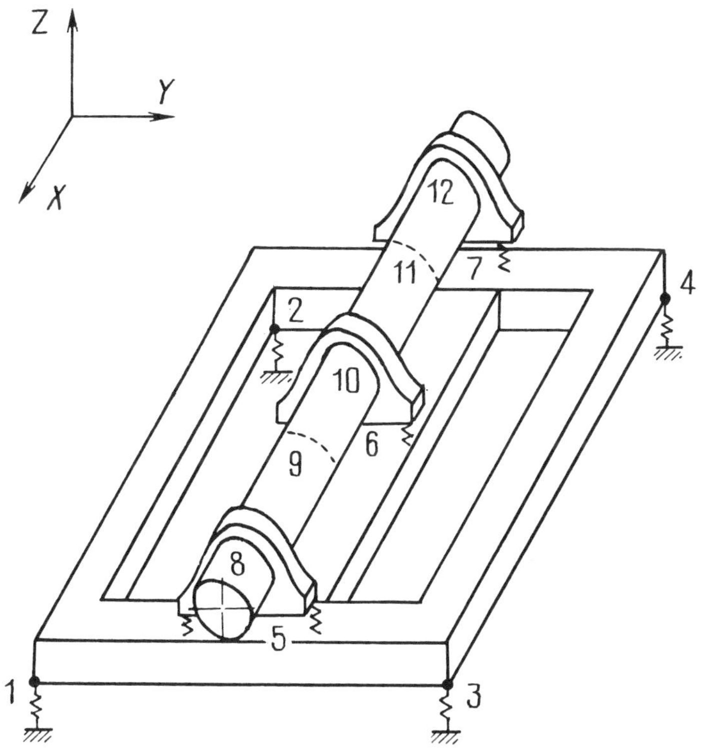

Figure 5–3 Rotor on an elastic foundation.

where $\mathbf{K}_f$ and $\mathbf{K}_r$ are stiffness matrices of the foundation and rotor, respectively; and $\mathbf{K}_{fr}$ is the stiffness matrix of elastic elements connecting the rotor and foundation. The 42×30-matrix $\mathbf{K}_{fr}$ has a block structure, with the blocks $\mathbf{K}_{fr}^{ij}$ ($i=1,\ldots,7$; $j=8,\ldots,12$). Only the 6×6-blocks $\mathbf{K}_{fr}^{5,8}=\mathbf{K}_{fr}^{6,10}=\mathbf{K}_{fr}^{7,12}$ are not equal to zero, nonzero elements of these blocks are $k_{11}=k_{22}=k_{33}=-10^3$N/m, $k_{15}=-k_{24}=-34.3\cdot 10^3$N/m. The first four natural frequencies for decoupled subsystems are 7.65, 8.36, 21.86, 39.8 Hz for the rotor, and 22.2, 25.76, 29.0, 79.8 Hz for the foundation. The oscillation shapes corresponding to these frequencies are $\mathbf{\Phi}_r^i$ and $\mathbf{\Phi}_f^i$ ($i=1,\ldots,4$). Having calculated the matrix $\mathbf{D}^*=\mathbf{\Phi}_f^T(\mathbf{K}-\lambda\mathbf{M})\mathbf{\Phi}_r$, according to (5-33), we find max $\alpha_{ij}^{ps}=\alpha_{ij}^{44}=0.11$, max $S_{ij}^{ps}=0.04$ ($\mu_i=\mu_j=50$). Hence, within the chosen frequency range, the system is weakly coupled, both in terms of energy and spectral coupling. For unknown oscillation modes with p, $s>4$, we can estimate α_{ij} by using (5-41). In the case in question, when λ_i, $\lambda_j>2\pi\cdot 80$Hz, we find that $|\alpha_{ij}| \leq ||\mathbf{M}_f^{-1/2}\,\mathbf{K}_{fr}\,\mathbf{M}_r^{-1/2}|| \cdot(\lambda_i\lambda_j\mu_i\mu_j)^{-1/2} \leq 0.2$. Taking into account (5-42) we conclude that the system is weakly coupled (in terms of energy coupling) within the whole frequency range and, hence, natural frequencies of the coupled system are close to the corresponding frequencies of the isolated subsystems. The calculations of natural frequencies confirm this conclusion.

The natural frequencies of the coupled system are 7.13, 7.96, 20.55, 22.8, 26.1, 29.4, 38.5, and 80.7 Hz.

The corrections for these frequencies do not exceed 6%, the frequencies of both the rotor and foundation sliding apart (the lesser frequencies decrease while the greater ones increase, approximately by the same amount) so that $\sum\lambda_{if}+\sum\lambda_{ir}=\sum\lambda_i$ (it follows from the Vi'ete theorem). The dynamic characteristics of subsystems obtained after partitioning the original system due to weak couplings are close to the respective characteristics of the original system.

6

Examples of Multicriteria Optimization of Machines and Other Complex Systems

At present, the parameter space investigation (PSI) method is widely used in various areas, such as pharmacy, petrophysics, nuclear physics, chemistry of polymers, geophysics, and nonlinear optics. However, of primary importance is its use in shipbuilding, machine tools, aircraft, railway cars, automobiles manufacture, etc. In this chapter we discuss some examples of effective applications of the method.

6-1. Vibration Machines Optimization

Resonant Table Vibrator Design (Sobol' and Statnikov 1981; Kryukov et al. 1980)

Resonant vibration machines are used in various industrial branches. They are created on the basis of nonlinear elastic systems ensuring their technological stability and allowing optimization of the laws of working parts vibration. Synthesis of vibration machines is a rather complicated problem, generally reducible to determination of the optimal dynamic structure and selection of the nonlinear system and drive parameters so as periodic motions of the working parts satisfy both the specifications and a number of design constraints in the best possible way.

Initial data

As the prototype of the vibration machine we have chosen a table vibrator used for moulding reinforced concrete products, whose load-carrying capacity is 8 t (Fig. 6-1). Its vibration is described by the following system of nonlinear differential equations:

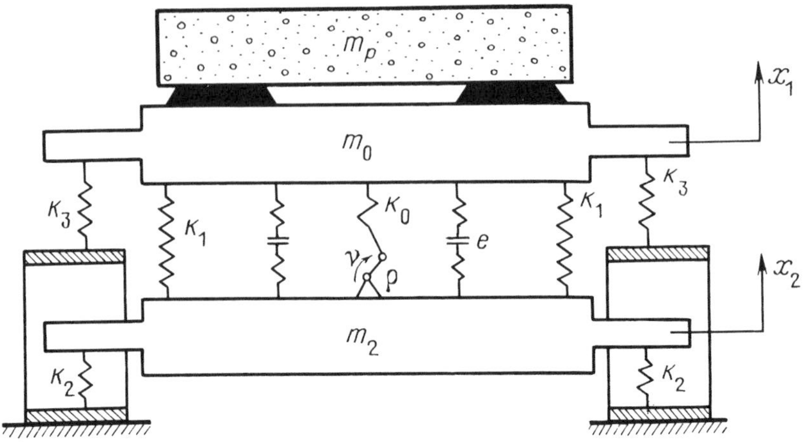

Figure 6–1 Schematic of an asymmetric resonant table vibrator.

$$\begin{aligned} m_1\ddot{x}_1+f(x)\dot{x}+\mu k_3\dot{x}_1+P(x)+k_3x_1= \\ k_0\rho\,(\sin\nu t+\mu\nu\cos\nu t) \\ m_2\ddot{x}_2-f(x)\dot{x}+\mu k_2\dot{x}_2-P(x)+k_2x_2= \\ -k_0\rho\,(\sin\nu t+\mu\nu\cos\nu t), \end{aligned} \tag{6-1}$$

where x_1 and x_2 are the displacements of masses m_0 and m_2 respectively; $x=x_1-x_2$, $P(x)=(k_1+k_0)x+\sigma(x)k_B(x+e)$, $f(x)=\mu_1[k_1+k_0+\sigma(x)k_B]$, $\sigma(x)=0$ for $x\geq e$, and $\sigma(x)=1$ for $x<e$. The physical meaning of the quantities appearing in (6-1) follows.

Design variables

There are 10 design variables: stiffness of the driver elastic links, $\alpha_1=k_0$ (measured in kN/cm); stiffness of the main linear elastic links, $\alpha_2=k_1$ (kN/cm); the buffers stiffness, $\alpha_3=k_B$ (kN/cm); stiffness of shock absorbers of the supports mounted under the frame, $\alpha_4=k_2$ (kN/cm); stiffness of shock absorbers of the supports mounted under the working part, $\alpha_5=k_3$ (kN/cm); drive eccentricity, $\alpha_6=\rho$ (cm); mass of the working part, $\alpha_7=m_0$ (t); balancing frame mass, $\alpha_8=m_2$ (t); initial clearance in buffers in the absence of technological loads, $\alpha_9=e_{in}$ (cm); operational frequency, $\alpha_{10}=\nu(\mathrm{s}^{-1})$.

The other quantities

The remaining quantities are: payload mass, m_p; coefficient of the payload mass addition, k_m; reduced mass of the working part of the system, $m_1=m_0+k_mm_p$; coefficient of internal resistance of the rubber elastic links, μ; reduced resistance coefficient, k_μ; reduced coefficient of internal resistance forces, $\mu_1=\mu+k_\mu m_p$; and buffers clearance, e.

The request for proposal specifies the product mass ($2\text{ t}\leq m_p\leq 8\text{ t}$), the operational frequency range ($50\text{ s}^{-1}\leq\nu\leq 100\text{ s}^{-1}$), and the maximum upward and downward accelerations of the working part w_{1u} and w_{1d}. Operational modes must correspond to the ascending branch of the frequency characteristic, that is, must belong to the subresonance region.

Approximate periodic solutions to system (6-1) were found using the Krylov-Bogolyubov method.

The following functional dependences are introduced:

$$f_1(\boldsymbol{\alpha})=w_{1u}, f_2(\boldsymbol{\alpha})=w_{1d}, f_3(\boldsymbol{\alpha})=k_0\rho,$$
$$f_4(\boldsymbol{\alpha})=m_0+M-(k_1+k_3)\,Q(gk')^{-1},$$
$$f_5(\boldsymbol{\alpha})=m_0+m_2+M-(k_2+k_3)Q(gk')^{-1},$$
$$f_6(\boldsymbol{\alpha})=\frac{k_2Mg-k_0\rho(k_2+k_3)}{k_1k_2+k_1k_3+k_2k_3}-e_{in},$$
$$f_7(\boldsymbol{\alpha})=\nu-\omega(\boldsymbol{\alpha})$$

where M is the maximum load (equal to 8 t); k' is the stiffness of a rubber shock absorber (measured in kN/cm); Q is the limiting load per shock absorber (kN); and $\omega(\boldsymbol{\alpha})$ is the linearized system's natural frequency (s^{-1}).

Functions $f_1(\boldsymbol{\alpha})$, $f_2(\boldsymbol{\alpha})$, and $f_7(\boldsymbol{\alpha})$ depend on the solution to system (6-1), while $f_3(\boldsymbol{\alpha})$–$f_6(\boldsymbol{\alpha})$ are expressed directly via $\alpha_1,\ldots,\alpha_{10}$.

The functional constraints are specified by the inequalities

$$9\leq f_1(\boldsymbol{\alpha})\leq 22,\ \ 40\leq f_2(\boldsymbol{\alpha})\leq 100,\ \ f_3(\boldsymbol{\alpha})\leq 120 \tag{6-2}$$

and

$$f_j(\boldsymbol{\alpha})\leq 0,\ \ j=4,\ldots,7. \tag{6-3}$$

The first two constraints ensure that the vibration of the working part corresponds to the design requirements; the third constraint limits the disturbing force; the fourth and the fifth ones limit the loads acting on the rubber shock absorbers; the sixth constraint is related to the load on the elastic suspension; and the seventh one ensures that the operational modes under consideration stay within the subresonance region.

The constraints imposed on f_1, f_2, and f_3, are not "rigid," and may be varied depending on the specific features of manufacture and the requirements to the reinforced concrete products. Conversely, the constraints imposed on f_4–f_7 cannot be violated.

Performance criteria

It is proposed to estimate the quality of the table vibrator using six criteria, all of which should be minimized. The first criterion (the mass of the machine) $\Phi_1=\alpha_7+\alpha_8$. The other five criteria are expressed through the solution to system (6-1), which depends on the product mass m_p. In line with the technique proposed in Kryukov et al. (1980), system (6-1) was solved four times for $m_p=2$, 4, 6, and 8 t. The max and min symbols employed in the subsequent formulas indicate that either the maximum or the minimum values obtained in the four series of calculations are used. Thus, we have the following performance criteria:

- The mass of the machine $\Phi_1(\boldsymbol{\alpha})=m_0+m_2$.
- The asymmetry of the law of the working part vibration, $\Phi_2(\boldsymbol{\alpha})=\max(w_{1u}/w_{1d})$.
- The dynamic force acting in the drive, $\Phi_3(\boldsymbol{\alpha})=\max(k_0a_0)$, where a_0 is the elastic deformation of driving links.
- The dynamic loads acting on the foundation, $\Phi_4(\boldsymbol{\alpha})=\max\left|\,|k_2a_2|-|k_1a_1|\,\right|$, where a_1 and a_2 are the vibration amplitudes of the working part and the balancing frame, respectively.
- The stability characteristics of the upward and downward accelerations, $\Phi_5(\boldsymbol{\alpha})=(\max w_{1u}/\min w_{1u})-1$ and $\Phi_6(\boldsymbol{\alpha})=(\max w_{1d}/\min w_{1d})-1$ respectively.

Problem 1: Analysis of the Potential for Modernizing the Initial Vibration Machine

We have to answer the question of whether comparatively small variations in the parameters of the existing vibration machine allow improvement without a cardinal change in the design.

To solve the problem the designer has indicated the boundaries of the design variable variation presented in Table 6-1. These boundaries define a nine-dimensional parallelepiped Π_1 whose center coincides with point $\boldsymbol{\alpha}^1$ representing the existing machine parameters (see Table 6-1). The criteria constraints were set equal to the values of the criteria at point $\boldsymbol{\alpha}^1$, that is, $\Phi_\nu^{**}=\Phi_\nu(\boldsymbol{\alpha}^1)$ for any $\nu=1,\ldots,6$.

Trial calculation and correction of the problem formulation

In parallelepiped Π_1, $N=256$ trials were conducted, subject to functional constraints (6-2) and (6-3). The number of models satisfying the functional constraints proved to be equal to $N'=24$, so that $\gamma\approx0.093$. Since the first three functional constraints were not rigid, functional dependences f_1, f_2, and f_3 were converted into pseudocriteria. Thus, we have

Table 6–1

	Problem 1					Problem 2			
α_j	α_j^*	α_j^1	α_j^{**}	α_j^{248}	α_j^{475}	α_j^*	α_j^{**}	α_j^{116}	α_j^{12}
1	44	48	52	44.96	50.98	20	100	31.79	48.78
2	92	96	100	99.03	99.42	30	150	59.04	83.12
3	1,400	1,600	1,800	1,758	1,490	400	3,000	1,449	2,396
4	20	25	30	22.07	20.09	20	60	20.42	20.00
5	20	25	30	29.49	26.54	20	60	48.64	46.87
6	0.5	0.7	0.9	0.848	0.830	0.5	2.0	0.971	1.065
7	3	3	3	3	3	3	7	3	3
8	3	5	7	4.02	5.43	3	7	3	3
9	0.0	0.2	0.4	0.398	0.110	−0.3	1.5	0.438	0.637
10	94	97	100	95.10	94.20	50	100	91.48	90.63

$$\Phi_7=f_3(\boldsymbol{\alpha}),\ \ \Phi_8=f_1(\boldsymbol{\alpha}),\ \ \Phi_9=f_2(\boldsymbol{\alpha}). \tag{6-4}$$

The second trial calculation

Again $N=256$ trials were conducted, 33 of which were included in the test tables. Models $\boldsymbol{\alpha}^{40}$ and $\boldsymbol{\alpha}^{248}$ proved to be advantageous, since they surpass $\boldsymbol{\alpha}^1$ in several important design variables, and this compensates for a certain deterioration in the rest criteria. Both models found their way into the test table solely due to transformation of functional dependences into pseudocriteria (6-4). The designer has decided that model $\boldsymbol{\alpha}^{248}$ was the most promising one (see Table 6-2).

Subsequently, the designer has tried to minimize criterion Φ_1.

Continuation of the calculations

The trials were continued, subject to constraints (6-3), within the same parallelepiped Π_1 for $N=1{,}024$. As a result, 108 models entered region G. At this stage, no criteria constraints were imposed. From among the 108 models included in the test tables, the Pareto optimal ones were selected. Taking into account all six criteria, there proved to be 59 such models, model $\boldsymbol{\alpha}^1$ included. Hence, the prototype (model $\boldsymbol{\alpha}^1$) could not be improved in all six criteria simultaneously. In a certain sense, this conclusion should be considered natural because the design subjected to analysis was very good. Subsequently, the designer has considered seven sets of criteria constraints of which three are described here.

The first designer-computer dialogue. Table 6-2 presents the trial data for all the performance criteria with the exclusion of pseudocriteria. The criteria constraints marked by horizontal lines define the feasible solutions set D containing only one solution represented by model $\boldsymbol{\alpha}^1$.

Table 6–2

i	$\Phi_1(\boldsymbol{\alpha}^i)$	i	$\Phi_2(\boldsymbol{\alpha}^i)$	i	$\Phi_3(\boldsymbol{\alpha}^i)$	i	$\Phi_4(\boldsymbol{\alpha}^i)$	i	$\Phi_5(\boldsymbol{\alpha}^i)$	i	$\Phi_6(\boldsymbol{\alpha}^i)$
248	7.02	445	0.233	319	28.25	475	3.70	588	0.047	214	0.051
353	7.04	475	0.241	248	28.34	1,022	3.74	418	0.106	295	0.120
132	7.17	141	0.241	794	28.46	957	4.10	910	0.133	853	0.120
673	7.30	40	0.241	47	28.94	593	4.15	863	0.144	910	0.122
47	7.31	78	0.242	935	29.38	253	4.37	214	0.155	231	0.130
528	7.32	831	0.242	176	29.52	768	4.52	134	0.170	807	0.163
905	7.37	183	0.243	498	29.59	143	4.55	853	0.176	819	0.177
733	7.39	667	0.243	13	30.15	643	4.62	295	0.180	445	0.183
379	7.41	925	0.243	40	30.34	116	4.66	637	0.197	40	0.186
836	7.42	433	0.243	528	30.62	204	4.77	947	0.214	116	0.188
141	7.42	655	0.244	922	30.82	588	4.91	347	0.233	637	0.214
690	7.43	396	0.245	475	30.94	552	4.93	396	0.236	588	0.217
620	7.47	319	0.246	905	31.62	176	5.16	203	0.243	406	0.221
421	7.49	913	0.246	682	31.68	733	5.20	231	0.256	475	0.227
922	7.50	819	0.248	141	31.69	667	5.27	925	0.256	790	0.243
498	7.52	682	0.248	819	31.70	661	5.42	406	0.256	655	0.245
787	7.57	794	0.248	1	31.72	379	5.43	116	0.271	362	0.260
218	7.58	498	0.249	1,022	31.83	224	5.54	819	0.281	203	0.266
78	7.59	176	0.250	831	32.00	368	5.63	619	0.283	36	0.266
181	7.61	935	0.250	433	32.41	619	5.68	153	0.294	619	0.274
552	7.63	619	0.250	421	32.54	406	5.69	913	0.298	78	0.279
990	7.65	134	0.252	379	32.80	433	5.70	648	0.298	224	0.281
319	7.70	922	0.252	445	33.62	794	5.77	807	0.303	149	0.282
768	7.70	615	0.252	181	33.95	682	5.78	224	0.303	667	0.285
336	7.73	879	0.252	368	34.77	533	5.81	40	0.305	48	0.286
879	7.73	115	0.255	353	34.79	13	5.81	445	0.305	628	0.286
13	7.75	1	0.256	218	35.01	408	5.87	628	0.310	925	0.287
831	7.76	243	0.256	787	35.06	874	5.89	183	0.316	418	0.294
605	7.79	600	0.258	874	35.18	498	5.92	520	0.322	706	0.304
153	7.80	996	0.260	643	35.34	859	5.96	996	0.335	615	0.306
40	7.81	143	0.261	768	35.64	637	5.96	862	0.336	847	0.313
293	7.82	103	0.261	688	35.70	913	5.97	78	0.342	1,022	0.313
794	7.82	643	0.262	836	35.75	922	6.03	115	0.343	347	0.315
643	7.86	368	0.262	341	36.43	293	6.13	48	0.346	513	0.324
433	7.87	874	0.262	847	36.43	787	6.18	1,022	0.352	336	0.332

continued

The second designer-computer dialogue. Having decided to make concessions in the less important criteria Φ_5 and Φ_6, the designer chose $\Phi_5^{**}=1$ and $\Phi_6^{**}=1$, while the first four criteria remained unaltered: $\Phi_\nu^{**}=\Phi_\nu(\boldsymbol{\alpha}^1)$, $1\leq\nu\leq4$.

In this case three models, $\boldsymbol{\alpha}^1$, $\boldsymbol{\alpha}^{794}$, and $\boldsymbol{\alpha}^{922}$, were included in the feasible solutions set. The latter two solutions were assumed to be approximately equivalent, since the best value of $\Phi_1(\boldsymbol{\alpha}^{922})$ was balanced by the best values of $\Phi_3(\boldsymbol{\alpha}^{794})$ and $\Phi_4(\boldsymbol{\alpha}^{794})$. It is noteworthy that model $\boldsymbol{\alpha}^1$ proved to be improvable in the four most important criteria.

Table 6–2 (Continued)

i	$\Phi_1(\boldsymbol{\alpha}^i)$	i	$\Phi_2(\boldsymbol{\alpha}^i)$	i	$\Phi_3(\boldsymbol{\alpha}^i)$	i	$\Phi_4(\boldsymbol{\alpha}^i)$	i	$\Phi_5(\boldsymbol{\alpha}^i)$	i	$\Phi_6(\boldsymbol{\alpha}^i)$
910	7.87	13	0.262	253	36.62	879	6.25	667	0.354	880	0.339
952	7.93	163	0.263	132	36.67	248	6.36	475	0.357	1	0.359
192	7.95	661	0.263	879	36.69	935	6.37	86	0.366	115	0.360
767	7.96	990	0.263	224	37.06	615	6.53	790	0.374	408	0.380
619	7.97	421	0.263	192	37.18	295	6.62	149	0.375	688	0.383
418	7.99	853	0.264	103	38.11	243	6.67	513	0.385	668	0.390
925	8.00	347	0.264	655	38.20	1	6.70	615	0.385	85	0.393
1	8.00	293	0.265	143	38.44	690	6.82	706	0.386	600	0.402
819	8.01	204	0.265	78	38.85	319	6.82	655	0.395	957	0.431
593	8.04	48	0.265	243	39.08	905	6.90	336	0.399	103	0.432
149	8.05	224	0.266	185	39.15	78	7.01	847	0.400	368	0.451
682	8.05	149	0.266	733	39.27	218	7.03	600	0.417	293	0.463
408	8.05	203	0.266	406	39.57	47	7.08	85	0.420	948	0.474
935	8.06	787	0.267	990	39.63	790	7.16	1	0.421	183	0.481
36	8.06	520	0.267	513	39.82	853	7.17	668	0.439	836	0.485
790	8.07	253	0.268	593	40.05	48	7.17	880	0.441	253	0.486
⋮	⋮	⋮	⋮	⋮	⋮	⋮	⋮	⋮	⋮	⋮	⋮
214	8.33	673	0.276	605	45.84	421	8.51	831	0.649	243	0.626
475	8.34	231	0.276	948	45.91	353	8.55	767	0.687	947	0.660
996	8.34	690	0.276	790	46.01	528	8.63	859	0.687	176	0.665
637	8.35	768	0.277	35	46.30	990	8.71	421	0.696	682	0.719
⋮	⋮	⋮	⋮	⋮	⋮	⋮	⋮	⋮	⋮	⋮	⋮
35	8.56	767	0.282	153	50.85	362	9.44	353	0.844	528	0.891
183	8.61	248	0.283	859	52.01	153	9.59	218	0.849	768	0.894
648	8.61	593	0.283	418	54.35	170	9.59	733	0.881	733	0.897
807	8.64	952	0.284	957	54.73	880	9.64	787	0.886	913	0.903
⋮	⋮	⋮	⋮	⋮	⋮	⋮	⋮	⋮	⋮	⋮	⋮
347	8.98	605	0.305	231	64.18	35	11.82	13	1.118	498	1.164
863	9.04	620	0.313	947	70.62	600	12.51	498	1.143	13	1.222

The third designer-computer dialogue. The designer has decided to look for the models that are somewhat worse than $\boldsymbol{\alpha}^1$ in one of the first four criteria, being at the same time notably better in all other criteria. To do this he has chosen the constraints

$$\Phi_1^{**}=\Phi_1(\boldsymbol{\alpha}^{996})=8.34,\quad \Phi_2^{**}=\Phi_2(\boldsymbol{\alpha}^{836})=0.291,$$

$$\Phi_3^{**}=\Phi_3(\boldsymbol{\alpha}^{445})=33.62,\quad \Phi_4^{**}=\Phi_4(\boldsymbol{\alpha}^{847})=7.61.$$

As defined earlier, $\Phi_5^{**}=1$ and $\Phi_6^{**}=1$.

The feasible solutions set contain 15 models of which 12 are Pareto optimal. Of the latter, the designer has preferred models $\boldsymbol{\alpha}^{248}$, $\boldsymbol{\alpha}^{475}$, and $\boldsymbol{\alpha}^{922}$.

The most intriguing proved to be model $\boldsymbol{\alpha}^{475}$, which is somewhat worse than $\boldsymbol{\alpha}^{1}$ in criterion Φ_1, but surpasses it in criteria Φ_2–Φ_6. The value of $\Phi_4(\boldsymbol{\alpha}^{475})$ is minimal and much better than $\Phi_4(\boldsymbol{\alpha}^{1})$. The designer's ideas related to model $\boldsymbol{\alpha}^{475}$ are summarized in the conclusions.

Subsequent designer-computer dialogues did not result in substantial improvements. Attempts to improve model $\boldsymbol{\alpha}^{475}$ by means of a local search in its neighborhood proved to be fruitless.

The issue of model $\boldsymbol{\alpha}^{475}$ stability also was analyzed. In doing so it was supposed that the model is stable if the parallelepiped whose center coincides with point $\boldsymbol{\alpha}^{475}$ and edges correspond to technological tolerances of the design variables does not contain points with corresponding performance criteria substantially differing from $\Phi_\nu(\boldsymbol{\alpha}^{475})$. Within this parallelepiped, 64 trials were conducted in which criteria proved to be close to $\Phi_\nu(\boldsymbol{\alpha}^{475})$.

Analysis of the criteria relations

The correlation coefficients of the criteria were calculated in region G containing 108 trial points. The results presented in Table 6-3 show that of the six criteria only Φ_5 and Φ_6 are strongly interdependent, the correlation coefficient $r_{5,6}$ being equal to 0.89. The analysis of the test tables has confirmed the conclusion that the groups of the best (and worst) models with respect to both the performance criteria consist mostly of the same models.

Table 6–3

μ \ ν	1	2	3	4	5	6
1	1	−0.14	0.60	0.00	−0.63	−0.56
2	−0.14	1	0.24	0.05	0.17	0.14
3	0.60	0.24	1	0.32	−0.64	−0.47
4	0.00	0.05	0.32	1	−0.32	−0.24
5	−0.63	0.17	−0.64	−0.32	1	0.89
6	−0.56	0.14	−0.47	−0.24	0.89	1

The results of solving Problem 1 can be summed up as follows.

1. The prototype machine (model $\boldsymbol{\alpha}^{1}$) cannot be improved in all six criteria simultaneously. However, models $\boldsymbol{\alpha}^{794}$ and $\boldsymbol{\alpha}^{922}$ surpass it in four of the most important criteria, Φ_1–Φ_4.
2. An advantageous model, $\boldsymbol{\alpha}^{248}$, has been found, which is optimal in criterion Φ_1. Its mass is less than that of model $\boldsymbol{\alpha}^{1}$ by approximately 1t, and the rest of the criteria are acceptable in the designer's opinion.

3. Model $\boldsymbol{\alpha}^{475}$ optimal in criterion Φ_4, was found. Though its mass is larger than that of the prototype by 0.34 t, it surpasses the latter in the remaining five criteria. For the newly found model the dynamic load on the foundation (criterion Φ_4) is almost twice as small as for the prototype! Since the newly designed plants, and the more so, advanced plants of the future, are supposed to be multistory buildings with vibrator tables installed not only on the ground floor but on the upper floors too, a reduction in the foundation load (criterion Φ_4) acquires major significance.

   At this point we would like to stress the usefulness of multicriteria analysis once again: At the start of the analysis criterion Φ_1 was assumed to be undoubtedly one of the most important criteria. However, it is absolutely clear that within the framework of the single-criterion problem of the criterion Φ_1 optimization we would fail to find the advantageous $\boldsymbol{\alpha}^{475}$ solution.
4. Generally, in the case of the previous design variable variations the possibilities of improving the prototype machine design are rather scarce.

Problem 2: Predesign of the Minimal-Mass Machine

It is necessary to improve criterion Φ_1 (the machine mass) considerably, improving at the same time, if possible, the values of the rest criteria. Since, according to the previous analysis the problem is not solvable within parallelepiped Π_1, it was decided to widen the region of search drastically.

Global analysis

The designer has constructed a new parallelepiped Π_2 using design-variable constraints α_j^* and α_j^{**} presented in Table 6-1. The limits of α_7 and α_8 in Π_2 have been determined from the dynamic strength conditions, as in Π_1.

Under the initial functional constraints $N=4{,}096$ trials were conducted in Π_2, of which only $N'=100$ trials were included into the test table. Hence, $N'/N \approx 0.025$. The reduction (as compared with $\gamma=0.093$ in the case of the first problem) is quite natural, since the preceding search was carried out over a rather limited volume.

Six designer-computer dialogues have been conducted, with the number of models entering the feasible solutions set varying from zero to 20. Upon analyzing all 100 models it was decided to continue analysis not in the whole of the parallelepiped, but in a certain portion of it.

Local analysis

The designer has selected the seven best models and constructed parallelepipeds centered in them. The local search was carried out within these parallelepipeds. Two best models were found, having minimum mass 6 t ($\Phi_1=6$ t) and acceptable values of all other criteria, namely, model $\boldsymbol{\alpha}^{116}$ found in the neighborhood of model $\boldsymbol{\alpha}^{1,452}$, and model $\boldsymbol{\alpha}^{12}$ in the neighborhood of $\boldsymbol{\alpha}^{2,406}$ (see Table 6-1).

Conclusion

The results of these calculations have been regarded by the designer as most promising. First, it was found that the mass of the machine can be reduced by 2 t, with the forces acting in the drive and the dynamic loads on the foundation being reduced by 10% to 20% (models $\boldsymbol{\alpha}^{116}$ and $\boldsymbol{\alpha}^{12}$ in Π_2). Second, it was shown that the reduction in the dynamic loads on the foundation may be accompanied by improvements in the rest of the criteria, if the machine mass is increased by less than 5% (model $\boldsymbol{\alpha}^{475}$ in Π_1).

These results have stimulated further improvements in the designs of resonant table vibrators used for moulding reinforced concrete products.

More Examples of Vibration Machine Optimization

Both the degree of perfection of vibration machines and their correspondence to the state of the art depend on how well they comply with numerous, often conflicting, requirements, such as small material consumption combined with high reliability and operability of the machine, small overall dimensions, high strength, operational stability, high efficiency, and ecological safety.

In Spivakov and Goncharevich (1983) the reader will find numerous interesting examples of multicriteria design of vibration-impact installations for handling ore, eccentric-drive vibration conveyers, double-screw crashers, and other vibration machines.

6-2. Truck Frame Design

As the major structure of a truck, a frame is subjected to the influences of both the road roughness and the units mounted on the truck itself. In turn, the properties of a frame strongly affect many significant characteristics of a truck, such as its controllability, smoothness of motion, vibration loads, stability, etc. Besides, the mass of a frame makes up a considerable portion of the overall mass of a truck.

A frame is designed subject to conflicting requirements: One has to decrease its mass, enhancing at the same time its strength and ensuring the specified level of a number of operational characteristics (Velikhov et al. 1986).

Here, the problem of designing an optimal truck frame is considered, formulated as follows: It is necessary to design a frame whose mass is smaller than that of the prototype, and the strength properties are improved as compared with the latter. Besides, the stiffness characteristics of the optimal frame must be close to those of the prototype whose dynamic properties are sufficiently high.

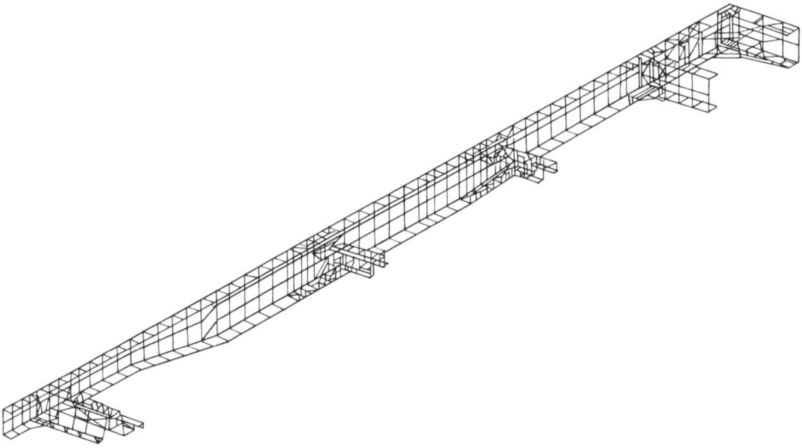

Figure 6–2 Finite element model of a truck frame.

Finite Element Model of a Truck Frame

Figure 6-2 shows a model composed of platelike elements possessing both the membrane and the flexural stiffness. By using these elements one can take into account the effects of stiffened torsion in the joints of a frame and in the zones where the cross pieces are fastened to the side rail in the most natural way, and analyze the stressed state of the structure under study in sufficient detail.

The adequacy of the model was confirmed by numerous bench and road tests. The calculated and experimental results were compared for the major loading modes resulting in torsion and bending in the vertical and horizontal planes. The loading modes were chosen taking into account the statistics of truck frame failures. The model allows estimating a stress-strained state taking into account the specific features of interaction of the frame's elements.[15] It proved to be highly efficient in determining the dynamic characteristics of a truck.

Optimization Criteria

In line with the objective of the study, seven criteria incorporating three pseudocriteria and four performance criteria were formulated. The torsional stiffness of a frame may be characterized by the overall twist angle ϕ. Since we are exploring the departures of the stiffness design variables of a frame subjected

[15]Later we had an opportunity to analyze a simplified model composed of beams. The comparison of the results calculated using the two models (as well as the comparison with experimental data) has shown that the beam model yields adequate results regarding displacements and stress distributions. The use of the beam model has allowed considerable reduction in time needed for conducting optimization calculation.

to optimization, from those of the prototype frame, the first pseudocriterion was represented in the form

$$\Phi_1(\boldsymbol{\alpha})=(\phi_i-\phi_p)\cdot\frac{100\%}{\phi_p}$$

where ϕ_i is the ith test twist angle (in the case of the PSI method $i=1,\ldots,N$), and ϕ_p is the prototype frame twist angle. The vertical-plane bending stiffness is characterized by the pseudocriterion

$$\Phi_2(\boldsymbol{\alpha})=(f^i_{max,\,V}-f^p_{max,\,V})\cdot\frac{100\%}{f^p_{max,\,V}}$$

where $f^i_{max,\,V}$ and $f^p_{max,\,V}$ are the maximum vertical-plane deflections for the ith test and the prototype frame, respectively. The horizontal-plane flexural rigidity of a frame was taken into account using the pseudocriterion

$$\Phi_3(\boldsymbol{\alpha})=(f^i_{max,\,H}-f^p_{max,\,H})\cdot\frac{100\%}{f^p_{max,\,H}}$$

where $f^i_{max,\,H}$ and $f^p_{max,\,H}$ are the maximum horizontal-plane deflections for the ith test and the prototype frame respectively.

The performance criterion $\Phi_4(\boldsymbol{\alpha})$ representing the side rail mass is defined as a sum of the web and the upper and lower flanges masses; and $\Phi_5(\boldsymbol{\alpha})$ is the sheet thickness. The latter criterion is also a design variable. Besides,

$$\Phi_6(\boldsymbol{\alpha})=(\max\sigma_{\text{tor}})\,\frac{\phi_{5^\circ}}{\phi_i}\quad\text{and}\quad\Phi_7(\boldsymbol{\alpha})=\max\sigma_{Hb}$$

where $\max\sigma_{\text{tor}}$ and $\max\sigma_{hb}$ are the maximum torsional and horizontal-bending stresses, respectively. The torsional stress was normalized by the twist angle 5°, which is known to be the average twist angle in moving over a road. This was done by introducing the coefficient ϕ_{5°/ϕ_i where $\phi_{5^\circ}=5^\circ$.

Design Variables

For solving the formulated problem, 21 design variables were chosen defining both the side rail geometry (see Fig. 6-3) and the cross piece rigidities (Table 6-4). Of special importance is thickness D of a sheet used for manufacturing the side rail. This design variable determines the latter's mass as well as rigidity characteristics and stresses. The geometry of a side rail is defined by a set of design variables (see Fig. 6-3). By H, B, and L we denote the height, width, and length of the side rail portions, respectively. The frame stiffness characteristics display a marked dependence on the torsional and vertical-bending stiffnesses

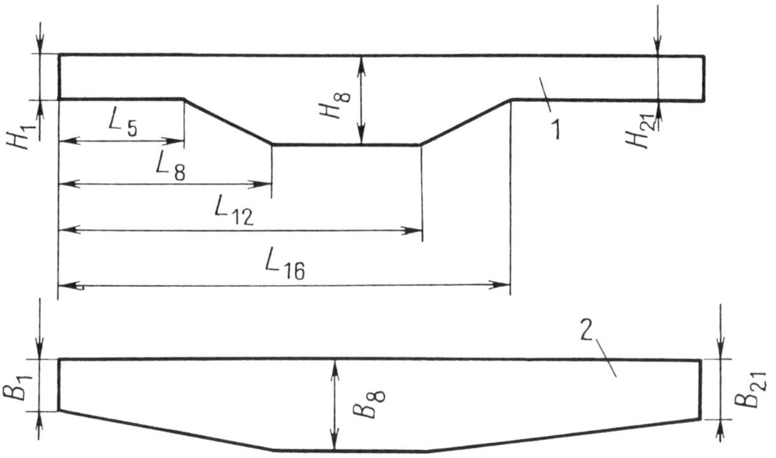

Figure 6–3 Side rail geometry (1 is a web; 2 is a flange).

of cross pieces. Therefore, the latter were included into the set of design variables: IT_i and IW_i are cross-section moments of inertia corresponding to the torsional and the horizontal-plane stiffnesses of cross pieces, respectively. Here $i=1,\ldots,5$, since the frame under consideration incorporates five cross pieces.

The boundaries of the design variables (the Π_1 parallelepiped) were chosen taking the design and technological potential into account.

Analysis in Parallelepiped Π_1

The number of trials was $N=1{,}024$. The large number of designer-computer dialogues has allowed detailed analysis of the results obtained for different criteria constraints, as well as detection of the design variations leading to a reduction in the mass and improvements in the strength properties of the truck frame. The results of the dialogues were used for compiling a test table representing the most noteworthy models. Table 6-5 is a fragment of the latter. The first row in Table 6-5 corresponds to the prototype, the values of its performance criteria being given. The first column presents the 15 best models: 836, 596,..., 824. The next three columns are occupied by pseudocriteria Φ_1, Φ_2, and Φ_3 (the "+" and "−" are the signs of deviations from the values corresponding to the prototype frame). Finally, the performance criteria Φ_4–Φ_7 are presented. We see that for the prototype frame the sheet thickness is $\Phi_5=6.35$ mm, and the side rail mass $\Phi_4=104$ kg. In Tables 6-5 and 6-6 Φ_5 corresponds to the rounded values of the sheet thickness.

Since in constructing the Pareto optimal set pseudocriteria were ignored, each dialogue resulted in construction of a feasible solutions set and determination of the Pareto optimal models in the criteria Φ_4–Φ_7.

Table 6–4

Design variables numbers	Design variables	Lower boundary	Upper boundary	Prototype	Optimal model
1	D (mm)	5	10	6.35	5.5
2	H_1 (cm)	11.4	15.4	13.4	14
3	B_1 (cm)	5.4	7.4	5.7	5.6
4	L_8 (cm)	185.6	205.6	185.6	196.5
5	H_8 (cm)	22.5	26.5	24	24.6
6	B_8 (cm)	5.4	9.4	7.7	7.9
7	L_{12} (cm)	354.1	384.1	374.1	362.5
8	H_{21} (cm)	15.4	19.4	17.4	16.7
9	B_{21} (cm)	5.4	7.4	5.7	6.8
10	L_5 (cm)	100.6	120.6	110.6	106.4
11	L_{16} (cm)	454.6	494.6	474.6	459.8
12	IW_1 (cm^4)	500	1,500	1,000	1,112
13	IT_1 (cm^4)	50	200	113	113
14	IW_2 (cm^4)	50	200	110	111
15	IT_2 (cm^4)	10	100	20.5	13
16	IW_3 (cm^4)	50	200	110	145
17	IT_3 (cm^4)	10	100	20.5	34
18	IW_4 (cm^4)	50	200	85.1	128
19	IT_4 (cm^4)	10	100	25.5	40
20	IW_5 (cm^4)	300	600	420	343
21	IT_5 (cm^4)	100	300	180	256

Table 6–5

Models \ Criteria	Φ_1 (%)	Φ_2 (%)	Φ_3 (%)	Φ_4 (kg)	Φ_5 (mm)	Φ_6 (kgf/cm^2)	Φ_7 (kgf/cm^2)
Prototype	—	—	—	104	6.35	1,000	2,220
836	+7.9	+6.8	−7.1	92.7	5.68	875	2,217
596	−8.8	+4.8	+1.5	97.5	5.83	958	1,928
716	−9.65	+7.3	−7.4	99.2	6.0	850	1,984
356	−3.42	+6.0	−9.7	94.9	5.75	817	2,356
504	−6.0	+6.0	−8.1	91.3	5.62	947	2,299
924	−12.9	+9.6	−6.7	100.8	6.13	856	2,089
436	−14.6	−0.8	−10.18	98.9	5.89	971	2,275
628	−14.5	+15.4	−11.94	93.9	5.9	866	2,314
708	+14.5	+15.3	+0.6	93.1	5.69	799	1,927
684	−17.9	+11.3	−15.7	96.2	6.04	932	2,354
56	+6.5	+6.7	−1.5	92.5	5.55	879	2,084
980	−7.82	+15.6	+18	94.1	5.85	916	2,036
20	−16.44	+3.5	−6.26	95.3	5.78	894	2,125
564	−16.23	+6.4	−19.7	96.9	5.86	910	2,381
824	−17.3	+6.1	−8.55	96.7	5.56	946	2,011

Table 6–6

D, mm	Models	Φ_1 (%)	Φ_2 (%)	Φ_3 (%)	Φ_4 (kg)	Φ_5 (mm)	Φ_6 (kgf/cm^2)	Φ_7 (kgf/cm^2)
5.5	244	−8.9	+5.2	−8.2	95.5	5.53	936	1,814
	168	−3.5	+2.7	+4	90	5.46	934	2,046
	176	−4.4	+9.32	−7.5	94	5.44	897	1,824
	20	−1.8	+4.7	+11	92	5.51	883	2,095
	56	+6.48	+7.16	+3.03	91.2	5.48	925	2,067
5.6	74	−6	+3.64	−4	95.3	5.62	933	1,817
	2	−5.7	+8.4	−6.9	94.7	5.57	903	1,947
	252	−4.6	+7.6	−0.7	92.4	5.57	932	1,955
	266	+3.2	+1.7	+7.5	93.4	5.59	870	2,018
5.7	62	−9.7	+8.2	+2.7	94.6	5.74	950	2,071
	90	−8.4	−0.5	−2.5	94.8	5.65	992	1,912
	142	−5.6	−8.2	−4.6	95.3	5.71	858	1,823
	238	−5.4	+7.7	−1.6	95.1	5.73	927	2,043
	166	−3.7	+5.4	+5.8	93.5	5.68	912	2,080
5.8	89	−9.6	+6.7	−13.4	98.6	5.82	886	1,698
	1	−9	+4.9	−1.3	95.9	5.75	886	2,009
	129	−7.3	+8.7	−14.44	97.7	5.75	892	1,945
	97	+5.03	+2.46	+1.2	97.1	5.77	823	1,969
5.9	13	−8.5	+8.67	−8	99	5.88	880	1,785
	67	−8.8	+6.41	−5.5	98.1	5.93	930	1,920
	37	−4	−1.5	+3.9	97.1	5.85	885	2,050
	53	−4	+7.9	+9.54	95.7	5.87	828	2,070
	45	−3	−0.85	+6.08	97.05	5.89	904	2,086
	195	+0.41	+3.52	+4.58	96.7	5.93	821	2,095
6.0	7	−6.5	3.84	−3.3	98.9	6.01	872	2,012

The first designer-computer dialogue

For the pseudocriteria accounting for torsional and vertical and horizontal bending deformations, the deviations from the prototype values Φ_ν^p were limited by 10%:

$$\Phi_\nu(\boldsymbol{\alpha})\leq 1.1\Phi_\nu^p,\quad \nu=1,2,3.$$

This allowed ensuring proximity of the optimal design and the prototype stiffness characteristics.

In what follows we consider the models whose mass is smaller than that of the prototype (104 kg).

The feasible solutions set contained 11 models, 10 of which were Pareto optimal.

The first dialogue resulted in selection of the five best models, 356, 504, 836,

596, and 716, whose masses are considerably less, and the strength characteristics better, than those of the prototype. However, the constraints on the frame stiffness characteristics introduced previously, are rather stringent. Making them somewhat looser may result in the appearance of additional, no less interesting, solutions.

Subsequently five more designer-computer dialogues were conducted, in which the constraints on $\Phi_1(\boldsymbol{\alpha})$, $\Phi_2(\boldsymbol{\alpha})$, and $\Phi_3(\boldsymbol{\alpha})$ were weakened by as much as 20%, and the torsional and horizontal-bending stresses did not exceed 1,200 kgf/cm^2 and 2,300 kgf/cm^2, respectively. As a result, the feasible solutions set was extended.

Conclusions Concerning the Results of Analyzing Parallelepiped Π_1

The main results are presented in Table 6-5 and have been analyzed from the viewpoints of the criteria and the design variables. Preference was given to models 836, 596, 716, 504, and 56, the first four of which proved to be the best in almost all the dialogues. Model 56 is the only one ensuring comparatively small stresses for a low side-rail mass.

For the remaining models (836, 596, 716, and 504) the sheet is thicker. Besides model 684, the thickest sheet is used in model 716. Naturally, its advantage in the mass, as compared with the prototype, is minimal.

The analysis of the best feasible models[16] has allowed construction of a new parallelepiped, Π_2, $\Pi_2 \subseteq \Pi_1$, in which 256 trials were conducted.

Analysis in Parallelepiped Π_2

Two dialogues were conducted. Since torsional stresses satisfied the formulated requirements for almost all the models, constraints were imposed on the horizontal-bending stresses.

In all the dialogues, the constraints were imposed that prevent the frame stiffness characteristics from exceeding by more than 10% those of the prototype. In order to achieve a more significant reduction in the mass, the side-rail mass was limited by 100 kg.

Conclusions Concerning the Results of Analyzing Parallelepiped Π_2

The search proved to be highly effective. Owing to the shrinkage of the search zone, the results were substantially improved. Of primary importance are the horizontal stresses reduction by 10% to 20% and an increase in the sheet thickness for a lower side-rail mass.

The results of the analysis allowed compiling a table of the best models, a

---

[16]Since the illustrative material pertaining to this chapter is extremely voluminous, only a small fragment of it is presented here.

fragment of which is presented in Table 6-6 where the models are arranged in the order of increasing sheet thickness D. As to this design variable the models were divided into six groups, with D=5.5 mm, 5.6 mm, 5.7 mm, 5.8 mm, 5.9 mm, and 6.0 mm (see the first column in Table 6-6). The second column contains the numbers of the models. Next follow pseudocriteria Φ_1, Φ_2, Φ_3, and criteria Φ_4–Φ_7.

It was found that the most promising models form three groups with the sheet thickness D=5.5 mm, 5.8 mm, and 6.0 mm. This circumstance has allowed choosing three models, 168, 1, and 7, subsequently subjected to further analysis. Compared to the prototype, the models are characterized by lower horizontal-bending stresses for acceptable side-rail masses and sheet thickness. The latter effect was attained by increasing the flange width in the middle and rear portions of the frame.

Model 168 was admitted to be the best one. Its design-variable vector as well as the prototype design variables are presented in Table 6-4.

The General Conclusion

The major performance criteria of the frame have been considerably improved. Among other things, the side-rail mass was reduced by 14 kg. The results of optimization have been confirmed by road tests.

Thus, the mass of the frame whose characteristics were being improved by traditional methods for 20 years, was reduced by 28 kg (Velikhov et al. 1986). Simultaneously, the stresses in the critical locations were reduced.

This example demonstrates the efficiency of multicriteria optimization in solving problems associated with mass production, as in the case of automotive industry.

6-3. Optimization of Metal-Cutting Machine Tools

Let us consider some aspects of searching for optimal solutions in designing metal-cutting machine tools.

Selection of the Optimal Design Variables of a Vertical Knee-Type Milling Machine

In Gorodetskii (1984) a closed dynamic model of the title machine tool is considered. The specifics of its general design make the machine prone to forced and self-sustained vibrations that limit its productivity and deteriorate the machining quality. This study is aimed at shortening the time needed for designing new millers characterized by improved vibration stability, machining accuracy, productivity, and other technological and economic indices.

Figure 6-4 shows schematically the structure of a vertical knee-type milling

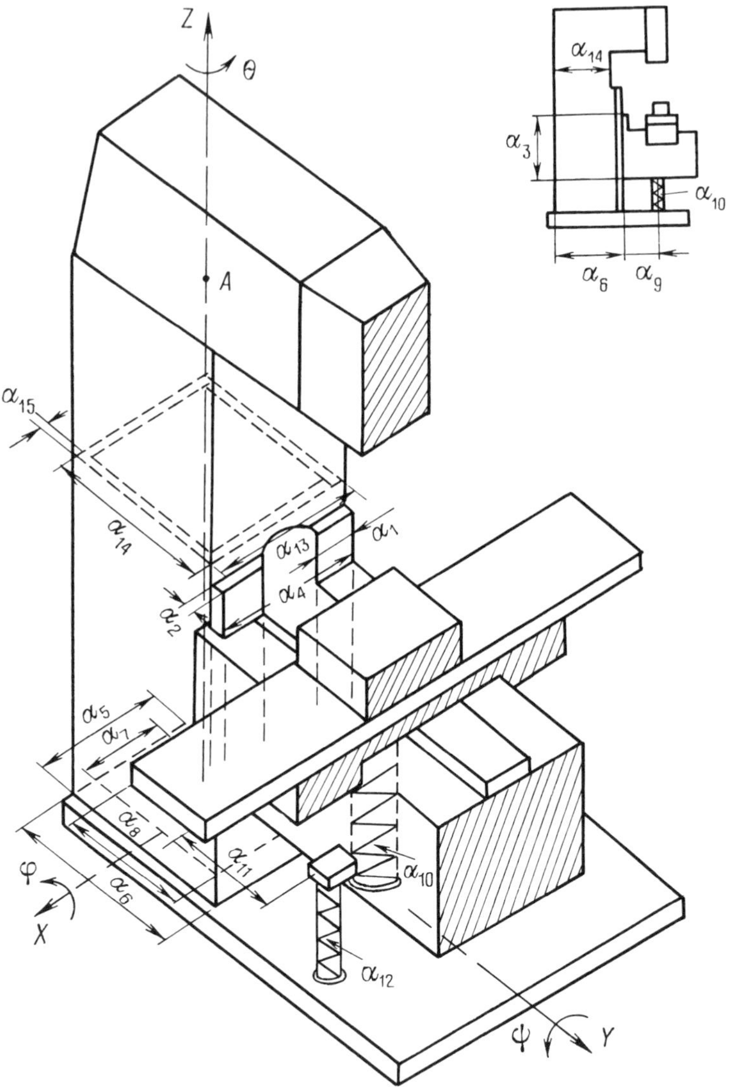

Figure 6–4 Schematic of the structure of a vertical knee-type milling machine.

machine. The knee, the slide, the table with a mounted part, and the swivel head with the upper portion of the column (whose end faces are dashed) were modeled by rigid (undeformable) bodies. Up to point A the column is considered a hollow rectangular beam. The flexural-torsional deformations of the column caused by low-frequency vibration are supposed to be qualitatively analogous to static deformations of the beam. This has allowed expressing the point A displacement via the beam's angular rotations and using the angular displacements as generalized coordinates. The column mass was reduced to point A.

The following factors were considered: the stiffnesses and the deformations in the column, the table and the slide drives, lifting mechanism, the knee supports, and the column base and the knee-column joints. The specifics of natural vibration in the latter two joints of the miller made it possible to take into account only the stiffness coefficients associated with angular displacements with respect to its axes.

In line with the adopted scheme of calculation the generalized mathematical model (the equations of motion) of the miller's closed dynamic system may be written in the following operator form:

$$\{[\mathbf{M}]p^2+[\mathbf{N}]p+[\mathbf{R}]+([\mathbf{\Gamma}^{(1)}_{(p)}]-e^{-p\tau}[\mathbf{\Gamma}^{(2)}_{(p)}])\}\,[\mathbf{\Lambda}]=[\tilde{\mathbf{M}}] \tag{6-5}$$

where $[\mathbf{M}]$, $[\mathbf{N}]$, and $[\mathbf{R}]$ are, respectively, the matrices of the inertial, dissipative, and stiffness coefficients of a miller; $[\mathbf{\Gamma}^{(1)}_{(p)}]$ and $[\mathbf{\Gamma}^{(2)}_{(p)}]$ are the matrices of averaged (over a period) dynamic characteristics (the transfer functions) of milling; $[\mathbf{\Lambda}]$ is the generalized coordinates vector; $[\tilde{\mathbf{M}}]$ is the vector of cutting force moments; $\tau=(nz)^{-1}$, n is the frequency of revolution; and z is the number of mill teeth.

The set of design variables included the dimensions of the equivalent cross sections of the column, the fixed column-base joint and the movable knee-column joint, as well as the stiffnesses of the lifting mechanism and the knee supports and the distances from the column face.

The quality of the machine tool was estimated using the following five criteria: the amplitudes of the mill-workpiece relative vibrations in the directions of the table, slide, and knee feeds, denoted by Φ_1, Φ_2, and Φ_3 respectively; the reciprocal of the limiting mill depth, Φ_4, and the metal consumption, Φ_5.

The algorithm of searching for the optimal solutions was based on the use of the PSI method and consecutive implementation of the following operations at each point of the design-variable space:

1. Calculation of natural frequencies ω_i with the help of the characteristic equation of an open-loop conservative system.

$$\det\{[\mathbf{M}]p^2+[\mathbf{R}]\}=0, \quad p=j\omega \tag{6-6}$$

2. Calculation of the natural modes of vibration $[\mathbf{\Lambda}_i]$ corresponding to natural frequencies ω_i, carried out using the conservative model ($[\mathbf{N}]=0$).
3. Calculation of forced vibration of the structure for each natural (resonance) frequency ω_i in the cutting zone, using the equation

$$\{[\mathbf{M}]p^2+[\mathbf{N}]p+[\mathbf{R}]\}\,[\mathbf{\Lambda}]=[\tilde{\mathbf{M}}], \quad p=j\omega \tag{6-7}$$

4. Calculation of the milling depth starting from which self-sustained vibration sets in (the depth is called limiting, and is calculated using Eq. (6-5))
5. Calculation of the metal consumption.

Next we present the results of optimization of a vertical knee-type milling machine, aimed at decreasing the latter's resonance amplitudes and increasing vibration stability (accompanied by a decrease in the metal consumption).

The analysis has been carried out for the case of symmetric longitudinal milling with a face milling cutter. The specimens of width $B=100$ mm were made of steel 45; the milling cutter frequency of revolution n equaled 160 rpm, and the feed $S_z=0.125$ mm per revolution. The problem was solved using mathematical model (6-5) of a closed dynamic system of a machine tool with a matrix of period-averaged dynamic characteristics of the milling process.

The set of design variables of the structure incorporated the dimensions of the movable knee-column joint, α_1, α_2, α_3, and α_4; the geometry of the fixed column-base joint, α_5, α_6, α_7, and α_8; the ordinate of the point of application of the knee feed mechanism force, α_9; the stiffness of the knee lifting mechanism, α_{10}; the knee supports ordinate α_{11}; the knee support stiffness α_{12}; and the geometry of the equivalent cross section of the column, α_{13}, α_{14}, and α_{15} (see Fig. 6-4).

The calculation of particular criteria Φ_1–Φ_5 at each trial point required a great deal of computer time. Therefore, first the spectrum of natural frequencies and the objective functions Φ_ν were analyzed.

Then the limiting milling depth was calculated for each natural frequency. The solution of the general problem was reduced to the analysis of the machine tool's dynamic quality over a limited set of potentially unstable modes of vibration.

Preliminary calculations have yielded the spectrum of natural frequencies and the limiting milling depths for the prototype machine tool over the range of frequencies from zero to 200 Hz. It was found that for 28 Hz, 87 Hz, and 147 Hz the machine tool had a substantial stability margin; however, for 58 Hz and 75 Hz milling became unstable for the depths of 10.8 mm and 3.3 mm, respectively.

The performance criteria were calculated using a system of modal equations derived by energy methods for the second and the third natural frequencies.

For the frequencies of 58 Hz and 75 Hz, 11 and 22 models were included into the feasible domain, respectively. The best models were found within the intersection of the optimal models sets obtained for the two frequencies. The values of their particular criteria are presented in Table 6-7.

Subsequent analysis has shown that the models correspond to substantially better values of the performance criteria as compared with the prototype. Thus,

Table 6–7

Models	Φ_1 (μm)	Φ_2 (μm)	Φ_3 (μm)	Φ_4 (m^{-1})	Φ_5 (kg)
17	6.74	23.4	16.7	39.6	3,340
26	15.5	61.8	42.8	84.3	3,330
59	18.7	19.3	13.7	60.7	3,380

for model 59 the limiting milling depth increased by approximately a factor of five, and the metal consumption decreased by 7%.

The analysis has allowed formulation of a number of general statements and concrete recommendations for improving vertical milling machines:

1. The oscillations were suppressed mainly owing to an increase in the flexural-torsional stiffness of the lower portion of the column and in the stiffnesses of the lifting mechanism, the knee support, and the column-base joint.
2. The metal consumption was reduced by optimizing the geometric variables of the basic units cross sections.
3. The dynamic quality indexes of vertical NC millers may be improved by passing from knee-type to compound-table milling machines.

Besides, the time needed for finding the optimal solution characterized by enhanced vibration stability, lower level of forced vibration, and lower metal consumption, has been considerably decreased.

Determination of Significant Design Variables in Optimizing the Structures of Lathes with Movable Workheads (Betin and Kaminskaya 1992)

Figure 6-5 shows the scheme used for calculation of the structure of a precision lathe. A finite element beam model of the machine tool was considered. The bed, spindle head, carriage, chuck, and pneumatic cylinder were assumed to be rigid bodies.

The scheme of the structure incorporates two subsystems: the carriage and the spindle head. By M_i we denote the ith inertial element possessing both mass and the moment of inertia. The joints and supports are denoted by K_i and J_i and are characterized by linear and angular stiffnesses of the corresponding structural elements of the machine tool. Within the framework of the finite element model of the machine tool under consideration the spindle is represented by beams.

Optimization criteria include the stresses in the guideways and the static and dynamic compliances.

Selection of significant design variables

As noted in Section 5-2, design variables are the variables that have the greatest effect on the performance criteria. The analysis allowed finding the distributions of kinetic and potential energies of vibration among the lathe elements for the natural frequencies equal to 181 Hz and 387 Hz (for the prototype). For the frequencies 181 Hz and 387 Hz the major kinetic energy component is determined by the spindle head (64%) and the chuck (68%) vibrations, respectively. For the frequency 181 Hz, 80% of the total potential energy correspond to the spindle head vibration (mainly the angular ones with respect to the axis Z), and only 6.5% to translational vibration of the carriage proceeding in the feed direction. For 387 Hz, 58% correspond to the chuck vibration. The contributions of the remaining elements into the kinetic and potential energies were insignificant.

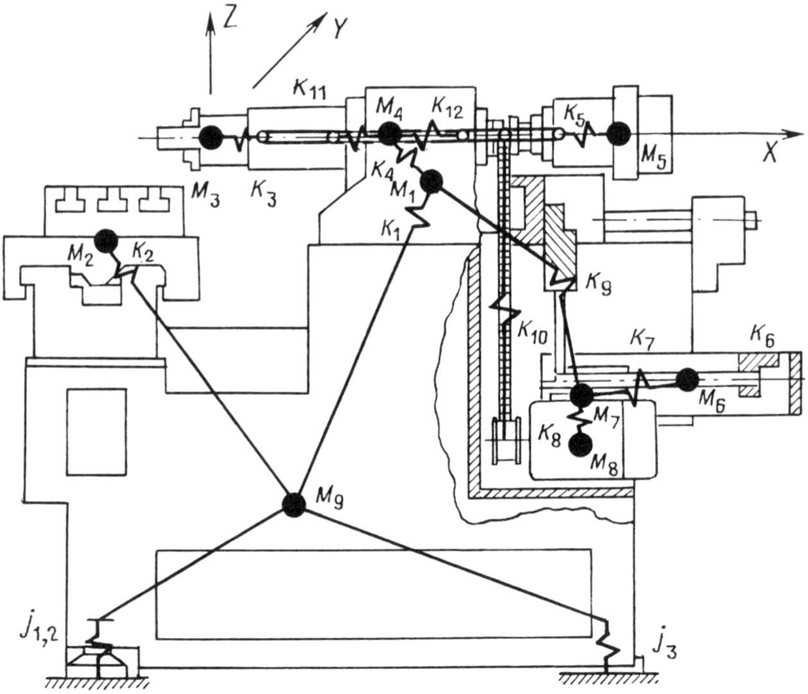

Figure 6–5 Scheme used for calculating the structure of a precision lathe. The elements and notation. Inertial elements  : M_1 is the spindle head; M_2 the carriage; M_3 the chuck; M_4 the spindle housing; M_5 the pneumatic cylinder; M_6 the bracket; M_7 the motor stator; M_8 the motor rotor; M_9 the bed. Stiffness elements  : k_1 refers to the spindle head-bed pair; k_2 to carriage-bed; k_3 to chuck-spindle; k_4 to spindle head–spindle housing; k_5 to spindle-pneumatic cylinder; k_6 to bracket-bed; k_7 to stator-bracket; k_8 to stator-rotor; k_9 to stator–spindle head; k_{10} to the belt transmission; k_{11} and k_{12} to the spindle bearings.  j_1, j_2, and j_3 are the machine tool supports. Finite element beam model: 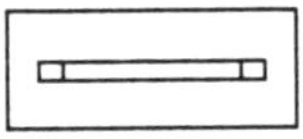 is the spindle.

Therefore, in solving the optimization problem the parameters related to these elements were not varied. Thus, it is clear that for the frequency 387 Hz the dynamic characteristics of the system may be primarily changed by varying the chuck parameters. Thus, in optimizing the design variables of the structure attention must be concentrated on improving the compliance determined by the spindle and carriage parameters. The analysis of the data on the kinetic and potential energy distribution, as well as of the shape of the structure's vibration at the frequency of 181 Hz, shows that the level of relative vibration within the cutting zone depends on the magnitude and direction of the spindle head angular vibratory displacements and translational vibratory displacements of the carriage.

In line with what was said previously, the following nine parameters were varied: the mass of the spindle head and carriage, the stiffness of the carriage feed drive, and the geometric variables of the main faces of the spindle guideways. The latter geometric variables determine the stiffness of the spindle head guideways.

The results of optimization (following $N = 1{,}024$ trials) are presented in Table 6-8 where the values of dynamic compliances are presented only for natural frequencies. For the model 70 and the prototype the natural frequencies lie outside the range of frequencies 270 Hz–350 Hz.

Let us summarize the major results of the study.

1. As regards the dynamic compliance, models 484 and 627 are the best ones over the entire frequency range. A still higher dynamic compliance of model 70 is explained by the proximity of the partial frequencies of the head and the carriage, while for model 302 it is explained by a low stiffness of the carriage feed drive.
2. For the most preferred models, 484 and 627, the design variable values are such that for the lower and the upper natural frequencies the vibrations of the carriage and the head dominate, respectively.

Table 6–8

	Performance criteria			
Models	Pressure in guideways (10^{-5} kgf/cm^2)	Static compliance (10^{-4} mm/N)	Dynamic compliance over the frequency range (10^{-3} mm/N) [170–240] Hz	[240–350] Hz
70	33,000	0.69	0.185	—
138	45,800	0.68	0.109	0.219
302	41,500	0.70	0.115	0.175
484	43,400	0.68	0.161	0.139
522	54,000	0.67	0.127	0.307
627	50,200	0.68	0.159	0.122
Prototype	77,900	0.75	0.212	—

The analysis has revealed a strong dependence of the dynamic compliance within the cutting zone on the relationship between the partial frequencies of the spindle head, the carriage, and the chuck vibrations. For the best solutions the frequencies differ by 30%–35%.

These results allow formulation of two general recommendations for decreasing the dynamic compliance of the structure:

1. The stiffnesses of the head guideways and the carriage feed drive must be made sufficiently large (within the fixed overall dimensions)
2. The inertial characteristics of the elements must ensure the corresponding relationships between the partial frequencies for the chosen values of the elements stiffnesses.

Other Studies in Machine Tools Optimization

In Sections 4-3 and 5-3 we discussed some studies aimed at optimization of slotting and grinding machines. Here we present some more interesting studies in the area, based on the PSI method.

Design of the structure of a multipurpose single-column vertical boring and turning machine

It was necessary to design a machine tool surpassing the prototype in the basic performance criteria. The two specific features of the problem are the high dimensionality of the design-variable vector and calculations taking a great deal of computer time. The latter was reduced by using multicriteria identification. The cross section of the machine's column had a complicated configuration and was described by a large number of design variables. In order to reduce the latter the column cross section was simplified, and the adequacy of the real and the simplified column cross sections in the basic geometric characteristics was demonstrated. The latter included the cross-sectional area of the column, its moments of inertia, and torsional stiffness. The adequacy was proved using the method of multicriteria identification discussed in Chapter 4. The differences in the geometry of the prototype and the column model with the simplified cross-section of the column were subjected to minimization. Minimization was implemented by varying the parameters of the cross section. As a result, the adequate vectors were found, the dimensionality of which was much smaller than that of the prototype. Subsequently, this allowed solving the problem of optimization of the machine tool's structure in an acceptable time. The following four criteria were optimized: vibration activity within the workpiece machining zone, vibration stability (the Nyquist criterion), static compliance, and the metal consumption per column (Kaminskaya 1984). By solving the problem the latter was reduced by 8%; the dynamic characteristics of the machine tool were improved by 12% compared to the prototype.

In Khomyakov and Yatskov (1984) the design variables of the structure of a heavy-duty single-column vertical boring and tuning machine were optimized with respect to the mass of the structure and the latter's total compliance, both of which are highly important for ensuring accuracy of workpiece machining. As a result, the stiffness of the structure increased in the direction normal to the surface of a large part being turned, and thus the turning accuracy and productivity were improved.

Work (Zinyukov et al. 1983) deals with optimizing the drives of various pipe-cutting machines for which 30 parameters were varied and five objective functions were optimized, which characterized the machine's vibration activity, mass, and dynamic loads acting on the spindle unit bearings and the foundation. As a result, the quality of the drives was improved in the basic criteria. Realization of the results has allowed designing compact machine tools characterized by smaller metal consumption and lower loads; simultaneously, the productivity of the machine tools increased.

Also, we would like to mention works (Debagyan and Khomyakov 1982) dealing with the general methodology of multicriteria optimization of machine tools.

6-4. Some Other Problems

Gear Unit Design

Since gear units are manufactured in large quantities, optimization of their design may result in a considerable economy. A continuous growth in operational loads and speeds of machines and mechanisms is to be accompanied by decreasing the overall dimensions and masses of their reduction gears, as well as by an increase in the load-carrying capacity, reliability, and durability of gearings.

The quality of a gear unit is primarily determined by the gearings. The characteristics of all the other components (shafts, bearings, body, joints, etc.) are defined by the dimensions and arrangement of the gearings and the loads acting on them.

Multicriteria optimization of the two-stage gear unit of a roller table rollers drive

The device (see Fig. 6-6) has been manufactured for some years already. Its operational output-shaft torque is T=4,905 N·m. The gear unit is reversible and is driven by a motor whose power is 22 kW and the rotational frequency $n_1=650$ rpm. The nominal gear ratio $u_0=9.92$ may vary by $\pm 1\%$.

In Fig. 6-6 superscripts 1 and 2 correspond to the first (input) and the second (output) stage, respectively; a_w^1 and a_w^2 are the center distances; β^1 and β^2 are the helix angles, b_w^1 and b_w^2 are the gears' working face widths; z_1^1 and z_1^2 are the numbers of pinion teeth; and z_2^1 and z_2^2 are the numbers of wheel teeth.

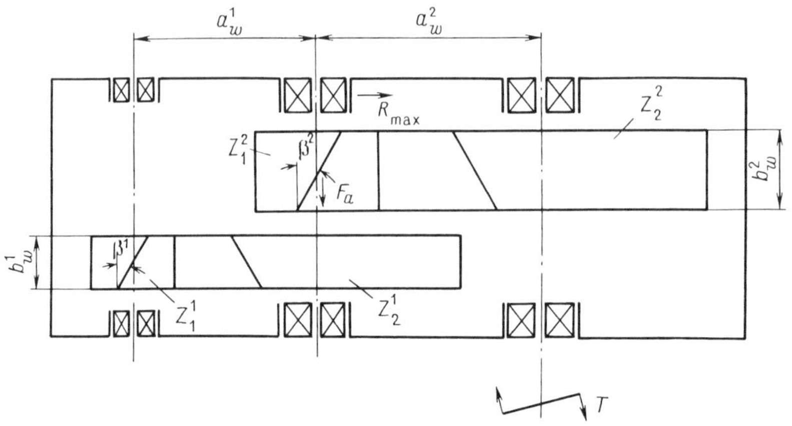

Figure 6–6 Schematic of a two-stage gear unit.

Problem formulation and solution

It is necessary to improve the prototype gear unit in the basic performance criteria.

Performance criteria. The volume Φ_1 of a gear unit determines the quantity of material needed for manufacturing it. Since the load-carrying capacity of a gear unit is often limited by the bearing units, the maximum loads on the supports are to be taken into account. This is done with the help of the criteria $\Phi_2=F_a^{(2)}$ and $\Phi_3=R_{max}^{(2)}$ representing the maximum axial load and the maximum reaction of the supports (in the case under consideration, on the second intermediate shaft).

The performance criteria of the first stage were represented by the contact endurance safety factors $\Phi_4^1=S_{H1}^1$ and $\overline{\Phi}_4^1=S_{H2}^1$ of the gear and the wheel, bending endurance safety factors $\Phi_5^1=S_{F1}^1$ and $\overline{\Phi}_5^1=S_{F2}^1$ of the gear and the wheel, the center distance $\Phi_6^1=a_w^1$, the gear working face width $\Phi_7^1=b_w^1$, the number of gear teeth $\Phi_8^1=z_1^1$, transmission ratio $\Phi_9^1=u^1$, and the wheel addendum circle diameter $\Phi_{10}^1=d_{a1}^1$.

Criteria Φ_4^2 through Φ_{10}^2 refer to the second stage of the gear unit. Table 6-9 presents the values of the most important criteria.

Functional constraints. The ratios of the center distances a_w^{i+1}/a_w^i and modules m^{i+1}/m^i of the ith and $(i+1)$th stages must satisfy the inequalities

$$c_{1,i}^{*}\leq\frac{a_w^{i+1}}{a_w^i}\leq c_{1,i}^{**},\quad c_{2,i}^{*}\leq\frac{m^{i+1}}{m^i}\leq c_{2,i}^{**}.$$

Table 6–9

	Performance criteria															
	Gear box			Stage 1						Stage 2						
Models	Φ_1 $V \cdot 10^{-7}$, (mm^3)	Φ_2 $F_a^{(2)}$, (N)	Φ_3 $R_{max}^{(2)}$, (N)	Φ_4^1 $S_{H_1}^1$	Φ_5^1 $S_{F_1}^1$	Φ_6^1 a_w^1, (mm)	Φ_7^1 b_w^1, (mm)	Φ_8^1 z_1^1	Φ_9^1 u^1	Φ_4^2 $S_{H_1}^2$	Φ_5^2 $S_{F_1}^2$	Φ_6^2 a_w^2, (mm)	Φ_7^2 b_w^2, (mm)	Φ_8^2 z_1^2	Φ_9^2 u^2	Φ_{10}^2 $d_{a_2}^2$, (mm)
Prototype	12.1	2,150	17,681	1.16	2.88	224	90	46	2.80	1.20	3.08	355	145	38	3.58	560
1,766	9.9	1,300	20,091	1.23	3.55	212	104	37	3.43	1.26	3.23	325	139	47	2.87	489

The requirements of equal contact and bending strength of interacting gears are represented by functional constraints of the form

$$c^*_{3,i} \leq \frac{m^i}{a^i_w} \leq c^{**}_{3,i}.$$

The values of $c^*_{j,i}$ and $c^{**}_{j,i}$, $j=1, 2, 3$, depend on the thermal treatment, gear hardness, and the stage transmission ratio.

Some other functional constraints, such as the conditions of teeth undercutting and interference absence, were also taken into account.

The following parameters were varied for the first and second stages of a gear unit: the center distances a^1_w and a^2_w, the relative face widths ψ^1_{bd} and ψ^2_{bd}, modules m^1 and m^2, and the helix angles β^1 and β^2. Also the transmission ratio of the first stage was varied. Characteristically, some of the determining variables vary continuously, while others (the modules and the numbers of teeth) vary in a discrete manner. Since the overall transmission ratio was specified, the last-stage transmission ratio was a dependent design variable. Some of the design variables played the role of criteria at the same time. The materials and thermal treatment were assumed to be the same for the prototype and the newly designed model.

Upon implementing the optimization calculations (4,096 trials, several seconds of computer time each) the initial boundaries of the design variables were corrected in line with Section 1-3.

It was found that the functional constraints are satisfied for 116 solutions. As a result, a feasible solutions set consisting of six solutions, was constructed. Tables 6-9 and 6-10 present the values of the performance criteria and design variables for the prototype and the most preferred model 1,766. The comparison of the latter two shows that:

1. The volume of the gear unit was reduced by 18%. Also, the contact endurance safety factors of the optimal solution were improved compared with the prototype.

Table 6–10

	Design variables								
	Stage 1					Stage 2			
	α_1	α_2	α_3	α_4	α_5	α_6	α_7	α_8	α_9
Models	a^1_w (mm)	ψ^1_{bd}	m^1 (mm)	β^1 (deg)	u^1	a^2_w (mm)	ψ^2_{bd}	m^2 (mm)	β^2 (deg)
Prototype	224	0.73	2.5	12°25′46″	2.80	355	0.95	4.0	11°23′49″
1,766	212	1.08	2.5	14°45′53″	3.43	325	0.83	3.5	11°28′42″

2. The second-stage wheel addendum circle diameter became substantially smaller (489 mm and 560 mm for the optimal solution and the prototype, respectively). This simplifies both the production technology and thermal treatment of the gear wheel.

The prototype was primarily improved owing to an increase in the transmission ratio and the relative width of the first stage.

This technique of multicriteria optimization may be used for any kind of multistage gear unit, gear boxes included.

Some other examples of gear units design. Multicriteria optimization of the design variables of machine tool kinematic chains composed of cylindrical gearings is discussed in Pluzhnikov (1983). It was necessary to determine the gearing design variables ensuring the best output accuracy (the minimal kinematic error), the minimal cost, mass, and overall volume of the mechanism. It was shown that the prototype may be considerably improved.

Multicriteria optimization of a gear unit is considered in Grinkevich et al. (1978). The study was aimed at improving vibroacoustic characteristics of submarines. The following were subjected to minimization: the amplitudes of displacements and accelerations of gear wheel bearings, the amplitudes of dynamic forces exerted by the bearings on the foundation, the mass of the parts, and the number of natural frequencies within the operational frequencies range 5–300 Hz. As compared with the prototype, the values of all the criteria were improved by 40–200%.

Optimization of Trunk Shakers (Chernikov 1986)

Trunk shakers are widely used in mechanized harvesting of fruit and nuts intended for industrial processing. The technological process involves gripping and squeezing the lower part of a tree trunk (or a branch) with shaker pads, and shaking it to remove the fruit.

Trunk shakers often incorporate linear inertial actuators that make a trunk vibrate owing to an oscillatory motion of a special mass in the direction perpendicular to the trunk. In what follows, the shaker is called a vibrator, since the latter term is widely used.

The kinematic, dynamic, and energy calculations were aimed at choosing the actuators' design variables guaranteeing the required completeness of the fruit removal, the minimal damage to a trunk, and minimal mass and energy characteristics.

The dynamic model of the "tree-vibrator" system takes into account the effects of the characteristics of a tree, the gripper pads, and the vibrator (Fig. 6-7).

In describing the model the following nomenclature was used: c and n are the coefficients characterizing the elastic and damping properties of a tree at the grip point; c_1 and n_1 are the analogous characteristics of the gripper pads; m_e is the

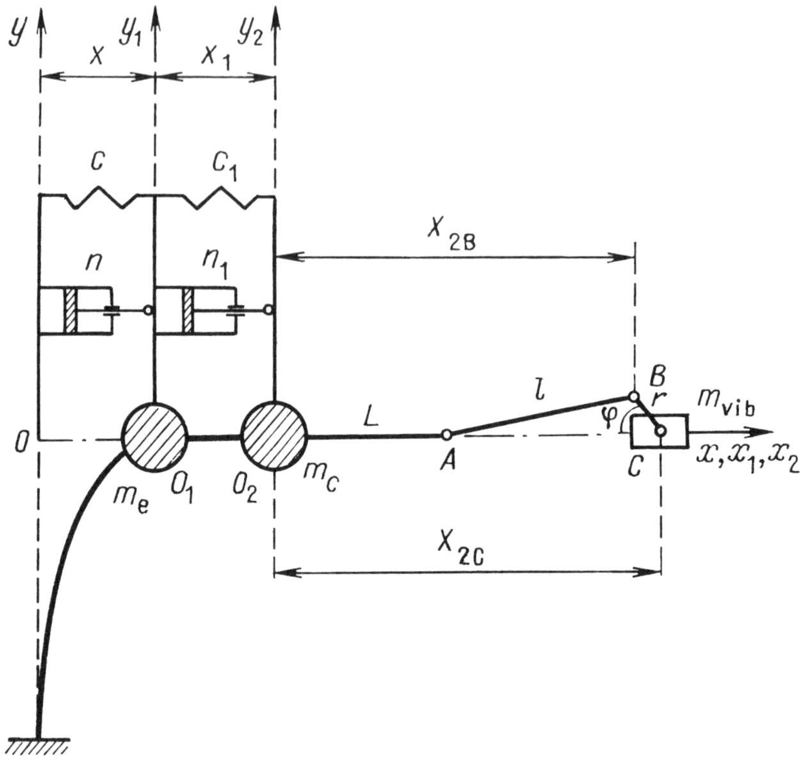

Figure 6–7 Dynamic model of the "tree-inertial virator" system.

equivalent mass of a tree reduced to the grip point; m_c is the mass of the vibrator part connected with the tree by means of the gripper; L is the distance from the plane where the pads are mounted to the connecting rod pin; l is the connecting rod length; r is the crank radius, m_{vib} is the mass of the movable parts of the vibrator; O_1 and O_2 are the origins of the coordinate systems $X_1O_1Y_1$ and $X_2O_2Y_2$, respectively; X_{2B} and X_{2C} are the coordinates of points B and C in coordinate system $X_2O_2Y_2$; X is the trunk deflection; and X_1 is the gripper pad deformation. Below, a crank vibrator is considered.

The mathematical model of the vibrator was described by the fourth-order system of differential equations. Upon solving the system of equations the necessary kinematic, dynamic, and energy calculations were carried out. The calculations yielded the amplitudes and phases of the trunk and pad displacements; forces Q_1 and Q_2 applied to the trunk and the mass m_{vib} of the movable parts of the vibrator, respectively; the energy spent during a period of vibration; the instantaneous energy-source power; mean power N_{m}; etc.

The parameters m_0, m_{vib}, n_1, r, l, ω, and c_1 were varied. Here m_0 is the total vibrator mass, and ω is the crank rotational frequency.

Functional constraints:

$$b_0^* \leq |X(t)| \leq b_0^{**},$$
$$|Q_1(\mathrm{t})| \leq Q_1^{**},$$
$$X_1(t) \leq b_1^{**}$$

where $X(t)$ is the trunk displacement at the grip point; $Q_1(t)$ is the force exerted on the trunk by the vibrator; and $X_1(t)$ is the gripper pad deformation.

The first constraint ensures the correspondence of the trunk deflection amplitude to the technological requirements, the second one prevents damaging the trunk bark, and the third constraint determines the limiting deformation of the gripper pads. The third functional dependence, $X_1(t)$, was converted into a pseudocriterion. The following are performance criteria to be minimized: the mass of the moving parts of a vibrator, m_{vib}; the force exerted on the crank, $Q_2(t)$; mean power, N_m; and the total vibrator mass, m_0.

The analysis has allowed finding the optimal model for which the total vibrator mass is 44% smaller than that of the prototype (whose mass is 140 kg (Chernikov 1986)). Simultaneously, the constraints imposed on the trunk deflection amplitude were satisfied and so were the constraints on the magnitude of the forces acting on a trunk. All other criteria were also improved.

Flexible Manufacturing Systems Design (Portman et al. 1992)

By definition a flexible manufacturing system (FMS) is a set of numerically controlled technological equipment and the systems for supporting the latter's automatic operation during a given time interval. Within the technological possibilities of its equipment an FMS may be readjusted for manufacturing diverse products.

Besides being highly automated, an FMS must allow rapid readjustment in response to variations in the production program.

The FMS for the electric discharge machining (EDM) of forming machinery parts at a die-making plant incorporates 23 EDM machines, a pallet loading station, and a stacker putting loaded pallets onto a transfer station positioned near a group of machine tools serviced by one mobile robot. The latter lifts the pallets from transfer stations and puts them onto the input tables of machine tools. Also a tunnel washer is installed to clean and dry the processed parts.

The route production process includes the following operations:

- The loading of blanks onto pallets at the loading station
- EDM
- Washing.

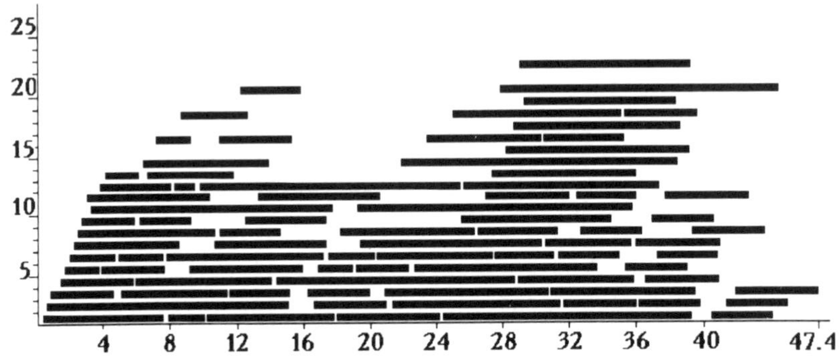

Figure 6–8 The cyclogram of the machine operation (the prototype).

In turn, the transporting-loading operations include:

- The transfer of a pallet from the loading station to storage
- The transfer of a pallet from a machine tool/washer to storage
- The transfer of a pallet from a machine tool to the washer (if the two machines are serviced by the same robot).

The processing of 84 batches of parts (an average fortnight production volume) was simulated moving through the technological and auxiliary equipment. Simulation of the prototype workshop has shown that (1) The time of processing of a complete set of parts was 47.4 h; and (2) the average machine-tool utilization factor equaled 0.5.

The general pattern of the workshop operation is clear from the cyclogram shown in Figure 6-8. This figure presents the time of the FMS workshop operation (shown on the abscissa axes), and the number of EDM machines (given on the ordinate axes). The time intervals during which the machine tools operate are blackened, blanks indicate machine down time. The simulation has shown that the machine tool down time was mainly caused by an inadequate number of loading stations and the operators servicing them, as well as by a limited central storage capacity (the number of cells).

In order to improve the major performance criteria of the prototype the problem of multicriteria optimization of the FMS parameters was formulated. The criteria to be optimized are listed in Table 6-11. These criteria determine the technological and economical efficiency of the FMS under consideration.

The following FMS parameters were varied: the number of electric discharge machines, the number of loading stations, the storage capacity, the average duration of the storage stacker cycle, the number of servicemen, and the form of the personnel work organization (the individual or team one).

Table 6–11

Criteria	Prototype	Optimal model
Duration of the production program execution, hours	47.4	34.0
Duty factor of EDM machines	0.5	0.67
Maximum loading of the storage	30	60
Stacker duty factor	0.19	0.17
Duty factor of an operator serving the loading station	0.44	0.47
Number of EDM machines	23	24
Number of servicemen	3	4
Workpiece storage capacity	30	60

The storage stacker working cycle includes the transfer of parts from a place of reloading to storage cells, the loading of containers, and the transfer from a cell to the point where the containers are reloaded and emptied.

The optimal solution criteria are compared with those of the prototype in Table 6-11. We see that though some of the former are worse than the latter (for instance, the number of servicemen and the number of EDM machines), the major criteria have been improved. The productivity of the workshop has increased by a factor of 1.4; also improved was the machine-tool utilization rate.

From the optimal machine tool operation cyclogram shown in Figure 6-9, it follows that the down time was markedly reduced compared with the prototype (Fig. 6-8). Figure 6-10 shows the current number of storage cells (K) loaded with blanks and semifinished articles during the workshop operation period (h). We see that for the optimal solution the time during which the storage is fully loaded, is considerably shorter than for the prototype.

Thus, the use of the PSI method has resulted both in a substantial improvement in the FMS design and a 12-fold reduction in the time needed to find the optimal solution.

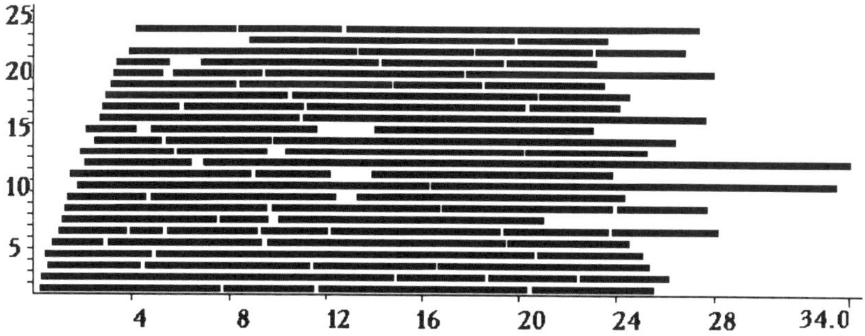

Figure 6–9 The cyclogram of the machine operation (the optimal solution).

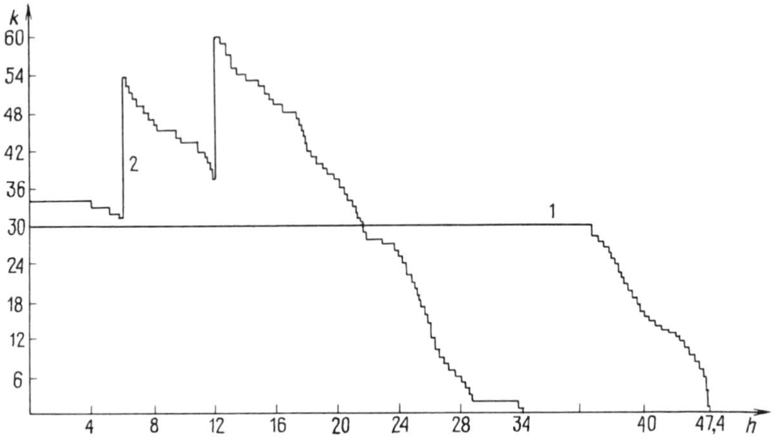

Figure 6–10 The diagram of storage capacity. 1 and 2 denote the prototype and the optimal solution, respectively.

Optimization of the Die-Casting Machine Locking Mechanism (Nogovitsin 1987)

Die-casting is a progressive and highly productive technology used for manufacturing nonferrous thin-walled castings requiring minimal additional processing. It is implemented using die-casting machines. Presently, widely used machines are universal machines with horizontal cold press chambers, whose locking and injection mechanisms are mounted on the same frame. The former mechanism serves for locking the half-molds within which a casting is formed, and is loaded by the forces arising during the filling of the mold with liquid metal and the latter's crystallization.

In designing a lever locking mechanism the determination of the slider stroke X_s and the force P_c acting on the cylinder rod for a fixed mold-locking force are of great importance. The latter is determined taking into account the friction torques in the lever system hinges and the mechanism's rigidity (Nogovitsin 1987). The functional constraints imposed on the lengths of the links and their angular positions were formulated taking into account the specific features of the mechanism.

The optimal design variables of the mechanism had to be obtained subject to the condition that they ensure the minimal slider stroke max $X_s \rightarrow \min$, and the maximum force acting on the cylinder rod, max $P_c \rightarrow \min$. (Since the force varies during the working cycle, its maximum value attained after the half-molds contact each other was minimized.) Also subjected to maximization was the movable plate travel, max $X_p \rightarrow \max$. (In doing so several prototype characteristics, such as the mold locking force equal to 8,000 kN, the diameters of the larger and the smaller hinge axes (0.14 m and 0.075 m, respectively), and the mechanism's stiffness coefficient equal to $2.4 \cdot 10^6$ kN/m, had to be preserved.)

Upon conducting 600 trials using the PSI method, an optimal solution was obtained, the criteria of which surpassed those of the prototype.

The analysis has allowed increasing the movable plate travel from 710 mm to 920 mm and decreasing the slider stroke without changing the maximum force acting on the locking cylinder rod.

Optimization of Freshly Harvested Haws Extraction (Murav'ev and Bredneva 1987)

Prevention and cure of cardiovascular diseases is one of the most pressing problems of the present-day medicine. Among the plants used for this purpose, hawthorn (*Crataegus gen.*) is of special importance. Its preparations produce cardiotropic, hypotensive, sedative, and hypocholesteremic effects, and are practically nontoxic.

The study was aimed, in particular, at analyzing the factors affecting the process of haws extraction. Factors subjected to analysis were the infusion time, ethanol concentration, the method used for chopping raw material, the raw material fineness, the raw material-to-extractor ratio, etc. It was supposed that extraction proceeded in three stages. The following eight parameters were varied: the character and the fineness of chopped haws (unchopped, milled, crushed, and rolled), the time of infusion in percolators, alcohol concentration, and the raw material-to-extractor ratio.

The experiment was planned using the PSI method. According to this method, 32 eight-dimensional points, $\boldsymbol{\alpha}^i=(\alpha_1^i,\ldots,\alpha_8^i)$, $i=\overline{1,32}$, were obtained. After that, 32 full-scale trials were conducted using these points. Then, a multiple-regression equation was compiled and analyzed. This allowed obtaining the expressions for the two optimization criteria: the dry residue output and the raw material exhaustibility factor. Next the feasible solutions and Pareto optimal sets were compiled, and the sensitivity of the criteria to the change of design variables was analyzed. It was found that the optimum value of the raw material-to-extractor ratio lay between 0.3 and 0.4. This corresponds to the raw material being exhausted by 90% and allows obtaining a fresh-haws tincture with the dry residue content in excess of 2%. The results of the study were used for developing the request for producing the tincture and were subjected to approbation at three pharmaceutical factories.

Ship Design (Berezanskii and Semenov 1988)

The ever-increasing scope of the support of off-shore oil extraction operations poses the problem of improving the efficiency of regional technological complexes of the oil and gas industry fleet. The boundaries of the ship design variables were determined by analyzing statistical data on the ships of a given type.

The quality of a ship being designed was characterized by its tonnage, speed, construction, capital investments, and operational costs. The study was aimed at improving the performance criteria of the UT-704 ship constructed by the Norwegian ULSTEJN firm for supporting oil drilling rig operation. The ship is used for transporting liquid, dry, and loose loads. In line with the previously formulated problem the following criteria constraints corresponding to the UT-704 performance criteria, were imposed:

$$P_t \geq 1{,}200 \text{ t}, V_s \geq 16 \text{ knots}, K \leq 5.01 \text{ million roubles (in 1984 prices), and } C \leq 0.62 \text{ million roubles}$$

Table 6-12 shows the design characteristics of supporting ships for the most advantageous Pareto optimal solutions and the prototype. The following notation is used: L is the length of the ship, B is its width, T is the draught, N_e is the power plant output, P_t is tonnage, and K and C are the capital investments and operational costs, respectively.

We see that the prototype has been substantially improved.

Selection of the Structure and Design Variables of Active Pneumatic and Hydropneumatic Vibration Isolators (Kreinin et al. 1986)

Vibration isolators of mechanisms are used for both reducing the loads transferred onto foundations over a wide range of frequencies and limiting the displacements due to low-frequency excitation in changing operational modes. Active pneumatic and hydropneumatic vibration isolators are widely used in transport engineering due to their reliability, economical efficiency, and structural simplicity. Such isolators develop low dynamic stiffness at frequencies ranging from 5 to 10^4 Hz, and high dynamic stiffness over a frequency range of 0.1–0.2 Hz.

Various designs of double-chamber active pneumatic and hydropneumatic vibration isolators were subjected to analysis. The domain of feasible design solutions was defined in the space of the following three performance criteria: damping at the resonance frequency f_0, the low-frequency dynamic stiffness for $f \leq 0.2 f_0$, and the working medium flow rate (the source intensity).

Table 6–12

Solutions	L (m)	B (m)	T (m)	N_e (kW)	P_t (10^{-3} t)	V_s (knots)	K (million roubles)	C (million roubles)
Pareto-optimal solutions	59.19	14.38	4.64	4,402	1.55	16.2	4.82	0.604
	59.22	14.69	4.75	4,402	1.62	16.3	4.86	0.607
	61.38	13.52	4.36	2,783	1.54	16.4	4.98	0.612
	56.24	13.67	4.47	2,989	1.23	16.0	4.56	0.540
UT-704 (prototype)	56.4	13.8	4.75	3,680	1.21	16.0	5.01	0.620

The stiffness, load-carrying capacity, and other parameters of a vibration isolator were assumed to be known, while the damping chamber volume, the flow rate through the interchamber capillary, and the regulator gain factor were varied. The functional constraints involved the damping chamber volume, the interchamber capillary tube dimensions (depending on these dimensions, the chambers can operate as if they were fully isolated through as if they were a single whole), the system's stability margin, etc.

The analysis has allowed determination of the optimal structure of an active vibration isolator possessing the maximum dynamic stiffness over a frequency range of 0.1–0.2 Hz and meeting the stability conditions. It has been shown that at a frequency of 0.2 Hz the dynamic stiffness of a double-chamber vibration isolator with a controlled damping chamber may be increased by a factor of 10–20 compared to a passive isolator with a natural frequency of 1.5 Hz.

Design variables of a passive double-chamber vibration isolator were also determined, ensuring maximum damping at the resonance frequency for a specified static stiffness. It has been shown that the introduction of a damping chamber results in the energy-absorption coefficient increasing by a factor of two to four at the resonance frequency, and the static stiffness decreasing by 10–20% depending on the type of the pneumatic isolator. Practical realization of these recommendations allowed decreasing amplitudes of the mechanism displacements at the first resonance frequencies by a factor of two to four.

In conclusion, we would like to mention some more practically important fields where the PSI method has been successfully used, such as an increase in the efficiency of mathematical simulation in petrophysics (Ingerman 1985), simulation in multicriteria problems of nonlinear adaptive optics (Karamzin et al. 1986), an increase in the efficiency of mechanization of subway construction on the basis of the module principle (Auerbakh et al. 1985), the design of transmissions of the main drives of rolling mills (Zhitomirskii, Rubanovich, and Filatov 1984), protection of the operators of open-cut machines against vibration (Tregubov 1983), an active vibration protection system (Zaikova and Yablonskii 1991), flying vehicles engines (Khronin et al. 1984), an electrohydraulic amplifier with nonunity feedback (Borovin et al. 1985), machines shafting (Godzhaev, Dmitrichenko, and Guberniev 1992), etc. Recently, studies on optimization of an automobile engine, a steering rack gear, the Macpherson wheel suspension, prospective truck cushioning systems, a high-speed centrifuge rotor, internal wheelheads mounted on ball bearings, low-power asynchronous machines, and damping devices for actuator suspension systems of a coal-loader have been fulfilled.

Finally, we draw the reader's attention to the problems of computer-aided design (CAD) and the role PSI plays in their solution (Gerasimov et al. 1985; Rizkin 1985; Popov 1986; Baturidi et al. 1991). At present, the PSI method is widely taught in higher educational establishments (see, for instance, textbooks (Odrin 1986; Reshetov et al. 1985)).

Conclusion

The world of engineering optimization problems is unbelievably vast.

Billions of dollars are spent on creation and operation of machines, mechanisms, instruments, and structures. That is why optimization in engineering, especially in designing, must be implemented at a qualitatively new level.

If we compare optimal design to a high building under construction, we are now on one of its first floors. Clearly, there is a colossal difference between an optimal design and the optimal machine. But one hardly may count on manufacturing an optimal machine without having first designed it in the optimal way.

Poor knowledge of the fundamentals and principles of engineering optimization is one of the reasons why the machines we are presently manufacturing are inferior compared to those we are capable of manufacturing. Such is the cost of our ignorance, and this is characteristic of all industrial countries without any exclusions. Someone may retort that there are and always have been good machines created without employing any science of optimal design. These are mainly exceptions. Therefore, the sooner we learn that an optimal design is not a whim but rather a must, that we are not so rich as to afford designing nonoptimal machines and structures, and also that the quantity of manufactured machines by no means compensates for their inferior quality, the sooner we will make decisive strides toward scientific and technological progress.

At present what hampers the search for the optimal solutions to engineering problems? The lack of adequate mathematical models and, no less important, the inability to formulate a problem of optimization correctly make the designers employ their intuition more often then is necessary. This would be acceptable if the cost of errors were not so high. In our opinion, such is the explanation (but not justification) of the present low level of solution of engineering optimization problems.

This book has been primarily aimed at perception of the essence of engineering optimization problems, starting from the formulation and finishing with the solution. Also we desired to demonstrate the efficiency of the new concept by considering numerous examples. In solving engineering optimization problems we prefer to be optimistic and believe that in the foreseeable future the term *optimal design* will be replaced by simply *design*. In practice, this will mean designing and manufacturing Pareto optimal automobiles and ships, machine tools and locomotives, machining centers and flexible manufacturing systems, robots and manipulators, tractors and combine harvesters, etc. Who would care for a less than optimal design?

Finally, Aristotle justly remarked that "one can't choose from the impossible." Through this book we are hoping to help make the choice possible.

Addendum

Uniformly Distributed Sequences and Nets in Multidimensional Parameter Space

A-1. Uniformity Characteristics of Distribution of Points in N-Dimensional Cube

In Section 1-2 the necessity of using uniformly distributed sequences (UDS) for solving the problems being considered was discussed. A definition has been given of the distribution uniformity for a sequence of points in a multidimensional space. The simplest properties of UDS have been stated therein. Here we shall attempt to study in more detail various properties of uniformity and consider new constructions of limited-length UDS (nets). The latter possess a number of interesting properties of uniformity that single out them from other known sequences.

Note that the well-known Weyl theorem was of great importance in developing the theory of uniform distribution of sequences[17] (Kuipers and Niederreiter 1974).

Theorem 1 (H.Weyl). Let $f(\mathbf{X})$ be a function integrable in the sense of Riemann. Then $\{\mathbf{X}^i\}$ is a uniformly distributed sequence $\Leftrightarrow$[18] the relationship

$$\lim_{N\to\infty} \sum_{i=1}^{N} f(\mathbf{X}^i) = \int_{K^n} f(\mathbf{X})d\mathbf{X}, \tag{A-1}$$

is satisfied where K^n is a unit n-dimensional cube.

This formula opened a new direction in the theory of quadrature formulas. The use of uniformly distributed sequences appeared to be rather promising for

[17]The results hereafter given without references were obtained by Matusov.

[18]Symbol $\Leftrightarrow$ means "in order to ... it is necessary and sufficient that."

the multidimensional integral calculations. In many cases it is more efficient than the use of other methods.

There exist various characteristics of uniformity for distribution of point sequences in multidimensional spaces.

Discrepancy

Let us consider in K^n an arbitrary net consisting of N points $\mathbf{X}^1,\ldots,\mathbf{X}^N$. An arbitrary point $\mathbf{X}$ within K^n will be bring in correspondence with a parallelepiped $\Pi_\mathbf{X}$ with its faces being in parallel with the coordinate planes and with its diagonal being OX (O denotes the origin of coordinates). Let us define $V_{\Pi_\mathbf{X}}$ as the volume of $\Pi_\mathbf{X}$.

Definition. A discrepancy of points $\mathbf{X}^1,\ldots,\mathbf{X}^N$ is an upper bound

$$D(\mathbf{X}^1,\ldots,\mathbf{X}^N) = \sup_{\mathbf{X}\in K^n} |S_N(\Pi_\mathbf{X}) - N \cdot V_{\Pi_\mathbf{X}}| \tag{A-2}$$

where $S_N(\Pi_\mathbf{X})$ is the number of points belonging to $\Pi_\mathbf{X}$.

The smaller D is, the more uniform the location of points $\mathbf{X}^1,\ldots,\mathbf{X}^N,\ldots$ in K^n should be considered. It can be easily seen that $D(\mathbf{X}^1,\ldots,\mathbf{X}^N) \leq N$ always.

There are also other characteristics of uniformity.

A-2. P_τ-Nets and LP_τ-Sequences

In this section we shall start to study the properties of uniformity of the known P_τ-nets and LP_τ-sequences (Sobol' 1969).

Let us introduce an equivalent definition of the distribution uniformity.

The sequence of points $\mathbf{X}^1,\ldots,\mathbf{X}^N$ is distributed uniformly in K^n if for any binary parallelepiped $\Pi_k \subset K^n$, the following condition

$$\lim_{N\to\infty} \frac{S_N(\Pi_k)}{N} = V_{\Pi_k}$$

is satisfied.

The aforementioned definition only allows such "good" domains as binary parallelepipeds. Let us give an appropriate definition.

Let us consider some intervals being binary if they can be obtained by dividing an interval $[0,1]$ into 2^m equal parts, where $m=0,1,2,\ldots$. For the sake of definiteness, consider so-called closed-open intervals, containing only the left endpoint, except in the case when the right endpoint is equal to 1. If the right endpoint is equal to 1, then the interval is closed (contains both endpoints). By this definition, the sum of all the binary intervals of the length 2^{-m} constitutes the interval [0,1].

For example, the following intervals are binary: [0,1], [0,1/2),[1/2,1], [0,1/4), [1/4,1/2), [1/2,3/4),... . Incidentally, intervals such as [1/4,3/4), or [5/8,7/8) are not considered binary.

Let us number all the binary intervals and denote them l_{ms}, that is, $s=1,2,\ldots,2^m$. Let $\mathbf{K}=(K_1,\ldots,K_n)$. We shall identify as a binary parallelepiped $\Pi_{\mathbf{K}}$ a set of points with coordinates $(x_1,\ldots,x_n)$ where $x_j \in l_{K_j^s}$ at $s=1,2,\ldots,2^{K_j}$.

It is obvious that any such binary parallelepipeds belong to a unit n-dimensional cube K^n.

P_τ-*Nets*

We shall refer to a net consisting of $N=2^\nu$ points of the cube K^n as a P_0-net if any binary parallelepiped with a volume of $1/N$ contains one point of this net.

It is not difficult to prove that the points of the P_0-net are uniformly distributed in K^n. Unfortunately, as has been shown in Sobol' (1969), such nets exist only at $n=1,2,3$. In a four-dimensional cube it is impossible to construct a P_0-net with a number of points $N\geq4$. Therefore, the requirement to the distribution of net points has to be relaxed so that a more general definition might be introduced.

Definition (Sobol' 1969). A net consisting of 2^ν points of the cube K^n is referred to as a P_τ-net if any binary parallelepiped $\Pi_{\mathbf{K}}$ with the volume $V_{\Pi_{\mathbf{K}}}=2^\tau/2^\nu$ contains 2^τ points of the net. (In this case, it is always supposed that $\nu>\tau$.)

P_τ-nets exist in K^n at any n, but the values of τ, however, grow invariably with the growth of n. The projection of the points consisting of P_τ-net in K^n into any s-dimensional face of the cube K^n forms an s-dimensional P_τ-net. Moreover, this s-dimensional net may appear to be a $P_{\tau'}$-net with $\tau'<\tau$.

How exactly can this definition reflect the process of the most uniform distribution of points? The following definition seems to be more appropriate.

We shall refer to an interval $|l|\subset[0,1]$ as quasi-binary if it consists of one or more binary intervals.

A parallelepiped $\Pi_{\mathbf{K}}$ is quasi-binary if its edges are quasi-binary intervals.

Definition. A net consisting of 2^ν points of the cube K^n is referred to as p-net if any quasi-binary parallelepiped $\Pi_{\mathbf{K}}\subset K^n$ with a volume $p/2^\nu$ contains p points of this net.

Theorem 2. For any dimensionality n of the cube K^n there exist p-nets with a length 2^ν, $2^\nu>p$ if and only if $p=p_1\cdot2^\tau$ (where p_1 is an odd number, and τ is a natural number).

Proof. Let us consider an arbitrary quasi-binary parallelepiped $\Pi_{\mathbf{K}}$ with a volume $p/2^\nu$, where 2^ν is length of a net in K^n. This parallelepiped is determined in the following way (in binary notation)

$$0.\ b_1^{(k)}\ldots b_{\mu_k}^{(k)} \leq x_k < 0.\ b_1^{(k)}\ldots b_{\mu_k}^{(k)} + 0.\ a_1^{(k)}\ldots a_{\mu_k}^{(k)} \leq 1$$

where $k=1,\ldots,n;$ $\mu_1+\ldots+\mu_n=\nu-l,$ $l\geq 0.$ Let $A^{(k)}=0, a_1^{(k)}\ldots a_{\mu_k}^{(k)}.$ Then $\prod_{k=1}^{n} A^{(k)}=p/2^\nu=V_{\Pi_\mathbf{K}}$. Any number $i\leq 2^\nu$ in binary notation is $i=e_te_{t-1}\ldots e_\nu\ldots e_1$. Let us denote the coordinates of point $\mathbf{P}(i)$ (with number i) within the net through $p^k(i)$. Let $p^k(i)=0, g_{i1}^{(k)}\ldots g_{ij}^{(k)}\ldots$is the appropriate binary number. Also, let

$$B_{1_k}^{(k)}=0.\ b_1^{(k)}\ldots b_{\mu_k}^{(k)}+0.\ \underbrace{0\ldots 1}_{\mu_k},$$

$$B_{m_k}^{(k)}=B_{m_{k-1}}^{(k)}+0.\ \underbrace{0\ldots 1}_{\mu_k},\quad 1\leq m_k\leq p,\quad \sum_{k=1}^{n} m_k=p$$

in the binary notation $B_{m_k}^{(k)} = 0.\ b_{1m_k}^{(k)}\ldots b_{\mu_k m_k}^{(k)}$.

The condition that $P(i)$ belongs to a given parallelepiped can be written as follows: $g_{ij}^{(k)}=b_{jm_k}^{(k)},\ 1\leq j\leq\mu_k,\ 1\leq k\leq n.$

It will be shown that the coordinates $p^k(i)$ of the point $\mathbf{P}(i)$ can be defined by means of a so-called direction matrix $[\upsilon_{ij}^{(k)}]$ in the following way

$$g_{ij}^{(k)}=e_1\upsilon_{1j}^{(k)} * e_2\upsilon_{2j}^{(k)} * \ldots * e_t\upsilon_{tj}^{(k)}$$

where $*$ is modulo 2 addition operation ($1*1=0$, $0*1=1*0=1$, $0*0=0$, and the result of addition is not transferred to the next digit. Thus, for example, $0.1011*0.1101=0.011$). Having substituted the latter expression to the left-hand side of the system, we shall obtain p systems of linear equations with ν unknowns

$$e_1\upsilon_{1j}^{(k)} * \ldots * e_\nu\upsilon_{\nu j}^{(k)}=b_{jm_k}^{(k)} * e_{\nu+1}\upsilon_{\nu+1j}^{(k)} * \ldots * e_t\upsilon_{tj}^{(k)},\qquad \text{(A-3)}$$
$$1\leq j\leq\mu_k,\ 1\leq k\leq n,\ 1\leq m_k\leq p.$$

Note that $\mathbf{P}(i)$ belongs to $\Pi_\mathbf{K}$ if and only if it belongs to one of the p binary parallelepipeds, that is, when one of the given systems of linear equations is satisfied.

The volume of $\Pi_\mathbf{K}$ might be equal either to $p/2^\nu$, or to $p_1/2^{\nu-1}$ (if $p=p_12^l$) where p_1 is an odd number.

In the first case, each of the systems (A-3) describes the pertaining of a point to a binary parallelepiped having a volume $(1/2)^\nu$, and consists of ν equations. Such a system has either a unique solution if the coefficients on the left-hand side of equations (A-3) form a nonsingular matrix, or else no solutions at all. If this matrix is nonsingular, then one point gets into each binary parallelepiped having a volume $(1/2)^\nu$. But, as was mentioned, this is impossible when $n\geq 4$.

In the second case, each of the systems (A-3) contains $\nu-l$ equations. Such a system has either no solutions or it has 2^l solutions if the matrix of the coefficients on the left-hand side of these equations has the rank $\nu-l$. If the latter condition is satisfied, it means that a P_l-net exists in K^n. In other words, the existence of p-nets is reduced to the existence of P_τ-nets where $\tau \leq l$, $p=p_1 2^l$.

LP_τ-*Sequence*

Let us consider an arbitrary sequence of points $\mathbf{Q}_1, \mathbf{Q}_2, \ldots, \mathbf{Q}_i, \ldots$, belonging to K^n. Let us identify as a binary part of the sequence a set of terms $\mathbf{Q}_i$, whose numbers i satisfy an inequality of the following form

$$k2^s \leq i < (k+1) \cdot 2^s \quad (k=0,1,2,\ldots;\ s=1,2,\ldots).$$

For example, parts $0 \leq i < 8$; $8 \leq i < 16$; $16 \leq i < 24, \ldots$ are binary, whereas part $4 \leq i < 16$ is not binary.

Definition (Sobol' 1969). A sequence of points $\mathbf{Q}_1, \mathbf{Q}_2, \ldots, \mathbf{Q}_i, \ldots$ within the cube K^n is referred to as an LP_τ-sequence if any of its binary parts containing at least $2^{\tau+1}$ points represents a P_τ-net.

Estimate of Discrepancy

Let us denote through $\begin{bmatrix} m \\ k \end{bmatrix}$, as usual, a binomial coefficient so that $\begin{bmatrix} m \\ k \end{bmatrix} = m(m-1)\ldots(m-k+1)/1 \cdot 2 \ldots k$.

Theorem 3 (Sobol' 1985). For any P_τ-net in K^n consisting of $N \geq 2^{n-1+\tau}$ points, the following estimate is valid:

$$D(\mathbf{Q}_1 \ldots, \mathbf{Q}_N) \leq 2^\tau \sum_{j=0}^{n-1} \begin{bmatrix} \nu-\tau \\ j \end{bmatrix}, \qquad \text{(A-4)}$$

where $\nu = \log_2 N$.

Theorem 4 (Sobol' 1985). For an arbitrary initial part of any LP_τ-sequence in K^n containing at least $2^{n-1+\tau}$ points, the following estimate holds:

$$D(\mathbf{Q}_1, \ldots, \mathbf{Q}_N) \leq 2^\tau \left[1 + \sum_{j=0}^{n-1} \begin{bmatrix} \nu_1-\tau+1 \\ j+1 \end{bmatrix} \right] - 1 \qquad \text{(A-5)}$$

where $\nu_1 = E(\log_2 N)$ is the integer part of the logarithm of N.

From the last theorem it follows that

$$D(\mathbf{Q}_1,\ldots,\mathbf{Q}_N)=O\,(ln^n N)$$

as $N\rightarrow\infty$. This means that $D/(ln^n N)<$const.

It follows from this theorem that the LP_τ-sequences are distributed uniformly in K^n.

Let us now turn our attention to the construction of LP_τ-sequences.

DP-Sequences

Let us select an infinite triangular matrix of the following form

$$(\boldsymbol{\upsilon}_{sj})=\begin{bmatrix} 1 & & & & 0 \\ & 1 & . & & \\ & & & . & \\ & & & . & \\ \boldsymbol{\upsilon}_{sj} & & & & 1 \end{bmatrix}$$

which will be referred to as a direction matrix. Elements υ_{sj} underlying the main diagonal may be zeroes or unities.

To specify a matrix (υ_{sj}) is equivalent to specifying the sequences of dyadic fractions (in binary system)

$$V_s = 0.\upsilon_{s1}\upsilon_{s2}\ldots\upsilon_{sj}\ldots$$

which are referred to as direction numbers.

What is termed as a sequence of dyadic type, or DP-sequence, is a sequence of numbers $r(0)$, $r(1),\ldots,$ $r(i),\ldots$ calculated in accordance with the following rules:

1. $r(0)=0$; $r(2^s) = V_{s+1}$
2. If $2^s\leq i<2^{s+1}$, then $r(i)=r(2^s) * r(i-2^s)$ where $*$ means a digit-by-digit modulo 2 addition in binary system.

It is not difficult to prove that rules 1 and 2 are equivalent to the following definition: If in a binary system $i=e_m\ldots e_2e_1$, then $r(i) = e_1V_1 * e_2V_2 *\ldots* e_mV_m$.

Lemma (Sobol' 1969). A DP-sequence $r(0)$, $r(1),\ldots,r(i),\ldots$corresponding to a direction matrix $(\boldsymbol{\upsilon}_{sj})$ is a one-dimensional LP_0-sequence consisting of various dyadic fractions.

Monocyclic Operators in a Field GF(2)

A field $GF(2)$ consists of two elements: 0 and 1. The rules of multiplication are usual and the rules of addition are as follows:

$$0+0=1+1=0, \quad 0+1=1+0=1.$$

Let us consider a linear difference equation of order m with constant coefficients:

$$Lu_i=0 \tag{A-6}$$

where the difference operator L is defined by the following expression

$$Lu_i= u_{i+m}+a_{m-1}u_{i+m-1}+\ldots+a_1u_{i+1}+u_i$$

Here, all u_i and a_i belong to $GF(2)$.

Let us define the solution of equation $Lu_i=0$ as an infinite sequence

$$\ldots, u_{-2}, u_{-1}, u_0, u_1, u_2, \ldots$$

definite at $-\infty<i<\infty$ and satisfying this equation at every i.

Each solution is determined uniquely by assigning the group $(u_1,\ldots,u_m)$, since all the values of $u_{m+1}, u_{m+2},\ldots$ and all the values of $u_0, u_{-1}, u_{-2},\ldots$ are calculated successively by means of equation (A-6):

$$u_{i+m} = a_{m-1}u_{i+m-1}+\ldots+a_1u_{i+1}+u_i, \quad i=1,2,\ldots$$
$$u_i = u_{i+m}+a_{m-1}u_{i+m-1}+\ldots+a_1u_{i+1}, \quad i=0,-1,-2,\ldots$$

Since only 2^m various groups $(u_1,\ldots,u_m)$ exist that consist of zeroes and unities, then only 2^m solutions exist, including a trivial one, $u_i\equiv 0$.

Let us consider the groups $(u_1,\ldots,u_m)$, $(u_{2,\ldots,}u_{m+1})$, $(u_3,\ldots,u_{m+2}),\ldots$. Among them, a group will certainly be found that coincides with one of the groups already considered. Therefore, any solution of equation (A-6) is periodical, its period not exceeding 2^m-1.

Definition. An operator L is referred to as monocyclic if equation $Lu_i=0$ has a solution with the least period 2^m-1.

Construction of LP$_\tau$*-Sequences*

Let us come to an agreement that the direction matrix $(\mathbf{v}_{sj})$ pertains to an operator L of mth order if the following two conditions are satisfied:

1. Every one of the first m columns of this matrix is a solution of the equation

$$L\upsilon_{ij}=0 \text{ at } 1\leq j\leq m.$$

2. Every one of the subsequent columns of the matrix is a solution of nonhomogeneous equation

$$L\upsilon_{ij}=\upsilon_{ij-m} \text{ at } m<j<\infty.$$

We shall also mention that a DP-sequence corresponding to such direction matrix ($\boldsymbol{\upsilon}_{sj}$) belongs to the operator L. Since the elements υ_{sj} underlying the main diagonal in the first m rows can be selected arbitrarily, a number of various DP-sequences can pertain to the same operator.

Theorem 5. (Sobol' 1969). Let $L_1,\ldots L_n$ be various monocyclic operators with their orders equal to $m_1,\ldots,m_n$ respectively. Also, let $q^j(1)$, $q^j(2),\ldots$, $q^j(i),\ldots$ be some DP-sequence pertaining to the operator L_j. Then, the sequence of points $\mathbf{Q}_1$, $\mathbf{Q}_2,\ldots,\mathbf{Q}_i,\ldots$ with coordinates

$$\mathbf{Q}_i = (q^1(i), q^2(i),\ldots,q^n(i)) \tag{A-7}$$

is an LP_τ-sequence in K^n with

$$\tau=\sum_{j=2}^{n} (m_j - 1).$$

A first coordinate may formally be defined as pertaining to the operator $Lu_i=u_{i+1}$.

Values of $\boldsymbol{\tau}$. It is seen from the last formula that in order to reduce τ, we should select the monocyclic operators L_j with orders m_j as low as possible. The last theorem has been used to construct the LP_τ-sequences for which $\tau=0$ at $n=1,2$ and $\tau=1$ at $n=3$. These values of τ have been proved to be minimum for the LP_τ-sequences. When $n=4$, we obtain $\tau=3$. In general, as $n\rightarrow\infty$

$$\tau=O\,(n\log_2 n).$$

The estimates obtained here show that asymptotically all the LP_τ-sequences can be ascribed to the best ones. However, in practice it would be important that the uniformity of location is set up rapidly and not only as $N\rightarrow\infty$. For this purpose some additional requirements have been formulated.

A-3. Additional Properties of Uniformity of LP_τ-Sequences

Most important in solving specific problems is to ensure the uniformity of initial parts of the sequences used. This section is devoted to consideration of the corresponding properties of uniformity.

Property A. Let us divide a cube K^n by planes $x_j = 1/2$ with $j=1,2,\ldots,n$ into $t=2^n$ octants that will be considered as binary parallelepipeds. Let us split the sequence of points $\mathbf{Q}_1$, $\mathbf{Q}_2$, ..., $\mathbf{Q}_i$, ... into binary parts that have their lengths equal to 2^n:

$$\mathbf{Q}_0,\ldots,\mathbf{Q}_{t-1},\mathbf{Q}_t,\ldots,\mathbf{Q}_{2t-1},\mathbf{Q}_{2t},\ldots,\mathbf{Q}_{3t-1},\ldots \qquad \text{(A-8)}$$

If for any of these parts all the points pertain to different octants, then we can say that the sequence possesses the Property A.

Theorem 6 (Sobol' 1976). The sequence $\mathbf{Q}_1$, $\mathbf{Q}_2$, ..., $\mathbf{Q}_i$,... constructed in the previous theorem possesses Property A if and only if the $(n\times n)$-determinant composed of the first columns of all direction matrices is equal to 1 (mod 2).

This theorem allows constuction of LP_τ-sequences possessing Property A. However, the following question remains open: Is it possible to use the operators of the lowest orders as the sequence of operators $\{L_k\}$? This question is justified by the fact that in precisely this case the value of τ will be minimum, and this will improve all the characteristics of uniformity. In order to answer the question, let us consider the following statements.

Let L_1, L_2,...,L_n be various monocyclic operators with their orders being m_1, m_2,...,m_n, respectively.

Lemma 1. Let $q^j(i)$ in (A-7) be a DP-sequence corresponding to all monocyclic operators L_1,...,L_n of the lowest orders. Let m be the order of operator L_n. Then, the following inequality holds

$$n+\frac{m(m+1)}{2}<2^m, \quad \text{at } m\geq 5 \qquad \text{(A-9)}$$

Proof. We shall prove this by induction with respect to m. When $m=5$, the inequality holds, since in this case the maximum value of n is equal to 13 (Sobol' 1969). Let for some value of m, inequality (A-9) take place for any n corresponding to this m. Let us prove that at any n corresponding to $m+1$, a similar inequality will be satisfied. Taking into account that the number of monocyclic operators of the $(m+1)$th order is equal to $\phi(2^m-1)/(m+1)$ (where $\phi(k)$ is the Euler number-theoretic function equal to the quantity of natural numbers less than k and relatively prime with k), we shall prove now that

$$n+\frac{\phi(2^m-1)}{m+1}+\frac{(m+1)(m+2)}{2}<2^{m+1}.$$

Let us represent this inequality in the following form

$$n+\frac{\phi(2^{m+1}-1)}{m+1}+\frac{(m+1)m}{2}+m+1<2^{m+1}$$

All that remains is to compare it with (A-9) to show that

$$\frac{\phi(2^{m+1}-1)}{m+1}+m+1<2^m,$$

i.e.,

$$\phi(2^{m+1}-1)+(m+1)^2<(m+1)2^m$$
$$\phi(2^{m+1}-1)+(m+1)^2<2^{m+1}-1+(m+1)^2<(m+1)2^m$$

Finally,. we have that $(m+1)^2-1<(m-1)2^m$.

Thus we have come to an obvious inequality, which is true at $m\geq 3$.

Let us denote through $m(n)$ the order of a monocyclic operator standing on the nth place in the sequence of all monocyclic operators taken in succession.

Lemma 2. Let us consider $n-1$ sequences of elements $\{u_i^{(k)}\}$, $k=\overline{1, n-1}$, $i>0$, of field $GF(2)$. Let a sequence consisting of unities belong to this set, and let the determinant be

$$\begin{vmatrix} u_1^{(1)}, \dots, u_{n-1}^{(1)} \\ \dots\dots\dots\dots\dots \\ u_1^{(n-1)}, \dots, u_{n-1}^{(n-1)} \end{vmatrix} = 1 \pmod 2. \tag{A-10}$$

Let us consider the monocyclic operator L_n with the order $m\geq m(n)$, $m\geq 5$. Then, such a solution $\{u_i^{(n)}\}$ of the operator L_n exists that the determinant is as follows

$$\begin{vmatrix} u_1^{(1)}, \dots, u_{n-1}^{(1)}, u_n^{(1)} \\ \dots\dots\dots\dots\dots\dots\dots\dots \\ u_1^{(n-1)}, \dots, u_{n-1}^{(n-1)}, u_n^{(n-1)} \\ 1, \dots, u_{n-1}^{(n)}, u_n^{(n)} \end{vmatrix} = 1 \pmod 2 \tag{A-11}$$

Proof. Let us assume the opposite: Whatever the first m values of the operator L_n solution are selected, the determinant in (A-11) is equal to 0 (mod 2). Since the columns in (A-10) are linearly independent, the first $n-1$ of the elements in the last column in (A-11) may be expressed uniquely through these columns. But by virtue of the aforementioned assumption, the last element of the nth column in (A-11) must be expressed in the same manner through the elements of the nth row. The latter belong to the appropriate columns in (A-11), no matter what first m values of the last row are selected in (A-11). That means we have for the last row

$$u_n^{(n)}=\sum_{k=1}^{n-1} c_k u_k^{(n)}, \ c_k \in \{0,1\}, \text{ or}$$

$$u_{i_0}^{(n)}=\sum_{k=1}^{n-1-i_0} c_{i_0+k}\, u_{i_0+k}^{(n)}+u_n^{(n)}=0, \ c_{i_0}=1, \qquad \text{(A-12)}$$

$$(c_i=0 \text{ when } i<i_0)$$

for arbitrary first m values on this row beginning with unity. This implies that any solution of monocyclic operator L_n is a solution of a linear difference operator with the order $n-i_0$

$$L^*(u_i)=u_i+\sum_{k=1}^{n-1-i_0} c_{i+k}u_{i+k}+u_{i+n-i_0} \qquad \text{(A-13)}$$

Indeed, let $\{u_i^{(n)}\}$ be a solution of monocyclic operator L_n. Then, if any part of this solution beginning with 1 is taken as the last row in (A-11), we shall obtain (A-12). Let us take an arbitrary part having a length of n and beginning with zero. Since the first m values can be selected arbitrarily, we shall present this part as a sum of two parts having the same length and beginning with 1. And because (A-12) is satisfied for these parts, then (A-12) is also true for the given part that begins with 0. This means that with any initial values of $u_1^{(n)}$, $u_2^{(n)}$, ... $u_m^{(n)}$ we shall have satisfied (A-12). And this also implies that any solution of monocyclic operator L_n of the mth order is a solution of a linear difference operator $L^*(u_i)$. Since it is so, the polynomial corresponding to $L^*(u_i)$ must be divisible by the polynomial corresponding to the operator L_n.

Let

$$x^{n-i_0}+a_{n-i_0-1}\, x^{n-i_0-1}+\ldots+a_1 x+1 \qquad \text{(A-14)}$$

be a polynomial, where all a_i and x pertain to the field $GF(2)$, corresponding to the operator L^*. We shall show now that if a polynomial corresponding to L_n, such as

$$x^m+a_{m-1}\, x^{m-1}+\ldots+a_1 x+1 \qquad \text{(A-15)}$$

is a divisor of the polynomial (A-14), then it will also be a divisor of some polynomial of a larger degree that has no monomials of x^k type, where $k\leq m$, excluding 1. Indeed, let (A-14) contain some monomial x^k, where $k\leq m$ (i.e., $a_k=1$). By multiplying (A-14) by (x^k+1) we obtain a polynomial of the $(n-i_0+k)$th degree, which has its coefficient equal to zero at x^k, that is, $a_k=0$.

We may treat similarly any term that has its degree not greater than m. So we shall obtain a polynomial that has the degree of its penultimate term higher than m, whereas the maximum degree of that polynomial will be equal to $n-i_0+m(m+1)/2$, that is,

$$x^{n-i_0+\frac{m(m+1)}{2}}+\ldots+a_{m+1}x^{m+1}+1. \qquad \text{(A-16)}$$

If the polynomial (A-15) is a devisor of the polynomial (A-14), then it must also be a divisor of the polynomial (A-16). On the other hand, the polynomial (A-16), like the polynomial (A-14), contains an even number of terms, since, in accordance with this lemma, one of the rows of the determinant (A-11) consists of unities. Therefore, the polynomial (A-16) can be represented as a sum of binomials

$$(x^{n-i_0+\frac{m(m+1)}{2}}+1)+\ldots+(x^{m+p}+1) \ (p>0) \qquad \text{(A-17)}$$

Let us divide each of these binomials by the polynomial (A-15). Since the polynomial (A-15) corresponds to the monocyclic operator L_n of the mth order, then the polynomial (A-15), according to Sobol' (1969), is also a divisor of the binomial $(x^{2m-1}+1)$, but at the same time it is not a divisor for any binomial (x^s+1), where $s<2^m-1$. But by the virtue of Lemma 1 the order of the polynomial (A-16) is less than 2^m-1. Therefore, having divided (A-17) term by term by the polynomial (A-15), we shall obtain the following:

$$\begin{aligned}&(x^m+\ldots+a_1x+1)(x^{p_1}+\ldots+x^{p_1-c_1})+x^{k_1}+\ldots+1+\ldots\\&+(x^m+\ldots+a_1x+1)(x^{p_k}+\ldots+x^{p_k-c_k})+x^{k_l}+\ldots+1\\&=(x^m+\ldots+a_1x+1)(x^c+\ldots+x^d)+x^k+\ldots+1\end{aligned}$$

where $k_i<m$, $k<m$. The remainder left after division will be $x^k+\ \ldots\ +1 \neq 0$ (mod 2), since the remainder left after the division of each one of the binomials (A-17) by the polynomial (A-15) will contain 1. There is an odd number of such binomials so that the total remainder will also contain 1. Thus, the polynomial (A-16) is not divisible by the polynomial (A-15). And we have obtained a contradiction.

Theorem 7. There exist such initial conditions for the first column of the direction matrices that the sequence of points $\mathbf{Q}_i = (q^1(i),\ldots,q^n(i))$ constructed from the sequence of all the monocyclic operators taken successively possesses Property A for any n.

Proof. We shall prove this by induction. It is easy to verify that this theorem is true up to $n=7$ ($m(8)=5$). Let it be true also at some $n>7$. This means that the conditions of Lemma 2 are satisfied. And if we consider any monocyclic operator L_{n+1} that has its order the same as the order of L_n or larger by unity, then a part having its length equal to $n+1$ and beginning with 1 will, by virtue of Lemma 2, be found within the solution of this operator, such that

$$\begin{vmatrix} 1, u_2^{(1)}, \ldots, u_n^{(1)}, u_{n+1}^{(1)} \\ \ldots\ldots\ldots\ldots\ldots\ldots \\ 1, u_2^{(n)}, \ldots, u_n^{(n)}, u_{n+1}^{(n)} \\ 1, u_2^{(n+1)}, \ldots, u_n^{(n+1)}, u_{n+1}^{(n+1)} \end{vmatrix} = 1 \pmod 2 \qquad \text{(A-18)}$$

With this, all the corner determinants (A-18) are also equal to 1 (mod 2). As follows from Sobol' (1976), this proves that the theorem is true also for $n+1$. Thus, the theorem is proved.

Property A'. Let us draw $3n$ planes, $x_k=1/4$, $1/2$, $3/4$, $k=\overline{1,n}$. They divide the cube K^n into 2^n smaller cubes. In accordance with Sobol' (1976), we shall say a sequence (A-7) possesses Property A' if within any of its binary parts having the length of 2^{2n} all the points will pertain to different smaller cubes.

So a sequence (A-7) possesses Property A' if and only if the determinant consisting of the first two columns of the direction matrices is equal to 1 (mod 2).

Just like in the case with Property A, a question arises as to whether all the monocyclic operators can be used in succession so that the sequence (A-7) of points corresponding to them would possess Property A' at any n.

Remark. By analogy with Lemma 1, it will not be difficult to show that the following inequality should take place

$$2n+\frac{m(m+1)}{2} < 2^m, \quad m\geq 6. \qquad \text{(A-19)}$$

Theorem 8. There exist such initial conditions for the first two columns of direction matrices that the sequence of points $\mathbf{Q}_i=(q^1(i),\ldots,q^n(i))$ that is constructed from the sequence of all the monocyclic operators taken in succession for any n possesses Property A'.

Proof. The theorem will be proved by induction by analogy with Theorem 7. We shall verify if the theorem is true for all values of n, such that $m=m(n)<6$. Now let it be also true for some value of $n(n>13, m(n)>5$ (Sobol' 1969).

Let us prove that this theorem is also true for $n+1$.

Thus, we suppose that

$$\begin{vmatrix} u_{1,1}^{(1)}, \ldots, u_{2n,1}^{(1)} \\ u_{1,2}^{(1)}, \ldots, u_{2n,2}^{(1)} \\ \ldots\ldots\ldots\ldots \\ u_{1,1}^{(n)}, \ldots, u_{2n,1}^{(n)} \\ u_{1,2}^{(n)}, \ldots, u_{2n,2}^{(n)} \end{vmatrix} = 1 \pmod 2 \qquad \text{(A-20)}$$

as well as that all the corner determinants of even dimensionality are equal to 1 (mod 2). Now we shall consider an arbitrary monocyclic operator L_{n+1} of the same order that the order of the operator L_n if $L_{n+1} \neq L_{n-k}$, where $k \geq 0$ (here, L_{n-k} is either one of the operators that has some parts of solution with length of $2n$, included in (A-20), or else an operator whose order is larger by 1 than that of L_n). In view of the remark given previously and by analogy with Lemma 2, it is possible to make a conclusion that for some parts of the solution of a monocyclic equation with length of $2n + 1$, corresponding to the operator L_{n+1} and beginning with 1, the following equality holds:

$$\begin{vmatrix} u_{1,1}^{(1)}, \ldots, u_{2n+1,1}^{(1)}, u_{2n+1,1}^{(1)} \\ u_{1,2}^{(1)}, \ldots, u_{2n,2}^{(1)}, u_{2n+1,2}^{(1)} \\ \cdots\cdots\cdots\cdots\cdots\cdots \\ u_{1,1}^{(n)}, \ldots, u_{2n,1}^{(n)}, u_{2n+1,1}^{(n)} \\ u_{1,2}^{(n)}, \ldots, u_{2n,2}^{(n)}, u_{2n+1,2}^{(n)} \\ u_{1,1}^{(n+1)}, \ldots, u_{2n,1}^{(n+1)}, u_{2n+1,1}^{(n+1)} \end{vmatrix} = 1 \pmod 2, \qquad \text{(A-21)}$$

Now we shall consider the following determinant:

$$\begin{vmatrix} u_{1,1}^{(1)}, u_{2,1}^{(1)}, \ldots, u_{2(n+1),1}^{(1)} \\ u_{1,2}^{(1)}, u_{2,2}^{(1)}, \ldots, u_{2(n+1),2}^{(1)} \\ \cdots\cdots\cdots\cdots\cdots\cdots \\ u_{1,1}^{(n+1)}, u_{2,1}^{(n+1)}, \ldots, u_{2(n+1),1}^{(n+1)} \\ 0, \quad 1, \ldots, u_{2(n+1),2}^{(n+1)} \end{vmatrix} \qquad \text{(A-22)}$$

Let us show that the initial conditions in the last row can be chosen in such a way that the determinant (A-22) would be equal to 1 (mod 2). In order to make sure we shall suppose the opposite, that the determinant (A-22) would be equal to 0 (mod 2) for any choice of the last row beginning with 0 or 1. Let us show that any solution of the operator L_{n+1} will be a solution of a linear difference operator L^* of the $[2(n+1) - l]$th order where $l>1$. In other respects this theorem is to be proved similarly to what has been done in the proof of Theorem 7.

Since the determinant (A-21) is equal to 1 (mod 2), whereas the determinant (A-22) is equal to 0 (mod 2) (as is supposed), this implies the following. The last row in the determinant (A-11) satisfies the solution of some linear difference operator L^* of the $[2(n+1)-l]$th order regardless of initial conditions beginning with 0 and 1. The last row of the determinant (A-22) is the solution of L_{n+1}. If we take any two parts of this solution that have their length equal to $2(n+1)$ and begin with 0 or 1, then their sum is a part satisfying the solution of the operator L^* as well. This means that any part of the solution of the operator L_{n+1} that begins with 0 and has a length of $2(n+1)$ is also the solution of the operator L^*.

This solution is satisfied by the penultimate row of the determinant (A-22) that begins with 1. The same row is also a part of the solution of the operator L_{n+1}. Also, that is why any part of the solution of L_{n+1} having a length of $2(n+1)$ and beginning with 1 will satisfy L^* as well. (This is due to the fact that it can be represented as a sum of the penultimate row and some part of the solution beginning with 0.) Thus, any solution of the operator L_{n+1} is also a solution of the operator L^*. So this theorem is thereby proved.

Now we shall consider another very important additional property of uniformity.

Property *B*. As has been assumed earlier, let K^n be a unit n-dimensional cube, where $0 \leq x_k \leq 1$, $k = \overline{1,n}$ We shall fix some i_0, such that $1 \leq i_0 \leq n$. Let us divide the cube K^n by planes $x_k = 1/2$, $k \neq i_0$ into 2^{n-1} smaller multidimensional cubes. Let us carry out the same operation for all other fixed i_0. In total, we shall receive $n\ 2^{n-1}$ smaller cubes.

Let us consider an infinite sequence of points belonging to the cube K^n:

$$\mathbf{Q}_1, \ldots, \mathbf{Q}_i, \ldots \tag{A-23}$$

Definition. The sequence (A-23) possesses Property B if each one of the $n\ 2^{n-1}$ smaller cubes considered contains one and only one point from a binary part of this sequence taken arbitrarily and having a length of 2^{n-1}. (This definition has been proposed by Sobol'.)

In a similar way, we may consider the division of the cube K^n into smaller cubes that have their two coordinates x_{i_0} and x_{j_0} varying so that $0 \leq x_{i_0} \leq 1$ and $0 \leq x_{j_0} \leq 1$, respectively, whereas all the other x_k belong to the interval $0 \leq x_k \leq 1/2$ or $1/2 \leq x_k \leq 1$.

With different i_0 and j_0 we shall obtain $n(n-1) \cdot 2^{n-2}$ smaller cubes. If each one of these smaller cubes possesses one and only one point belonging to any binary part of the sequence (A-23), which has its length equal to 2^{n-2}, then we shall say that the sequence (A-23) possesses Property $B^{(2)}$. In a similar way we can define Property $B^{(k)}$, where $k \leq n-1$.

Let us assume that $Q_i = (q^1(i), \ldots, q^n(i))$ is an infinite sequence of points whose coordinates $q^k(i)$ are DP-sequences, whereas $\mathbf{\upsilon}_{sj}^{(k)}$ are the direction matrices that generate them.

It is easy to prove next statement.

Statement. There exist such sequences $\mathbf{Q}_i$ that possess both Properties B and $B^{(n-1)}$ for any n. Property $B^{(k)}$, where $2 \leq k < n-1$ is not met by any sequence $\mathbf{Q}_i$.

If $\upsilon_{1,1}^{(k)} = 1$, then it can be easily seen that Property $B^{(n-1)}$ is met.

Corollary. If $\upsilon_{1,1}^{(k)} = 1$ (in particular, when the sequence of $\mathbf{Q}_i$ is an LP_τ-sequence), then Property B is met, provided that n is an even number.

Also, it is not difficult to see that there is such a sequence of direction matrices that the corresponding sequence of $\mathbf{Q}_i$ will possess Property B at every value of n. Indeed, we shall put for every n

$$v_{i,1}^{(n)} = \sum_{k=1}^{n-1} v_{i,1}^{(k)} \text{ where } 1 \leq i < n-1, \tag{A-24}$$

$$\text{whereas} \qquad v_{n,1}^{(n)} = \sum_{k=1}^{n} v_{n,1}^{(k)} + 1.$$

By analogy with introduction of Property A', we can introduce Property B', $B' = (B^1)'$ or $(B^k)'$, $1 \leq k \leq n-1$. In this case, as it was with Property B, we deal with distribution of the points taken from a binary part of the sequence $\mathbf{Q}_i$ with a length $2^{2(n-k)}$. Incidentally, it is not difficult to prove the following statement.

An n-dimensional sequence of points $\mathbf{Q}_i$ that possesses Property $(B^k)'$ exists if and only if

$$n \geq \frac{3}{2k} + 1 \tag{A-25}$$

Do multidimensional sequences of points exist that possess Properties A and B simultaneously?

The answer to the question is in the following statement for which proof is not presented because of its simplicity.

If all $v_{1,1}^{(k)}$ are not identically equal to 1, then there exists such a sequence of direction matrices that the sequence of Q_i possesses Properties A and B simultaneously at every value of n. In this case, for any n,

$$v_{j,1}^{(n)} = 0, \;\; 1 \leq j < n-1, \;\; v_{n-1,1}^{(n)} = 1, \;\; v_{n,1}^{(n)} = \sum_{j=1}^{n-1} v_{n,1}^{(j)} + 1 \tag{A-26}$$

Otherwise, if $v_{1,1}^{(k)} = 1$ at all values of k, then it is easy to show that such a sequence of Q_i can be found that at every n it possesses Property A and at every even n it possesses Property B. In this case $q^{(k)}(i)$ can be selected so that it will belong to some monocyclic operator L_k of the mth order.

A question arises as to whether this statement can be proven if m is the order of operator L_k, where $m = m(k)$, that is, if all the monocyclic operators are taken in succession. The following statement answers this question.

Let $Q_i = (q^{(1)}(i), \ldots, q^{(n)}(i))$ be an infinite sequence of points in the cube K^n, $q^{(k)}(i)$ corresponding to all the monocyclic operators of the lowest orders. Then the sequence of Q_i cannot possess Properties B and A simultaneously at even $n \geq 12$.

The proof can be made by the reader without any difficulties.

Construction of New LP$_\tau$-*Sequences*

If a direction matrix ($\mathbf{v}_{ij}$) is such that all of its corner determinants are equal to 1 (mod 2), then the DP-sequence determined by this matrix will be an LP_0-sequence.

Can these sequences determining the values of various coordinates of multidimensional points be used to construct an LP_τ-sequence?

The following results enable us to extend Theorem 4 of Sobol' (1969), where the construction of an LP_τ-sequence is given, to our case.

Let L be a monocyclic operator of the mth order. Now we shall consider the sequence of operator degrees, that is, L, $L^2,\ldots,L^n,\ldots$. To each operator L^n in this sequence we put in accordance its m linearly independent solutions

$$\{u_{11}^{(n)}, \ldots, u_{k1}^{(n)},\ldots\}, \ldots, \{u_{1m}^{(n)}, \ldots, u_{km}^{(n)},\ldots\}, \tag{A-27}$$

where any solution of operator L^n is not a solution of operator L^{n-1}. In addition, the first nm, $1\leq n\leq\infty$, elements in each of the sequences (A-27) should be selected so that in the following matrix

$$\left|\begin{array}{l} u_{11}^{(1)},\ldots,u_{1m}^{(1)},\ldots,u_{11}^{(n)},\ldots,u_{1m}^{(n)} \\ \ldots\ldots\ldots\ldots\ldots\ldots\ldots\ldots\ldots\ldots \\ u_{k1}^{(1)},\ldots,u_{km}^{(1)},\ldots,u_{k1}^{(n)},\ldots,u_{km}^{(n)} \\ \ldots\ldots\ldots\ldots\ldots\ldots\ldots\ldots\ldots\ldots \end{array}\right| \tag{A-28}$$

all the corner determinants would be equal to 1 (mod 2). Let $\hat{L}^p$ be an arbitrary degree of a monocyclic operator $\hat{L}$ different from L.

Lemma 1. The operators $\hat{L}^p$ and L^k do not possess common solutions.

Proof. Indeed, let $\{u_i\}$ be one of the sequences in (A-27) such that $L^*(u_i)=0$. So, if $\hat{L}^p(u_i)=0$ would take place, then $L^{k-1}\hat{L}^p(u_i)=\hat{L}^pL^{k-1}$ $(u_i)=0$. But it is impossible because $L^{k-1}(u_i)$ is a nontrivial solution of monocyclic operator L. Thus, by Lemma 4 of Sobol' (1969), $\{u_i\}$ cannot be a solution of operator $\hat{L}^p$.

Lemma 2. If the operator $L\hat{L}^p$ is applied to any of the $n(n\leq m)$ linearly independent solutions of operator L^k, then we shall obtain n linearly independent solutions of operator L^{k-1}.

Proof. Let us at first show that $\hat{L}^p$ preserves the linear independence of the solutions of operator L^k. We shall apply the operator $\hat{L}^p$ to n solutions of L^k. Since $\hat{L}^p$ is a linear operator, then the linear dependence of the resulting solutions implies that $\hat{L}^p[\sum_{j=1}^{n} u_{ij}^{(k)}]=0$, where $i=\overline{1,\infty}$. But $\sum_{j=1}^{n} u_{ij}^{(k)}$ is not a trivial solution of L^k, since n solutions selected are linearly independent. And since the solution of L^k cannot be a solution of $\hat{L}^p$, the considered solutions are linearly independent.

Now we shall apply the operator L to these solutions of operator L^k. It is clear that in this way we shall obtain n solutions of operator L^{k-1} that are linearly independent. Actually, the opposite implies that $\sum_{j=1}^{n} L(u_{ij}^{(k)})=L(\sum_{j=1}^{n} u_{ij}^{(k)})=0$. The sum $\sum u_{ij}^{(k)}$ is not a solution of operator L since the matrix (A-28) contains all of the m linearly independent solutions of operator L. If $\sum_{j=1}^{n} u_{ij}^{(k)}$ would be a solution of L, then it should be expressed linearly through these m solutions. But this is in contradiction with the condition that all the corner determinants of the matrix (A-28) are equal to 1 (mod 2).

Theorem 9. Let $L_1,\ldots,L_n$ be various monocyclic operators that have orders equal to $m_1,\ldots,m_n$, respectively. Bring a matrix $(\mathbf{v}_{ij}^{(k,l)})$ similar to that of (A-33) in correspondence with every operator L_k and let $\mathbf{Q}_i$ be a sequence of points with the coordinates defined previously by these matrices. Then $\mathbf{Q}_i = (q^1(i),\ldots,q^n(i))$ is an LP_τ-sequence in K^n with the value of $\tau = \sum_{k=2}^{n} (m_k-1)$.

Proof. In accordance with the proof of the aforementioned Theorem 4 of Sobol' (1969), an arbitrary binary part of sequence with a length of 2^ν is a P_τ-net $\Leftrightarrow$ the rank of the matrix

$$[\mathbf{W}_{i,j}]=\begin{bmatrix} \upsilon_{1,1}^{(1,1)},\ldots,\upsilon_{1,\mu_1}^{(1,k_1)},\ldots,\upsilon_{1,1}^{(n,1)},\ldots,\upsilon_{1,\mu_n}^{(n,k_n)} \\ \cdots\cdots\cdots\cdots\cdots\cdots\cdots\cdots \\ \upsilon_{\nu,1}^{(1,1)},\ldots,\upsilon_{\nu,\mu_1}^{(1,k_1)},\ldots,\upsilon_{\nu,1}^{(n,1)},\ldots,\upsilon_{\nu,\mu_n}^{(n,k_n)} \end{bmatrix}, \qquad \text{(A-29)}$$

consisting of $\nu-\tau$ columns of direction matrices corresponding to various monocyclic operators is equal to $\nu-\tau$.

We shall use μ_k' to denote the remainder of the division of μ_k by m_k so that $\mu_k=\rho_k m_k+\mu'_k$ ($\mu_1+\ldots+\mu_n=\nu-\tau$, see Theorem 2).

Let $\rho_1\geq 1$. We shall keep in the matrix (A-29) the first m_1 rows without changes, whereas all the others will be replaced by linear combinations of rows in accordance with the application of the operator L_1 to rows. Thereby we shall obtain an equivalent matrix

$$\left[\begin{array}{c|c|c} \begin{array}{c} \upsilon_{1,1}^{(1,1)},\ldots,\upsilon_{1,m_1}^{(1,l_1)} \\ \cdots\cdots\cdots \\ \upsilon_{m_1,1}^{(1,1)},\ldots,\upsilon_{m_1,m_1}^{(1,l_1)} \end{array} & \multicolumn{2}{c}{\mathbf{W}_{i,j}} \\ \hline 0 & \begin{array}{c} \upsilon_{1,1}^{\prime(1,1)},\ldots,\upsilon_{1,\mu_1-m_1}^{\prime(1,d_1)} \\ \cdots\cdots\cdots \\ \upsilon_{\nu-m_1,1}^{\prime(1,1)},\ldots,\upsilon_{\nu-m_1,\mu_1-m_1}^{\prime(1,d_1)} \end{array} & \mathbf{L}_1\mathbf{W}_{i,j} \end{array}\right] \tag{A-30}$$

In the upper left corner, according to the condition of the theorem mentioned, there is a nonsingular matrix. It is not difficult to see that the columns $\upsilon_{i,j}^{\prime(l,k)}$, where $k \leq d_1$, $j \leq \mu_1 - m_1$, that are parts of the solutions of operator L_1, are linearly independent. If $\rho_1 > 1$, then in the lower portion of matrix (A-30) we repeat the same transformation for ρ_1 times. If $\mu_1' > 0$, then after these transformations there will be μ_1' columns with elements $\upsilon_{ij}^{\prime(1,l)}$. We shall move them to the end of this matrix. Thus, we obtain the following equivalent matrix

$$\left[\begin{array}{c|c|c} \begin{array}{cc} \mathbf{B}_1 & \\ & \mathbf{B}_1 \end{array} & \multicolumn{2}{c}{\mathbf{W}_{i,j}} \\ \hline 0 & \begin{array}{c} \dot{\upsilon}_{1,1}^{(2,1)},\ldots,\dot{\upsilon}_{1,\mu_2}^{(2,k_2)},\ldots,\dot{\upsilon}_{1,1}^{(n,1)},\ldots,\dot{\upsilon}_{1,\mu_n}^{(n,k_n)} \\ \cdots\cdots\cdots \end{array} & \begin{array}{c} \upsilon_1^{\prime(1,1)},\ldots,\upsilon_{1,\mu_1'}^{\prime(1,l_1)} \\ \cdots\cdots\cdots \end{array} \end{array}\right] \tag{A-31}$$

According to Lemma 2, it is so that $\dot{\upsilon}_{i,j}^{(p,b)}$ $(2 \leq p \leq n,\ 2 \leq b \leq k_p)$ are parts of the solutions of operator L_p^b. By the aforementioned Lemma 4 of Sobol' (1969), the columns $\dot{\upsilon}_{i,j}^{(2,1)}$, $1 \leq j \leq m_2$, form the following matrix

$$\mathbf{B}_2 = \begin{bmatrix} \dot{\upsilon}_{1,1}^{(2,1)},\ldots,\dot{\upsilon}_{1,m_2}^{(2,1)} \\ \cdots\cdots\cdots \\ \dot{\upsilon}_{m_2,1}^{(2,1)},\ldots,\dot{\upsilon}_{m_2,m_2}^{(2,1)} \end{bmatrix}$$

with determinant equal to 1.

Applying L_2 to a part of matrix (A-31) disposed under the line, we carry out transformations similar to the previous ones. We shall move the remaining μ_2' columns to the end of the matrix. Having made the same transformations with all L_k we shall obtain the following matrix

$$\left[\begin{array}{c|c} \begin{array}{cccc} \mathbf{B}_1 & & & \\ & \mathbf{B}_1 & & \\ & & \mathbf{B}_2 & \\ & & & \mathbf{B}_n \end{array} & \mathbf{W}_{i,j} \\ \hline 0 & \begin{array}{c} \upsilon_{1,1}^{\prime(1,1)},\ldots,\upsilon_{1,\mu_1'}^{\prime(1,l_1)},\ldots,\upsilon_{1,1}^{\prime(n,1)},\ldots,\upsilon_{1,\mu_n'}^{\prime(n,l_n)} \\ \cdots\cdots\cdots \\ \upsilon_{\nu',1}^{\prime(1,1)},\ldots,\upsilon_{\nu',\mu_1'}^{\prime(1,l_1)},\ldots,\upsilon_{\nu',1}^{\prime(n,1)},\ldots,\upsilon_{\nu',\mu_n'}^{\prime(n,l_n)} \end{array} \end{array}\right] \tag{A-32}$$

where $\nu' = \mu'_1 + \ldots + \mu'_n + \tau$, and the elements of matrix in the lower right corner are:

$$\upsilon'^{(k,l)}_{i,j} = L_1^{\rho_1} \ldots L_{k-1}^{\rho_{k-1}} L_{k+1}^{\rho_{k+1}} \ldots L_n^{\rho_n} (\upsilon^{(k,l)}_{i,j}) \qquad \text{(A-33)}$$

If all ρ_k were equal to 0, then the initial matrix would have already had the form of (A-32).

Since $j \leq \mu'_k < m_k$ in (A-33), then $L_k \upsilon'^{(k,l)}_{i,j} = 0$. This means that the columns of the right-corner matrix are parts of solutions of monocyclic equations.

Let among μ'_k be only s numbers different from zero. We shall designate them as μ'_f, where $1 \leq f \leq s \leq n$. All the solutions $\upsilon'^{(k,l)}_{i,j}$ corresponding to a given f are linearly independent by Lemma 2. Since

$\nu' = \sum_{k=1}^{m} (m_k - 1) + \sum_{f=1}^{s} \mu'_f \geq \sum_{f=1}^{s} (m_f + \mu'_f - 1) \geq \sum_{f=1}^{s} m_f$, from Lemma 7 Sobol' (1969), it follows that all the columns of the right-corner matrix are linearly independent, that is, its rank is equal to $\mu'_1 + \ldots + \mu'_n$. Hence, the rank of the entire matrix is equal to $\nu - \tau$.

The result of this theorem allows construction of LP_τ-sequences of a wider class than in Sobol' (1969). This positively influences the additional properties of uniformity. These properties must affect the results obtained when these sequences are used to solve practical problems.

Method for Calculating the Coordinates of Points of LP_τ-Sequences

Sobol' and Statnikov (1981) provide a table of numerators $r_j^{(l)}$, $1 \leq j \leq 51$, $1 \leq l \leq 20$. This table can be used to calculate the points $\mathbf{Q}_t$ with numbers within $1 \leq i \leq 2^{20}$ in the cube K^n, when $j \leq n \leq 51$. If points are required that have their dimensionality equal to n, then only the first n rows of the table should be used; if the number of points to be used is $j \leq 2^l$, where $l \leq 20$, then use should be actually made of the first l columns of that Table (see Table A-1, where $l \leq 16$, $j \leq 20$).

Algorithm. Prior to starting the calculations, one needs to replace table $r_j^{(l)}$ by table $V_j^{(l)}$, $V_j^{(l)} = r_j^{(l)} \cdot 2^{-l}$.

Then, if in binary notation the number of point i is represented as $i = e_m \ldots e_1$, all the Cartesian coordinates of the point $\mathbf{Q}_\mathrm{i} = (q_{i,1}, \ldots, q_{i,n})$ are to be calculated using the same formula

$$q_{i,j} = e_1 V_j^{(1)} * e_2 V_j^{(2)} * \ldots * e_m V_j^{(m)},\ j = \overline{1,n} \qquad \text{(A-34)}$$

where * denotes a digit-by-digit modulo two addition in binary notation. It was discussed previously in greater detail.

In Formula (A-34), there is no need to multiply e_j by V_j: if $e_j = 1$, then the corresponding value of $V_j^{(l)}$ will be included in (A-34), whereas if $e_j = 0$, then

Table A–1

j \ l	1	2	3	4	5	6	7	8	9	10	11	12	13	14	15	16
1	1	1	1	1	1	1	1	1	1	1	1	1	1	1	1	1
2	1	3	5	15	17	51	85	255	257	771	1,285	3,855	4,369	13,107	21,845	65,535
3	1	1	7	11	13	61	67	79	465	721	823	4,091	4,125	4,141	28,723	45,311
4	1	3	7	5	7	43	49	147	439	1,013	727	987	5,889	6,915	16,647	49,925
5	1	1	5	3	15	51	125	141	177	759	267	1,839	6,929	16,241	16,565	17,139
6	1	3	1	1	9	59	25	89	321	835	833	4,033	3,913	11,643	18,777	35,225
7	1	1	3	7	31	47	109	173	181	949	471	2,515	6,211	2,147	3,169	35,873
8	1	3	3	9	9	57	43	43	225	113	1,601	579	1,731	11,977	7,241	63,609
9	1	3	7	13	3	35	89	9	235	929	1,341	3,863	1,347	4,417	5.087	12,631
10	1	1	5	11	27	53	69	25	103	615	913	977	6,197	14,651	2,507	27,109
11	1	3	5	1	15	19	113	115	411	157	1,725	3,463	2,817	9,997	7,451	12,055
12	1	1	7	3	29	51	47	97	233	39	2,021	2,909	5,459	2,615	13,329	35,887
13	1	3	7	7	21	61	55	19	59	761	1,905	3,379	8,119	13,207	8,965	9,997
14	1	1	1	9	23	37	97	97	353	169	375	1,349	5,121	13,313	19,457	1,033
15	1	3	3	5	19	33	3	197	329	983	893	3,379	7,669	2,671	18,391	31,161
16	1	1	3	13	11	7	37	101	463	657	1,599	347	2,481	5,201	3,123	32.253
17	1	1	7	13	25	5	83	255	385	647	415	387	7,101	11,469	11,699	15,865
18	1	3	5	11	7	11	103	29	111	581	605	2,381	2,677	14,855	721	26,903
19	1	1	1	3	13	39	27	203	475	505	819	2,821	1,405	12,165	709	41,543
20	1	3	1	15	17	63	13	65	451	833	975	1,873	7,423	5,837	20,481	12,291

the corresponding value of $V_j^{(l)}$ should be omitted. Thus, in order to carry out the calculations according to Formula (A-34), only logical operations are needed.

Two Examples

Now we shall show how this rule can be used in computing the values of coordinates for a point in a five-dimensional cube, $\mathbf{Q}_{22}=(q_{22,1},\ldots,q_{22,5})$. In binary notation, 22 is written as 10110. This means that $q_{22,j}=V_j^{(2)} * V_j^{(3)} * V_j^{(5)}$. Using Table A-1, we shall obtain the following:

$$q_{22.1}=1/4 * 1/8 * \ 1/32=0.01 * 0.001 * 0.00001=0.01101$$
$$q_{22.2}=3/4 * 5/8 * 17/32=0.11 * 0.101 * 0.10001=0.11101$$
$$q_{22.3}=1/4 * 7/8 * 13/32=0.01 * 0.111 * 0.01101=0.11001$$
$$q_{22.4}=3/4 * 1/8 * 31/32=0.11 * 0.001 * 0.11111=0.00011$$
$$q_{22.5}=1/4 * 5/8 * 15/32=0.01 * 0.101 * 0.01111=0.10011$$

Thus, $\mathbf{Q}_{22}=(13/32;\ 29/32;\ 25/32;\ 3/32;\ 19/32)$.

As another example, we shall present the values of coordinates of point Q_{30} in the four-dimensional space: $Q_{30} = (15/32;\ 3/32;\ 15/32;\ 9/32)$.

As to well-known uniformly distributed sequences (nets), the following remarks should be made. Cubic nets (Sobol' 1969) are close to the worst ones, regarding the discrepancy D growth order. This indicates that these nets have poor distribution uniformity. Hammersley nets (Hammersley 1960) and Halton sequences (Halton 1960) are also well-known. The estimate of D for the latter is similar to that for LP_τ-sequences. In addition, it is worth noting parallelepiped nets (Korobov 1959; Hlawka 1962) and perfectly distributed sequences introduced by Hlawka. It is especially worth mentioning the work of Faure (1982) in which r-nary LP_0-sequences that have better asymptotic behavior than LP_τ-sequences are constructed.

It is interesting to mention the research (Bretley et al. 1993) in which the construction of sequences is given that generalize the r-nary LP_0-sequences as well as the LP_τ-sequences. In this case, the sequences proposed are asymptotically of the same order as the r-nary Faure's LP_0-sequences. These sequences slightly improve the value of index τ, and it is well-known that the smaller the value of τ, the better the sequence is distributed. However, it should be mentioned that the initial parts are not distributed better in Faure's LP_0-sequences nor in sequences from (Bretley et al. 1993) than those of the LP_τ-sequences.

In the next section, the constructions of P_τ-nets will be given in which the value of index τ is much smaller than that of the LP_τ-sequences. These also have initial parts with better uniformity as compared to the LP_τ-sequences, at least for the spaces that have their dimensionalities in excess of 10–15.

A-4. Nets with Improved Uniformity Characteristics

The purpose of this section is to construct the P_τ-nets (Statnikov and Matusov 1989) that have a substantially smaller value of index τ than the LP_τ-sequences.

Let n be the dimensionality of a unit cube K^n; $n(\tau)$ being the maximum dimensionality of K^n in which a P_τ-net can be constructed.

Theorem 10. For an arbitrary n there exist P_τ-nets having a length of $2^{\tau+3}$, for which $n(\tau)=2^{\nu-1}-1$ ($\tau=\log_2(n(\tau)+1)-2$) where 2^ν is the length of the net, $\nu=\tau+3$.

Proof. Let us consider the correctness of this statement for the nets having the length of $2^{\tau+3}$. As has been done earlier, the matrices of direction numbers ($\mathbf{v}_{i,j}^{(k)}$) are used to calculate the coordinates $q^k(i)$ of points Q_i of the net under consideration. As follows from Theorem 4 of Sobol' (1969) in order to have a net with a length of 2^ν, ($\tau<\nu$), being also a P_τ-net, it is necessary and sufficient that the rank of the following matrix

$$\begin{bmatrix} v_{1,1}^{(1)},\ldots,v_{1,\mu_1}^{(1)},\ldots,v_{1,1}^{(n)},\ldots,v_{1,\mu_n}^{(n)} \\ \cdots\cdots\cdots\cdots\cdots\cdots\cdots\cdots \\ v_{\nu,1}^{(1)},\ldots,v_{\nu,\mu_1}^{(1)},\ldots,v_{\nu,1}^{(n)},\ldots,v_{\nu,\mu_n}^{(n)} \end{bmatrix}, \tag{A-35}$$

is equal to $\nu-\tau$. (This matrix has ν rows , $\mu_1+\ldots+\mu_n=\nu-\tau$ columns, where μ_i are nonnegative numbers). In other words, it must be ensured that any $\nu-\tau$ vectors of dimensionality ν that are the first columns of the matrices ($\mathbf{v}_{i,j}^{(k)}$) would be linearly independent. In doing so, a set of $\nu-\tau$ vectors may be made up arbitrarily of the first columns of only one, several, or all of the matrices ($\mathbf{v}_{i,j}^{(k)}$).

Let us use the numbers $1,2,\ldots,\nu$ to designate an arbitrary basis of a ν-dimensional vector space of the field $GF(2)$. In this case, $\nu-\tau=3$.

Now we shall show how linear combination of these vectors can be used to construct a matrix of such a type as (A-35) with the aforementioned property. Let us consider Table A-2 for the sets of linearly independent vectors: The nth row in the table corresponds to the nth dimensionality of the cube K^n. We shall assume that the number i designates an n-dimensional vector that has its ith coordinate equal to 1, whereas all other coordinates are equal to 0. The sum of the numbers designates a vector that is the sum of the vectors corresponding to the numbers included in this sum.

Thus, it is necessary to make sure that any triad of the first vectors taken in succession from the rows of the aforementioned table are linearly independent. (It is obvious that all triads of vectors being in the same row in Table A-2 are linearly independent.) For this purpose, it would be sufficient to consider either

Table A–2

Dimension of cube K^n	Binary vectors for determining coordinates of points
1	1, 2, 3
2	3, 2, 1
3	3+2+1, 2, 1
4	4, 2, 1
5	4+2+1, 2, 1
6	4+3+1, 2, 1
7	4+3+2, 2, 1
8	5, 2, 1
9	5+2+1, 2, 1
10	5+3+2, 2, 1
11	5+3+1, 2, 1
12	5+4+1, 2, 1
13	5+4+2, 2, 1
14	5+4+3+2+1, 2, 1
15	5+4+3, 2, 1
•••	•••••••••••••••••
n	$\boldsymbol{K}$, 2, 1
$n+1$	$\boldsymbol{K}$+2+1, 2, 1
$n+2$	$\boldsymbol{K}$+3+2, 2, 1
$n+3$	$\boldsymbol{K}$+3+1, 2, 1
$n+4$	$\boldsymbol{K}$+4+3+2, 2, 1
$n+5$	$\boldsymbol{K}$+4+1, 2, 1
•••	•••••••••••••••••
$n+m$	$\boldsymbol{K}+\boldsymbol{K}-1+\boldsymbol{K}-2+$ ..., 2, 1
•••	•••••••••••••••••

the first vectors of any three rows from the table or the first two vectors of any one row and the first vector of another row.

Table A-2 has been constructed according to the following principle. In all rows the second position is taken by the same number, 2. This is possible since we are interested here only in the triads composed of the first vectors of the rows.

It is not difficult to see that the initial three rows satisfy the requirement formulated before. And all the rows, beginning with the fourth, are constructed in accordance with the same rule. The first position in the fourth row is taken by number 4 (it is impossible to use any sum of the first three numbers in this position (Sobol' 1969)). Further, on the first position in the following rows we shall put a vector that is coded in the following way: It is equal to the sum of number 4 and the sum of the initial two numbers standing in each of the previous rows. After this, number 5 is put in the first position, since it is already impossible to use the sums consisting only of the first four numbers, etc. The table thus constructed will be in conformity with our condition.

Actually, we shall denote the first vector in the ith row through $\mathbf{k}+\mathbf{a}_1^j+2$, where $\mathbf{k}$ is the largest number in the sum of the numbers that designates the given vector, $\mathbf{a}_1^j$ is the first vector in the jth row, $j<i$, and $\mathbf{a}_1^j \neq \mathbf{k}$.

If we represent now this vector as $\mathbf{k}+\mathbf{a}_1^n+2$ where $n<i$, then we obtain that $\mathbf{a}_1^j=\mathbf{a}_1^n$, but this is impossible on the assumption of induction. And if $\mathbf{k}+\mathbf{a}_1^j+2=\mathbf{k}+\mathbf{a}_1^n+2+\mathbf{a}_1^m$, then we have that $\mathbf{a}_1^j=\mathbf{a}_1^n+\mathbf{a}_1^m$, but this is also impossible.

In addition, it is impossible to construct a row that has its first vector coded by the sum of number $\mathbf{k}$ with other terms different from those as defined previously. Actually, one can easily be convinced of this for the seven rows of Table A-2. Let us assume that the considered statement is fulfilled for all rows having their numbers smaller than some fixed value of i. Then, this vector is $\mathbf{k}+\mathbf{a}=\mathbf{k}+\mathbf{a}_1^j+2$ or $\mathbf{k}+\mathbf{a}=\mathbf{k}+\mathbf{a}_1^j$, $j<i$, that is, $\mathbf{k}+\mathbf{a}$ stands either at the first position in some row preceding the ith row or else it is the sum of the first two vectors and $\mathbf{a}_1^j$.

Thus, if a number appears in the record of the first vector of the new row that is larger than all the previous ones, we obtain that quantity of rows in which it is used is larger by unity than all the preceding ones. It is not difficult to see that $n(\tau)=2^{\nu-1}-1$, where ν is the largest number used in the table for the given dimensionality (2^ν is the number of points of the P_τ-net in the space of the given dimensionality). Hence, we have that $\log_2(n(\tau)+1)=\nu-1$ or $\tau=\nu-3=\log_2(n(\tau)+1)-2$.

Corollary. When $\nu-\tau=3$, there do not exist P_τ-nets with $\tau < \log_2(n(\tau)+1)-2$.

Theorem 11. In the cube K^n there exist P_τ-nets with a length of $2^{\tau+2}$ for which $n(\tau)=2^\nu-1$, $\nu=\tau+2$, ($\tau=\log_2(n(\tau)+1)-2$).

Proof. Let us consider Table A-3. Here, we are interested only in pairs of linearly independent vectors composed of the first taken in succession elements of the rows of the given table. Therefore, in order that the linear independence of the pairs of vectors would be fulfilled, it is necessary to ensure that any vectors not equal to each other are placed at the first position of the rows. The second places can be occupied by any vector not coinciding with the first vector, such as, for instance, 1. Therefore, with a fixed value of ν the first positions of the rows are occupied by the $2^\nu-1$ vectors, that is, we have $2^\nu-1$ rows. The latter means that by the equality of $\nu-\tau=2$, we obtain

$$\tau=\log_2(n(\tau)+1)-2.$$

Corollary. In K^n there do not exist P_τ-nets with a length of $2^{\tau+2}$ for which $\tau < \log_2(n(\tau)+1)-2$.

Theorem 12. When $\nu-\tau=4$, there exist P_τ-nets with

Table A–3

Dimension of cube K^n	Binary vectors for determining coordinates of points
1	1, 2
2	2, 1
3	2+1, 1
4	3, 1
5	3+1, 1
6	3+2+1, 1
7	3+2, 1
8	4, 1
9	4+1, 1
10	4+2, 1
11	4+3, 1
12	4+2+1, 1
13	4+3+1, 1
14	4+3+2, 1
15	4+3+2+1, 1
16	5, 1
17	5+1, 1
18	5+2, 1
19	5+1+2, 1
20	5+3, 1
21	5+3+1, 1
22	5+3+2, 1
23	5+3+2+1, 1
24	5+4, 1
25	5+4+1, 1
26	5+4+2, 1
27	5+4+3, 1
28	5+4+2+1, 1
29	5+4+3+1, 1
30	5+4+3+2, 1
31	5+4+3+2+1, 1
...	....
n	$\boldsymbol{K}$, 1
$n+1$	$\boldsymbol{K}+1$, 1
$n+2$	$\boldsymbol{K}+2$, 1
...	....
$n+2^{k-1}-1$	$\boldsymbol{K}+(\boldsymbol{K}-1)+(\boldsymbol{K}-2)+\ldots+2+1,1$
...	....

$$\tau=\begin{cases}\dfrac{n-3}{2}, & n \text{ is odd},\\[2mm] \dfrac{n-2}{2}, & n \text{ is even},\end{cases}$$

where n is an arbitrary dimensionality of K^n.

Proof. In order to prove this statement, it is necessary to make sure that the tetrads of vectors compiled arbitrarily from the initial elements of the rows taken in succession of Table A-4 are linearly independent. In other words, there must be some linearly independent vectors of at least one from the following sets: the first three vectors of an arbitrary row and the initial vector taken from some other row; the first two vectors taken from some two rows; the tetrad of initial vectors taken from various rows; the first two vectors taken from one row and the first two vectors taken from some other two rows. It is not difficult to ascertain directly the correctness of this statement for dimensionalities not exceeding ten. Let us consider the Table A-4. In this table, beginning from the tenth row, the following regularity is observed. For each couple of neighboring rows in which first numbers are odd, we have vectors:

$$2\mathbf{K}-1,\ (2\mathbf{K}-2)+2,\ 3,\ 1;$$
$$(2\mathbf{K}-1)+(2\mathbf{K}-2)+3+1,\ (2\mathbf{K}-3)+2+1,\ 2,\ 1$$

For each couple of neighboring rows in which first numbers are even, we have vectors:

$$2\mathbf{K},\ (2\mathbf{K}-1)+2,\ 3,\ 1;$$
$$2\mathbf{K}+(2\mathbf{K}-1)+3+2+1,\ (2\mathbf{K}-2)+2+1,\ 2,\ 1.$$

Now, relying on the aforementioned regularity it is easy to prove that all the tetrads of vectors are linearly independent.

For P_τ-nets with lengths of 2^{n+2}, 2^{n+3}, and 2^{n+4}, where n is the dimensionality of the cube K^n, the constructions of the matrices consisting of direction numbers have been studied. Common for these nets is that in all three cases, $\tau=n-3$. We shall prove the theorem for the case where the length is 2^{n+4}. The other cases are proved by analogy.

Theorem 13. For any n in the cube K^n there exist P_τ-nets of length equal to 2^{n+2}, 2^{n+3}, 2^{n+4}, for which $\tau = n-3$.

Proof. As was mentioned previously, we shall prove this theorem for the net with a length of 2^{n+4}. In this case, $\nu-\tau=7$. This means that it is necessary to prove that for given dimensionality n we can make up n rows formed by sevens of linearly independent vectors with the binary dimensionality of $n+4$, which possess the following property. Any sevens of vectors composed arbitrarily of the first vectors of these rows taken in succession are linearly independent. Such sets of vectors we will call correct.

It is not difficult to make sure (see Table A-5) that for $n=3$ the property of the sevens of vectors as formulated here is fulfilled. Table A-5 represents in parentheses in the nth row those vectors that should be used only in the case of the dimensionality n of a unit cube. Besides, a direct check shows that the vector

Table A–4

Dimension of cube K^n	Binary vectors for determining coordinates of points
1	1, 2, 3, 4
2	4+3, 3+2, 2+1, 1
3	4, 1+2+3, 2, 1
4	5, 3+1, 2, 1
5	6, 5+2, 2, 1
6	6+5+2+3, 4+2+1, 2, 1
7	6+4+2, 5+3+2+1, 2, 1
8	7, 6+3+2+1, 3, 1
9	7+5+4+2+1, 6+2+1, 2, 1
10	8, 7+2, 3, 1
11	8+7+3+2+1, 7+6+2+1, 2, 1
12	9, 8+2, 3, 1
13	9+8+3+1, 7+2+1, 2, 1
14	10, 9+2, 3, 1
15	10+9+3+2+1, 8+2+1, 2, 1
•••	•••••••••••••••••••••••••••••••
n	$2K-1$, $(2K-2)+2$, 3, 1
$n+1$	$(2K-1)+(2K-2)+3+1$, $(2K-3)+2+1$, 2, 1
$n+2$	$2K$, $(2K-1)+2$, 3, 1
$n+3$	$2K+(2K-1)+3+2+1$, $(2K-2)+2+1$, 2, 1
•••	•••••••••••••••••••••••••••••••

5+3+2 is not contained in the linear combination of any correct four vectors taken from the first three rows, that is, in the four made up arbitrarily of the first vectors taken in succession from the first three rows of Table A-5. In a similar manner, the vector 6+4+3+1 is not contained in any correct five of vectors taken from the first three rows, and the sum of the corresponding elements, the vector 6+5+4+2+1, is not contained in any correct four. Further, the vector indicated by number 8 is not contained in any of the correct sixes taken from the first three rows. Therefore, the vector 8 can be used as the first element of the fourth row; and as the second element we use the vector 6+4+3+1, since it is not a linear combination of any correct five. If we want to construct a net only in a four-dimensional space, we use as the third element the vector 5+3+2 since it is not contained in any correct four. And the vector 5+3+2+6+4+3+1=6+5+4+2+1 is not contained in any correct four. By analogy, one can make sure that the elements in the parentheses in the fourth row can be used for the construction of a P_1-net.

If Table A-5 is not completed on the fourth row, then the elements presented in this row without parentheses should be used as the remaining elements. Actually, the vector 9+4+2+1+9+3+1=4+3+2 is not a linear combination of any correct triad made up of the elements taken from the first three rows and the vectors 8 and 6+4+3+1. The vector 5+3+1 cannot be obtained as a linear

Table A–5

Dimension of cube K^n	Binary vectors for determining coordinates of points
1	1, 2, 3, 4, 5, 6, 7
2	7+6, 6+5, 5+4, 4+3, 3+2, 2+1, 1
3	7, 1+2+3+4+5+6, 8+4+2, 8+3+2, 5+3+2, 6+4+3+1, 1
4	8, 6+4+3+1, 9+4+2+1(5+3+2), 9+3+1(4), 5+3+1(7+2), 6+5+4+2+1(3), 1
5	9, 8+7+6+5+4+2+1, 10+4+2(5+3+1), 10+3+2(4+3), 5+3+2(8+6), 6+3+1(3), 1
6	10, 9+8+7+6+3+1, 11+4+2+1(5+3+2), 11+3+2+1(4+3+2+1), 5+3+1(9+7+6+1), 6+5+4+1(3+1), 1
7	11, 10+9+8+6+5+4+1, 12+4+2(5+3+1), 12+3+2(4+3), 5+3+2(9+8+7), 6+3(10+9), 1
8	12, 11+10+9+6+3, 13+4+2+1(5+3+2), 13+3+2+1(4+3+2+1), 5+3+1(10+9+8), 6+5+4(11+10), 1
9	13, 12+11+10+6+5+4, 14+4+2(5+3+1), 14+3+2(4+3), 5+3+2(11+10+9), 6+3+1(12+11), 1
10	14, 13+12+11+6+3+1, 15+4+2+1(5+3+2), 15+3+2+1(4+3+2+1), 5+3+1(12+11+10), 6+5+4+1(13+12), 1
...	....
2*K*	**n**+4, **n**+3+**n**+2+**n**+1+6+3(+1), **n**+5+4+2+1(5+3+2), **n**+5+3+2+1(4+3+2+1), 5+3+1(**n**+2+**n**+1+**n**), 6+5+4(+1) (**n**+3+**n**+2), 1
2*K*+1	**n**+4, **n**+3+**n**+2+**n**+1+6+5+4(+1), **n**+5+4+2(5+3+1), **n**+5+3+2(4+3), 5+3+2(**n**+2+**n**+1+**n**), 6+3(+1) (**n**+3+**n**+2), 1
...	....

combination of the first four vectors taken from the fourth row and any correct two taken from the previous rows. This vector could have been obtained only by attaching to the given four a correct two taken either from the third or fourth row. But, as is plain to see, such attachments do not lead to the vector 5+3+1. In addition, it can be noted that the vector 5+3+1 does not belong to any correct four made up of the elements from the first four rows.

The vector 6+5+4+2+1 cannot be formed as a linear combination of five of vectors from the fourth row and the first vector taken from any one of the previous rows. Out of all the first vectors that are possible, it is only the initial vector takén from the first row, 1, that can be attached to this five.

Let us join 1 to the given five. It is easy to see that no one of the linear combinations of vectors in the resulting six can give 6+5+4+2+1. Besides, the vector 6+5+4+2+1 does not belong to any correct four made up of the vectors from the first four rows.

The sum of vectors 6+5+4+2+1+5+3+1=6+4+3+2 does not belong to any correct triad made up of the first four rows.

The theorem for the dimensionalities of the cube K^n when n is 5, 6, or 7, is proved similarly to what has been given herein. This proof is not presented because it is rather cumbersome. The reader can reconstruct it without any difficulties using Table A-5.

Note, that the elements from Table A-5 starting from the eighth row are subjected to the following regularity. If n is the number of a row and n is an even number, then this row takes the following form: **n**+4, **n**+3+**n**+2+**n**+1+6+3(+1), 12+5+4+2+1 (5+3+2), **n**+5+3+2+1 (4+3+2+1), 5+3+1 (**n**+2+**n**+1+**n**), 6+5+4(+1) (**n**+3+**n**+2) where the designation (+1) shows that the second and the sixth places are taken either by the vectors given here (without 1), or else 1 is to be simultaneously added to corresponding vectors. These additions are to take place on the even dimensionalities only, every second one, starting from n=6.

If n is an odd number, then the nth row takes the following form: **n**+4, **n**+3+**n**+2+**n**+1+6+5+4(+1), **n**+5+4+2(5+3+1), **n**+5+3+2(4+3), 5+3+2 (**n**+2+**n**+1+**n**), 6+3(+1) (**n**+3+**n**+2), where the vector 1 is added on the odd dimensionalities, every second one, starting from n=9.

Therefore, the proof that the linear independence conditions are fulfilled for all the correct sevens as is necessary for existence of the n-dimensional P_{n-3}-net having a length of 2^{n+4}, is similar to the proof for $n \leq 7$.

Corollary. If τ is an index of an n-dimensional LP_τ-sequence equal to $\sum_{i=1}^{n} m_i - n > n-3+k$ for some natural k, then there exists an n-dimensional P_τ-net with $\tau < \sum_{i=1}^{n} m_i - n$ having a length of 2^{n+k+4}.

Here, m_i is the order of the monocyclic operator corresponding to the ith dimensionality of the unit cube.

Proof. The validity of this formulation of the corollary is obvious. Indeed, let us project the points of the P_{k+n-3}-net having length of 2^{k+n+4} from the space of the dimensionality $n+k$ into the n-dimensional space. In doing so, we obtain an n-dimensional net with a length of 2^{k+n+4} for which $\tau=n+k-3$.

From what was said previously, in particular from the method used in the proof, the following statement can be read quite naturally.

For an arbitrary dimensionality of the cube K^n and for any ν, there exist P_τ-nets with length of 2^ν where $\tau=n-3$.

Remarks

1. It is not difficult to show that the direction matrices for the nets given here can be selected so that the next condition would be satisfied: Different points of the net have different projections into all coordinate axes.
2. As far as the use of the nets being constructed is concerned, the following can be said. They have a substantially smaller value of index τ than those of LP_τ-sequences. This improves essentially all the characteristics defining the uniformity of their distribution. The results thus obtained allow construction of the P_τ-nets with the number of points acceptable in practice even in the spaces of relatively large dimensionalities ($n>50$). In this case it is appropriate to use one of the nets corresponding to Tables A-2 or A-4. When the dimensionalities are large enough, a preferable net corresponds to Table A-2.
3. The algorithm used to calculate the coordinates of the points for these nets is the same as that used for the LP_τ-sequences. The difference exists only in the direction matrices. Here, they are defined by the matrices formed by the binary vectors taken from Tables A-2–A-5. One such matrix is represented in Table A-6.

According to the table of numerators $r_j^{(l)}$, the coordinates of the direction numbers $V_j^{(l)}$ are calculated as

$$V_j^{(l)} = r_j^{(l)} \cdot 2^{-l}.$$

Here, j is the number of the coordinate of an appropriate point included in an P_τ-net and l corresponds to the point number equal to 2^l. The calculations of the coordinates q_{ij} for the ith point are to be carried out in accordance with (A-34).

Example

Let us show how Table A-6 can be used to obtain the coordinates of some specific point in K^n. Let $n=5$, and the point number is 22. As has been already

Table A–6

j \ l	1	2	3	4	5	6	7	8	9	10	11	12
1	1.00	1	1	1	1	1	1	1	1	1	1	1
2	1.25	3	4	1	1	1	1	1	1	1	1	1
3	0.25	1	0	9	1	1	1	1	1	1	1	1
4	1.25	3	0	9	1	1	1	1	1	1	1	1
5	1.25	1	4	9	1	1	1	1	1	1	1	1
6	0.25	3	4	9	1	1	1	1	1	1	1	1
7	0.25	1	0	1	17	1	1	1	1	1	1	1
8	1.25	3	0	1	17	1	1	1	1	1	1	1
9	1.25	1	4	1	17	1	1	1	1	1	1	1
10	0.25	3	4	1	17	1	1	1	1	1	1	1
11	0.25	3	0	9	17	1	1	1	1	1	1	1
12	1.00	1	1	9	17	1	1	1	1	1	1	1
13	1.25	3	4	9	17	1	1	1	1	1	1	1
14	1.25	3	4	1	1	33	65	129	257	1	1	1
15	0.25	1	4	1	1	33	65	129	257	1	1	1
16	0.00	1	1	9	1	33	65	129	257	1	1	1
17	1.25	3	0	9	1	33	65	129	257	1	1	1
18	1.25	1	4	9	1	33	65	129	257	1	1	1
19	0.25	3	4	9	1	33	65	129	257	1	1	1
20	0.00	1	1	1	17	33	65	129	257	1	1	1

said, 22 corresponds to 10110 in binary notation, $q_{22,j}=V_j^{(2)} * V_j^{(3)} * V_j^{(5)}, j=\overline{1,5}$.

Finding the values in Table A-6 at the intersections of the jth row with the second, third, and fifth columns, we shall obtain, respectively:

$$
\begin{aligned}
q_{22,1}&=1/4 * 1/8 * 1/32=0.01 * 0.001 * 0.00001=0.01101\\
q_{22,2}&=3/4 * 1/2 * 1/32=0.11 * 0.1 \quad * 0.00001=0.01001\\
q_{22,3}&=1/4 * 1/2 * 1/32=0.01 * 0.1 \quad * 0.00001=0.11001\\
q_{22,4}&=3/4 * 1/2 * 1/32=0.11 * 0.1 \quad * 0.00001=0.01001\\
q_{22,5}&=1/4 * 1/8 * 1/32=0.01 * 0.001 * 0.00001=0.01101
\end{aligned}
$$

Thus, $\mathbf{Q}_{22}$ = (13/32; 9/32; 25/32 9/32; 13/32).

When using a P_τ-net, the values of the coordinates of the ith point depend not on the dimension of the design-variable vector, r, only, but also on the number of trials, $N=2^k$ specified beforehand. The coordinates of the given point are defined for $N=32$.

References

Aivazyan, S. A., I. S. Yenyukov, and L. D. Meshalkin. 1983. *Applied Statistics. Principles of Modelling and Initial Data Processing*. Moscow: Finansy i Statistika (in Russian).

Aivazyan, S. A., I. S. Yenyukov, and L. D. Meshalkin. 1986. *Applied Statistics. Study of Relationships*. Moscow: Finansy i Statistika (in Russian).

Artobolevskii, I. I., M. D. Genkin, V. I. Sergeev, and R. B. Statnikov. 1974. "Search for a trade-off solution in choosing machines parameters." *Doklady AN SSSR*. Vol. 219, No. 1: 53–57 (in Russian).

Auerbakh, V. M., I. Z. Manevich, M. O. Volya, and I. I. Zelikovich. 1985. "Increasing efficiency of the underground railway construction mechanization on the basis of the modular principle." *Transportnoye Stroitel'stvo* No. 10: 24–28 (in Russian).

Banach, L. Y. 1988. "Weak energy and spectral couplings in mechanical oscillatory systems." *Izvestiya AN SSSR. Mekhanika Tverdogo Tela*. No. 2: 38–43 (in Russian).

Banach, L. Y., and M. D. Perminov. 1972. "Investigation of complex dynamic systems taking into account weak couplings between subsystems." *Mashinovedeniye,* No. 4: 3–9 (in Russian).

Bartel, D. L., and R. W. Marks. 1974. "The optimum design of mechanical systems with competing design objectives." *ASME Transactions-Journal of Engineering for Industry*. Vol. 96, Ser. B., No. 1: 171–179.

Baturidi, A. I., V. M. Burlakov, V. I. Molotkov, M. I. Osin, V. I. Senozatskii, S. I. Starkov, K. A. Tuskayev, G. G. Usov, and I. P. Fel'dman. 1991. *Automation of Designing Machine Industry Articles*. Moscow: Znaniye (in Russian).

Bekey, G. A. 1970. "System identification—An introduction and a survey." *Simulation*. Vol. 15, No. 4: 151–166.

Belsley, D. A., E. Kuh, and L. E. Welsch. 1980. *Regression Diagnostics: Identifying Influential Data and Sources of Collinearity*. New York: John Wiley.

Benayoun, R., J. De Montgolfier, J. Tergny, and O. Larichev. 1971. "Linear programming with multiple objective functions: Step method (STEM)." *Mathematical Programming* Vol. 1, No. 3: 366–375.

Benson, H. P. 1992. "A finite, non-adjacent extreme point search algorithm for optimization over the efficient set." *Journal of Optimization Theory and Applications*. Vol. 73, N1: 47–64.

Berezanskii, O. M., and Y. N. Semenov. 1988. "Solution of ships design problems on the basis of multicriteria optimization methods." *Sudostroitel'naya Promyshlennost': Ser. Sistemy Avtomatizatsii Proektirovaniya, Proizvodstva i Upravleniya*. No. 9: 78–85 (in Russian).

Betin, A. V., and V. V. Kaminskaya. 1992. "Parametric optimization of structures of lathes with a movable workhead." In *Integrated CAD Systems for Automated Manufactures,* ed. V. V. Kaminskaya, pp. 102–113. Moscow: Experimental Research and Development Institute for Metal-Cutting Machine-Tools (in Russian).

Bezier, P. 1987. "Courbes et surfaces." In *Mathematiques et CAO,* Vol. 4. Paris: Hermes Publishing.

Bode, H. W. 1945. *Network Analysis and Feedback Amplifier Design*. New York: Van Nostrand Reinhold.

Bondarenko, M. I., A. Y. Nazemkin, A. A. Pozhalostin, R. B. Statnikov, and V. S. Shenfel'd, 1994. "Construction of consistent solutions in multicriteria problems of optimization of large systems.", *Physics-Doklady* Vol. 39, No. 4: 274–279. Translated from *Doklady Rossiyskoy Akademii Nauk,* Vol. 335, No. 6: 719–724.

Borovin, G. K., I. A. Kuz'min, D. N. Popov, and P. S. Romashkin. 1985. *Optimization of Parameters of an Electrohydraulic Amplifier with Indirect (Nonunity) Feedback*. Moscow: Keldysh Institute of Applied Mathematics (in Russian).

Breiman, L., and J. H. Friedman, 1985. "Estimating optimal transformations for multiple regression and correlation." *Journal of the American Statistical Association,* Vol. 80, No. 391: 580–598.

Bretley, P., B. L. Fox, and H. Niederreiter. 1993. "Implementation and tests of low discrepancy sequences." *Association of Computing Machinery. Transactions on modeling and computer simulation* Vol. 2, No. 2: 195–213.

Chernikov, V. A. 1986. "Multicriteria optimization of actuators of fruit harvesters." *Traktory i Sel'khozmashiny* No. 6: 36–39 (in Russian).

Cohon, J. L., G. Scavone, and R. Solanki. 1988. "Multicriteria optimization in resources planning." In *Multicriteria Optimization in Engineering and in the Sciences,* ed. W. Stadler, pp. 117–160. New York: Plenum Press.

Craig, R. R., and M. C. C. Bampton. 1968. "Coupling of substructures for dynamic analysis." *AIAA Journal* Vol. 6, No. 7: 113–121.

Da Cunha, N. O., and E. Polak. 1967. "Constrained minimization under vector valued criteria in finite dimensional spaces." *Journal of Mathematical Analysis and Applications* Vol. 19, No. 1: 103–124.

Dauer, J. P., and O. A. Saleh. 1992. "A representation of the set of feasible objectives in multiple objective linear programs." *Linear Algebra and Its Applications* Vol. 166: 261–275.

Debagyan, O. V., and V. S. Khomyakov. 1982. "Parametric optimization of the universal milling machine structure on the basis of LP-search." *Izvestiya AN Armyanskoy SSR* Vol. 35, No. 5: 3–10 (in Russian).

Den Hartog, J. P. 1956. *Mechanical Vibrations*. New York: McGraw-Hill.

Dokukin, A. V., Y. D. Krasnikov, and Z. Y. Khurgin. 1978. *Statistical Dynamics of Mining Machines*. Moscow: Mashinostroyeniye (in Russian).

Dol'berg, M. D., and N. N. Jasnitskaya. 1973. "Estimates from below for frequencies of elastic system oscillations. Generalized Donkerlay-Papkovich estimates." *Doklady AN SSSR,* Vol. 212, No. 6: 1317–1319 (in Russian).

Draper, N., and H. Smit. 1966. *Applied Regression Analysis.* New York: John Wiley.

Dubov, Y. A., S. I. Travkin, and V. N. Yakimets. 1986. *Multicriteria Models for Formation and Selection of Systems Variants.* Moscow: Nauka (in Russian).

Dyer, J. S., P. C. Fishburn, R. E. Steuer, J. Wallenius, and S. Zionts. 1992. "Multiple criteria decision making, multiattribute utility theory: The next ten years." *Management Science* Vol. 38, No. 5: 645–654.

Eschenauer, H. A. 1988. "Multicriteria optimization techniques for highly accurate focusing systems." In *Multicriteria Optimization in Engineering and in the Sciences,* ed. W. Stadler, pp. 309–354. New York: Plenum Press.

Ester, J. 1987. "Some applications of MCDM on engineering problems." *Operations Research Spectrum* Vol. 9, No. 2: 59–80 (in German).

Faure, H. 1982. "Discrépance de suites associées á un systeme de numeration (en dimension S)." *Acta Arithmetica.* Vol. 41, No. 4: 337–351.

Fishburn, P. C. 1970. *Utility Theory for Decision Making.* New York: John Wiley.

Friedman, J. H., and W. Stuetzle. 1981. "Projection pursuit regression." *Journal of American Statistic Association* Vol. 76, No. 376: 817–823.

Gass, S., and T. Saaty. 1955. "The computational algorithm for the parametric objective function." *Naval Research Logistics Quarterly* Vol. 2, No. 1: 39–45.

Gearhart, W. B. 1979. "Compromise solutions and estimation of the noninferior set." *Journal of Optimization Theory and Applications.* Vol. 28, No. 1: 29–47.

Genkin, M. D., and R. B. Statnikov. 1987. "Basic problems of machines optimal design." *Vestnik AN SSSR.* No. 4: 28–39 (in Russian).

Genkin, M. D., L. V. Korchemnyi, I. B. Matusov, L. N. Sinel'nikov, A. I. Stavitskii, and R. B. Statnikov. 1983. "Multicriteria selection of optimal parameters for the automobile engine valve gear." *Mashinovedeniye* No. 3: 60–68 (in Russian).

Genkin, M. D., R. B. Statnikov, I. B. Matusov, and M. D. Perminov. 1987. "On the adequacy of a mathematical model to a real object. Vector identification." *Doklady AN SSSR* Vol. 294, No. 3: 549–552 (in Russian).

Gerasimov, Y. N., Y. M. Pochtman, and V. V. Skalozub. 1985. *Multicriteria Optimal Structural Design.* Kiev, Donetsk: Vishcha Shkola (in Russian).

Godzhaev, Z. A., S. S. Dmitrichenko, and A. Y. Guberniev. 1992. "Optimal design of machines shaftings." *Vestnik Mashinostroyeniya* No. 1: 3–5 (in Russian).

Goldman, R. L. 1969. "Vibration analysis by dynamic partitioning." *AIAA Journal* Vol. 7, No. 6: 1152–1154.

Gorodetskii, Y. I. 1984. "Development of an object-oriented CAD system for searching for optimal parameters of vertical milling machines with respect to dynamic quality criteria." In *Mathematical Modelling and Software for CAD Systems,* ed Y. I. Gorodetskii; pp. 48–66. Gorky: Gorky State University (in Russian).

Graupe, D. 1976. *Identification of Systems.* New York: Robert E. Krieger Publishing Company.

Grinkevich, V. K., P. I. Zinyukov, R. B. Statnikov, L. V. Sukhorukov, and S. I. Fridman. 1978. "Determination of optimal parameters of a mechanism according to a number of local quality criteria." In *Methods of Developing Low-Noise Machines,* pp. 44–49, Moscow: Nauka (in Russian).

Gurychev, S. E., A. V. Gringlaz, and G. S. Bolotin. 1985. "Investigation of dynamic characteristics of a multi-objective machine tool." *Stanki i Instrumenty*. No. 1: 22–24 (in Russian).

Halton, J. H. 1960. "On the efficiency of certain quasi-random sequences of points in evaluating multi-dimensional integrals." *Numerical Mathematics* Vol. 2, No. 2: 84–90.

Hammersley, J. M. 1960. "Monte Carlo methods for solving multivariable problems." *Annals of the New-York Academy of Science,* Vol. 86, No. 3: 844–874.

Hlawka, E. 1962. "Zur angenäherten Berechnung mehrfacher Integrale." *Monatshefte für Mathematik* Vol. 66, No. 2: 140–151.

Ingerman, V. G. 1985. "Increasing efficiency of mathematical modelling in petrophysics." *Neftyanaya Promyshlennost': Ser. Neftegazovaya Geologiya, Geofizika i Bureniye*. No. 10: 15–18 (in Russian).

Isermann, H. 1977. "The enumeration of the set of all efficient solutions for a linear multiple objective program." *Operational Research Quarterly* Vol. 28, No. 3: 711–725.

Kaminskaya, V. V. 1984. "Analysis of dynamics of heavy-duty vertical turning lathes." *Stanki i Instrumenty* No. 12: 8–12 (in Russian).

Kaminskaya, V. V., and A. V. Gringlaz. 1989. "Computational analysis of dynamic characteristics of machine tools structures." *Stanki i Instrumenty* No. 2: 10–13 (in Russian).

Karamzin, Y. N., A. P. Sukhorukov, and V. A. Trofilov. 1986. *Optimal Control of Light Beams in Nonlinear Medium*. Moscow: Znaniye (in Russian).

Karlin, S. 1959. *Mathematical Methods and Theory in Games, Programming, and Economics*. London-Paris: Pergamon Press.

Kato, T. 1966. *Perturbation Theory for Linear Operators*. New York: Springer-Verlag.

Keeney, R. L. 1972. "Utility functions for multiattributed consequences." *Management Science* Vol. 18, No. 5: 276–287.

Kelley, J. L. 1957. *General Topology*. New York: Van Nostrand Reinhold.

Khachaturov, A. A. 1976. *Dynamics of the Road-Tire-Automobile-Driver System*. Moscow: Mashinostroyeniye (in Russian).

Khomyakov, V. S., and A. I. Yatskov. 1984. "The structure optimization of a heavy-duty single-column vertical boring and turning machine." *Stanki i Instrumenty* No. 5: 14–16 (in Russian).

Khronin, D. V., V. I. Baulin, Y. P. Kirpikin, and M. K. Leontiev. 1984. *Basic Principles of Computer Aided Design of Flying Vehicles Engines*. Moscow: Mashinostroyeniye (in Russian).

Korchemnyi, L. V. 1981. *Automobile Engine Valve Gear*. Moscow: Mashinostroyeniye (in Russian).

Korobov, N. M. 1959. "On the approximate computation of multiple integrals." *Doklady AN SSSR* Vol. 124, No. 6: 1207–1210 (in Russian).

Kornbluth, J. S. H. 1974. "Accounting in multiple objective linear programming". *Accounting Review* Vol. 49, No. 2: 284–295.

Koski, J. 1988. "Multicriteria truss optimization." In *Multicriteria Optimization in Engineering and in the Sciences,* ed. W. Stadler, pp. 263–308. New York: Plenum Press.

Krasnoshchokov, P. S., A. A. Petrov, and V. V. Fedorov. 1986. *Computer Science and Design*. Moscow: Znaniye (in Russian).

Kreinin, G. V., N. M. Ostapishin, and G. V. Tarkhanov. 1986. "The choice of design variables of active pneumatic vibration isolators." In *Vibrations and Vibroaconstical Activity of Machines and Structures,* ed. Y. I. Bobrovnitskii, pp. 20–25. Moscow: Nauka (in Russian).

Kron, G. 1963. *Diakoptics*. London: Macdonald.

Kryukov, B. I., L. M. Litvin, I. M. Sobol', and R. B. Statnikov. 1980. "Optimal design of resonance-type vibration machines." *Mashinovedeniye*. No. 5: 31–39 (in Russian).

Kuhn, H. W., and A. W. Tucker. 1951. "Nonlinear programming." In *Proceedings of the Second Berkeley Symposium on Mathematical Statistics and Probability,* edited by J. Neyman, pp. 481–492. Berkeley: University of California Press.

Kuipers, L., and H. Niederreiter. 1974. *Uniform Distribution of Sequences*. New York: John Wiley.

Larichev, O. I. 1987. *Objective Models and Subjective Decisions*. Moscow: Nauka (in Russian).

Lieberman, E. R. 1991. *Multi-Objective Programming in the USSR*. New York: Academic Press.

Ljung, L. 1987. *System Identification: Theory for the User*. Englewood Cliffs, N.J.: Prentice-Hall.

Lukyanov, N. K. 1981. "Aggregation in simulation models of ecological systems." *Izvestiya AN SSSR. Tekhnicheskaya Kibernetika*. No. 5: 30–35 (in Russian).

Masataka, Y. 1977. "Study on optimum design of machine structures with respect to dynamic characteristics. (Approach to optimum design of machine tool structures with respect to regenerative chalter)." *Bulletin of JSME*, Vol. 20, No. 145: 811–818.

Matusov, I. B., and R. B. Statnikov. 1981. "On the choice of a metric in the space of criteria for determining an optimal machine model." *Soviet Math. Dokl.* Vol. 24, No. 2: 434–437.

Matusov, I. B., and R. B. Statnikov. 1985. "Approximation and regularization in vector optimization problems" In *Problems and Methods of Decision Making in Managerial Systems,* pp. 56–62. Moscow: The Institute for Systems Science (VNIISI) (in Russian).

Matusov, I. B., and R. B. Statnikov. 1987. "Approximation and vector optimization of large systems." *Doklady AN SSSR,* Vol. 296, No. 3: 532–536 (in Russian).

Merkur'ev, V. V., and M. A. Moldavskii. 1979. "A family of convolutions of a vector-valued criterion for finding points in the Pareto optimal set." *Avtomatika i Telemekhanika*. No. 1: 110–121 (in Russian).

Molodtsov, D. A., and V. V. Fedorov. 1979. "Stability of optimality principles." In *Modern State of Operations Research Theory,* ed. N. N. Moiseyev, pp. 236–262. Moscow: Nauka (in Russian).

Murav'ev, I. A., and N. D. Bredneva. 1987. "Optimization of freshly harvested haws fruit extraction process." *Farmatsiya* No. 1: 18–21 (in Russian).

Nogovitsin, B. F. 1987. *Basic Principles of Calculation and Design of Die-Casting Machines*. Irkutsk: Irkutsk University Publishing House (in Russian).

Nyquist, H. 1932. "Regeneration theory." *Bell System Technical Journal* Vol. 11, No. 1: 126–147.

Odrin, V. M. 1986. *Morphological Synthesis of Systems: Statement of the Problem, Classification of Methods, Morphological Design Methods*. Kiev, Ukraine: Institute of Cybernetics of the Ukrainean Academy of Sciences (in Russian).

Ozernoy, V. M. 1988. "Multiple criteria decision making in the USSR: A survey." *Naval Research Logistics* Vol. 35: 543–566.

Parlett, B. N. 1980. *The Symmetric Eigenvalue Problem*. *Englewood Cliffs,* N.J.: Prentice-Hall.

Pel'tsverger, B. V. 1984. "Construction of a special basis in the state space for decomposition of nonlinear multiconnected systems." *Izvestiya AN SSSR*. *Tekhnicheskaya Kibernetika* No. 2: 45–57 (in Russian).

Perminov, M. D., and R. B. Statnikov. 1987. "Multicriteria approach to the problem of identification of structurally-complex dynamical systems." In *Automation of Experiment in Machine Dynamics,* pp. 53–64. Moscow: Nauka (in Russian).

Pervozanskii, A. A. and V. G. Gaitsgori. 1979. *Decomposition, Aggregation, Approximate Optimization*. Moscow: Nauka (in Russian).

Pluzhnikov, A. I. 1983. *Accuracy and Optimization of Machine-Tools Kinematic Chains*. Moscow: Mashinostroyeniye (in Russian).

Popov, D. N. 1986. "Efficiency estimation and optimal design of hydraulic drives." *Vestnik Mashinostroyeniya* No. 9: 20–23 (in Russian).

Popov, N. M. 1981. "Approximation of the set of semi-effective points in the design problems decomposition." *Vestnik MGU. ser. Vychislitel'naya Matematika i Kibernetika* No. 1: 44–48 (in Russian).

Portman, V. T., Y. I. Sklyarevskaya, and A. Y. Yakovlev. 1992. "Special-purpose FMS simulation system." *Stanki i Instrumenty* No. 7: 2–4 (in Russian).

Raybman, N. S. 1970. *Identification: What Is This?* Moscow: Nauka (in Russian).

Red'ko, S. F., V. F. Ushkalov, and V. P. Shabel'skii. 1971. "Identification of some mechanical systems." In *Engineering Cybernetics,* pp. 69–82. Kiev, Ukraine: Institute of Cybernetics of the Ukrainian SSR Academy of Sciences (in Russian).

Red'ko, S. F., V. F. Ushkalov, and V. P. Yakovlev, 1985. *Identification of Mechanical Systems*. Kiev, Ukraine: Naukova Dumka (in Russian).

Reshetov, D. N., S. A. Shuvalov, V. D. Dudko, A. V. Klypin, and O. P. Lelikov. 1985. *Computer Aided Calculation of Machines Components*. Moscow: Vysshaya Shkola (in Russian).

Rizkin, I. K. 1985. *Computer Aided Analysis and Design of Engineering Systems*. Moscow: Nauka (in Russian).

Seber, G. 1977. *Linear Regression Analysis*. New York: John Wiley.

Sobol', I. M. 1969. *Multidimensional Quadrature Formulas and Haar Functions*. Moscow: Nauka (in Russian).

Sobol', I. M. 1976. "Uniformly distributed sequences with an additional uniformity property." *Zhurnal Vychislitel'noy Matematiki i Matematicheskoy Fiziki* Vol. 16, No. 5: 1332–1337 (in Russian).

Sobol', I. M. 1985. *Points Uniformly Distributed over a Multidimensional Cube*. Moscow: Znaniye (in Russian).

Sobol', I. M. 1987. "On functions satisfying the Lipschitz condition in multidimensional problems of computational mathematics." *Doklady AN SSSR,* Vol. 293, No. 6: 1314–1319 (in Russian).

Sobol', I. M., and R. B. Statnikov. 1977. *Statement of Some Problems of Computer Aided Optimal Design*. Moscow: Keldysh Institute of Applied Mathematics (in Russian).

Sobol', I. M., and R. B. Statnikov. 1981. *The Choice of Optimal Parameters in Multicriteria Problems*. Moscow: Nauka (in Russian).

Sobol', I. M., and R. B. Statnikov. 1982. *The Best Solutions: Where They May Be Found*. Moscow: Znaniye (in Russian).

Spivakov, A. O., and I. F. Goncharevich. 1983. *Vibration and Wave-Type Transporting Machines*. Moscow: Nauka (in Russian).

Sprague, C. H., and R. H. Kohr. 1969. "The use of piecewise continuous expansions in the identification of nonlinear systems." *Transactions of the ASME* Vol. 91, Series D, No. 2: 179–184.

Stadler, W. (ed.). 1988. *Multicriteria Optimization in Engineering and in Science*. New York: Plenum Press.

Stadler, W. and J. P. Dauer. 1992. "Multicriteria optimization in engineering: A tutorial and survey." In *Structural Optimization: Status and Promise*, ed. Manohar P. Kamat, pp. 209–249, vol. 150. Washington: American Institute of Aeronautics and Astronautics, Inc.

Statnikov, R. B. 1978. "Solution of multicriteria machines design problems on the basis of parameters space investigation." In *Multicriteria Decision-Making Problems*, ed. J. M. Gvishiani and S. V. Yemelyanov, pp. 148–155. Moscow: Mashinostroyeniye (in Russian).

Statnikov, R. B., and I. B. Matusov. 1989. *Multicriteria Machines Design*. Moscow: Znaniye (in Russian).

Statnikov, R. B. and I. B. Matusov. 1994. "General-purpose finite-element programs in search for optimal solutions." *Physics-Doklady*. Vol. 39, No. 6, pp. 441–443, *Translated from Doklady Akademii Nauk* Vol. 336, No. 4: 481–484.

Statnikov, R. B., I. B. Matusov, P. V. Miodushevskii, Y. Y. Uzvolok, D. S. Fel'dman, Y. A. Shevchenko, and V. S. Shenfel'd. 1993. "Parameter space investigation method and multicriteria optimization of objects using finite element models." *Doklady Rossiyskoy Akademii Nauk* Vol. 329, No. 1: 17–21 (in Russian).

Statnikov, R. B., and Y. Y. Uzvolok. 1990. "Determination of parameters boundaries in problems of optimal design and vector identification." *Doklady AN SSSR* Vol. 315, No. 5: 1057–1061 (in Russian).

Steuer, R. E. 1986. *Multiple Criteria Optimization: Theory, Computation and Application*. New York: John Wiley.

Steuer, R. E., and E. U. Choo. 1983. "An interactive weighted tchebycheff procedure for multiple objective programming." *Mathematic Programming* Vol. 26, No. 1: 326–344.

Strobel, H. 1968. *Systemanalyse mit Determinierten Test Signalen*. Berlin: VEB-Verlag Techn.

Sukharev, A. G. 1971. "Optimal search for an extremum." *Zhurnal Vychislitel'noy Matematiki i Matematicheskoy Fiziki* Vol. 11, No. 4: 265–269 (in Russian).

Tanino, T., and Y. Sawaragi. 1980. "Stability and nondominated solutions in multicriteria decision-making." *Journal of Optimization Theory and Applications*. Vol. 30, No. 2: 229–253.

Tikhonov, A. N. 1952. "Systems of differential equations containing small parameters as factors of derivatives." *Matematicheskiy Sbornik* Vol. 31/73 No. 3: 576–586 (in Russian).

Tregubov, V. A. 1983. "Multicriteria choice of operators vibration isolation parameters." *Mashinovedeniye* No. 2: 34–45 (in Russian).

Tsurkov, V. I. 1988. *Dynamic Problems of Large Dimension*. Moscow: Nauka (in Russian).

Tsypkin, Y. Z. 1982. "Optimal quality critéria in identification problems." *Avtomatika i Telemekhanika* No. 11: 5–24 (in Russian).

Tumanov, Y. A., B. Y. Lavrov, and Y. G. Markov. 1981. "On the issue of identification of nonlinear mechanical systems." *Prikladnaya Mekhanika* No. 9: 106–110 (in Russian).

Vasil'ev, F. P. 1981. *Methods of Solving Extremum Problems*. Moscow: Nauka (in Russian).

Velikhov, E. P., V. B. Betelin, and A. I. Stavitskii. 1986. "The use of computers in mechanical engineering." *Mashinovedeniye* No. 5: 3–11 (in Russian).

Voevodenko, S. M., and Y. M. Pevzner. 1985. "Approximate semigraphical method for calculating the vehicle vibrations in road conditions." *Avtomobil'naya Promyshlennost'* No. 7: 12–18 (in Russian).

White, D. J. 1990. "A bibliography on the applications of mathematical programming multiple-objective methods." *Journal of the Operational Research Society* Vol. 8: 669–691.

Woodside, C. M. 1971. "Estimation of the order of linear systems." *Automatica*. Vol. 7, No. 6: 727–733.

Yevtushenko, Y. G. 1971. "Numerical method of searching for the global extremum of a function (exhaustive search on a nonuniform grid)." *Zhurnal Vychislitel'noy Matematiki i Matematicheskoy Fiziki* Vol. 11, No. 6: 1390–1403 (in Russian).

Yevtushenko, Y. G., and V. P. Mazurik. 1989. *Software for Optimization Systems*. Moscow: Znaniye (in Russian).

Zaikova, I. G., and V. V. Yablonskii. 1991. "Parametric optimization of an active vibration isolation system with controlled damping for single- and six-degree-of-freedom objects." *Problemy Mashinostroyeniya i Nadyozhnost' Mashin* No. 1: 16–20 (in Russian).

Zeleny, M. 1974. *Linear Multiobjective Programming*. Lecture Notes in Economics and Mathematical Systems, No. 95, Springer-Verlag, Berlin-New York, New York.

Zhitomirskii, B. E., Y. A. Rubanovich, and A. A. Filatov. 1984. "The use of the multicriteria optimization method in designing transmissions of main drives of rolling mills." *Mashinovedeniye,* No. 1: 33–39 (in Russian).

Zinyukov, P. I., K. S. Samidov, and S. I. Fridman. 1983. "The choice of optimal parameters of machines according to vibration activity criteria." In *Mechanics of Machines,* pp. 78–91. Tbilisi: Metsniereba (in Russian).

Zionts, S., and J. Wallenius. 1980. "Identifying efficient vectors: some theory and computational results." *Operations Research* Vol. 28, No. 3: 788–793.

Index